T0100746

The ISDN Subscriber Loop

TELECOMMUNICATIONS TECHNOLOGY AND APPLICATIONS SERIES

Series editor:
Stuart Sharrock, Consultant,
The Barn, Sugworth Lane, Radley, Abingdon,
Oxon, OX14 2HX, UK

This series covers research into and the development and application of a wide range of techniques and methods used in telecommunications. The industry is undergoing fundamental change under the combined impact of new technologies, deregulation and liberalisation, and the shift towards a service oriented philosophy. The field of communications continues to converge, encompassing all of the associated technologies of computing, networking, software, broadcasting and consumer electronics. The series presents this material in a practical and applications-based manner which equips the reader with the knowledge and tools essential for an engineer working in the industry.

Titles available

1. **Coherent Lightwave Communications Technology**
   Edited by Sadakuni Shimada

2. **Network Management**
   Concepts and tools
   Edited by ARPEGE Group

3. **The Informatics Handbook**
   A guide to multimedia communications and broadcasting
   Stewart Fist

4. **Mobile Communications Safety**
   Edited by N. Kuster, Q. Balzano and J.C. Lin

5. **The ISDN Subscriber Loop**
   N. Burd

The ISDN Subscriber Loop

Nick Burd
National Semiconductor GmbH
Fürstenfeldbruck
Germany

CHAPMAN & HALL
London · Weinheim · New York · Tokyo · Melbourne · Madras

Published by Chapman & Hall, 2-6 Boundary Row, London SE1 8HN, UK

Chapman & Hall, 2-6 Boundary Row, London SE1 8HN, UK

Chapman & Hall GmbH, Pappelallee 3, 69469 Weinheim, Germany

Chapman & Hall USA, 115 Fifth Avenue, New York, NY 10003, USA

Chapman & Hall Japan, ITP-Japan, Kyowa Building, 3F, 2-2-1 Hirakawacho, Chiyoda-ku, Tokyo 102, Japan

Chapman & Hall Australia, 102 Dodds Street, South Melbourne, Victoria 3205, Australia

Chapman & Hall India, R. Seshadri, 32 Second Main Road, CIT East, Madras 600 035, India

First edition 1997

© 1997 Nick Burd

Printed in Great Britain by Cambridge University Press

ISBN 0 412 49730 1

A catalogue record for this book is available from the British Library

∞ Printed on permanent acid-free text paper, manufactured in accordance with ANSI/NISO Z39.48-1992 and ANSI/NISO Z39.48-1984 (Permanence of Paper).

To Juliet, Tristan and Rosie

Contents

Preface

The motivation for writing a book on the integrated services digital network (ISDN) was founded some seven years ago when, as a technical support engineer with National Semiconductor, I took on the responsibility of supporting customer's designs with National's ISDN products. At this time, ISDN was in its infancy, with only a handful of commercial services in operation, and most of these running as pilot schemes. Invariably I found myself in a position of having to educate people about ISDN as well as the application of our products. As a newcomer to ISDN myself, this required me to embark on a rapid learning curve to keep pace with the demands made from all quarters. A survey of existing books on the subject and commercial courses revealed a lack of material that dealt with the issue of how you turn a collection of standards into a functional piece of equipment. From this, the idea started to formulate of a book that would attempt to combine a working knowledge of standards, how those standards are embodied in functional building blocks such as integrated circuits, and how these circuits are combined with software into systems that are connected to the ISDN.

To this end, the material in this book is focused primarily on subscriber loop technology and the terminals that subscribers connect to the ISDN. The subscriber loop in this context is considered to be the entire connection from the network to the user's terminal.

The book is aimed primarily at engineers with an interest in design and development of telecommunications equipment, particularly that related to the ISDN subscriber loop. It may also provide useful supplementary reading for undergraduate and postgraduate students taking courses which specialize in telecommunications and ISDN.

Throughout the book, there are several instances where a particular device is taken and used as an example of how a standard interface specification is translated into an IC component, and how that component may be used in an application. In most cases, the examples chosen are devices that have been designed and manufactured by National Semiconductor, for the simple reason that these devices are ones with which I have had most experience in working with in the past. All information relating to these parts, and others mentioned or described in this book, has been taken from readily available sources such as databooks and datasheets, magazine articles or journal papers.

As well as using specific devices as examples, some tables are provided throughout the book that list a selection of alternative devices from different manufacturers, and are used to compare specific features and functions in support of their explanation provided in the text. Again, this information has been gleaned from manufacturer's data sheets. For more detailed information concerning these parts, the reader is directed to these data sheets. Although specific references to the data sheets for devices that appear in these tables are normally not given, the manufacturer's name and device number provided should be sufficient to locate the data sheet. In addition, it should be noted that while every attempt has been made to provide accurate details concerning the devices mentioned, this and the availability of the devices cannot be guaranteed.

Finally, much information concerning ISDN can be found on the World Wide Web. A short list of sites that may be of interest can be found at the end of the Bibliography.

Contents of the book

The book has a total of ten chapters. Chapter 1 discusses the justification for ISDN in terms of the service integration it provides for users of public telecommunications networks. The title of the chapter, *ISDN – the dawning of a new era* relates to the fact that although ISDN as a service has been around for ten years now, it has only just begun to make an impact on users, not because it is 'neat' technology, but rather that it adds sufficient value to the way people communicate as to provide compelling reasons for why people should use it. Much of ISDN technology is derived from similar technologies that have been in use in public networks for some time. Indeed, ISDN can be viewed as a stage in the evolution of the existing networks.

Chapter 2 introduces ISDN from a standards viewpoint and discusses both the organizations responsible for defining the standards as well as the basic standards documents that cover the architecture, interface and service definitions of ISDN. Chapters 3 and 4 describe in detail the S and U interfaces of a basic rate access ISDN subscriber loop, illustrating each with an example of how the standard is translated into a transceiver integrated circuit. Chapter 5 then goes on to describe the signalling protocols used for call control between a terminal and the network.

These four chapters provide the basics concerning subscriber loop technology in terms of the physical interfaces and the signalling required to establish an end-to-end connection across the ISDN. The following three chapters then build on this foundation by discussing application specific areas focused around the terminal. Chapter 6 discusses hardware and software design aspects of telephone terminals, while in Chapter 7 we examine the compression techniques and their implementation for audio and video. Chapter 8 then looks at the different techniques and protocols for interfacing and transporting data across the ISDN.

Within a large organization, users will most likely be connected to the public network through a private branch exchange, or PBX. The PBX provides the framework around which Chapter 9 is written due to its common use of primary rate access connections to the network, and the use of alternative subscriber connections to user's terminals that are capable of delivering ISDN–like services.

Both these topics are discussed in Chapter 9 along with an overview of extensions to the ISDN signalling protocol that allow PBXs to be networked together.

The final chapter in the book, Chapter 10, is devoted to discussing future technologies for wired subscriber loops and the services and applications they will deliver to the user.

Nick Burd
Fürstenfeldbruck, Germany
June 1996

Acknowledgments

My thanks go to National Semiconductor for permission to publish this book and to numerous friends and colleagues that have either directly or indirectly helped me in formulating ideas for it. My thanks also go to Loren Dooley at Telenetworks for permission to use the Telenetworks ISDN software package as an example in Chapter 6.

Many names appear in the text, tables or illustrations of this book, which have either been trademarked (indicated with the symbol ™) or registered (indicated with the symbol ®) by the respective organizations that own those names.

The opinions expressed in this book are those solely of the author's and are not intended to represent in any way those of National Semiconductor.

1

ISDN – the dawning of a new era in telecommunications

Since the invention of the telephone by Alexander Graham Bell in 1876, and the subsequent installation and operation of the first commercial telephone exchange in 1878 in New Haven, USA, telephone networks have been deployed in countries all over the world, and have rapidly grown in size over the past 20 years to meet the popular demand for voice communications from both private and business users. Today, we take for granted the ability to pick up a telephone handset, dial, and within a few seconds be able to talk with someone on the other side of the world, or someone just down the road. Indeed, with the world's nations now connected by international telephone exchanges, the telephone network appears to the user as a single global network.

When we make a telephone call to someone, we are invoking the primary service for which the telephone system was originally designed; to be able to talk with anyone else who also has a telephone connected to the telephone system. However, although speech is by far the predominant form of communication between human beings, in today's world there is an increasing need to communicate in other ways which lead to higher levels of effectiveness. For example, it is known that the ability of a person to retain information is enhanced if that information is communicated visually as well as verbally. Furthermore, the systems which provide the communications must also offer features and services which improve the productivity of its users.

The integrated services digital network (ISDN) is the next phase in the evolution of the world's public communication networks that will provide a user with access to a multitude of new and existing services which lead to more effective and productive communications, with the ability to support voice as well as video and data communications from a single access connection to the network. This book deals with the technical aspects of this connection between the user and the network; the ISDN subscriber loop.

1.1 PUBLIC NETWORKS

The operators of telecommunications networks base their business on the provision of telecommunications services to subscribers. On subscription to a service, the subscriber or user is connected to a network capable of providing the service requested. Telephony, or the plain old telephone service, also popularly known as POTS, is by far the most widespread service in use today with more than an estimated 700 million connections worldwide.

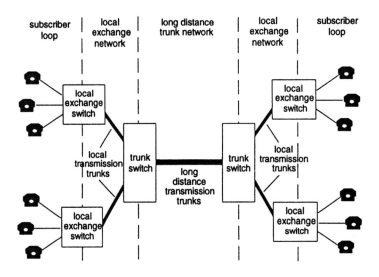

Fig. 1.1 A simplified view of a public network.

Services are provided by public switching networks that allow users to be interconnected with one another, or with equipment that provide a particular service. A switching network consists of three parts, as depicted in Fig. 1.1: the subscriber interface, also referred to as a subscriber loop; switching exchanges; and transmission systems. The subscriber interface connects the users to the switching exchanges which are responsible for routing calls to their destinations, and may involve the call passing through one or more exchanges in its route to its destination. The switching exchanges themselves are interconnected by local or trunk networks which employ transmission systems responsible for multiplexing the many individual calls into a smaller number of high-speed physical connections between the exchanges.

A call made between users of a switching network involves:

- the control of the switching function of the network in order to route the call to its destination, and

- if the call is established and a connection is made, the transmission and reception of information through the network between the connected parties.
- Once the dialogue is complete, the call is terminated by either user replacing the telephone handset which signals to the network that connection is no longer required. The control function is referred to as **signalling**.

Today, a number of different networks exist to provide the services already mentioned. Amongst these, there are two fundamental techniques used in the conveyance of information between the two end-points of a connection across the network. These techniques are known as **circuit switching** and **packet switching**.

1.1.1 Switching in the public networks

Different switching techniques are used as a direct result of the characteristics of the information carried by different networks. One of the first networks to be developed was the telephone network, also known today as the public switched telephone network (PSTN). The PSTN employs circuit switching, which switches a physical circuit between two users that is then dedicated to that use for the duration of the call. Once the call is terminated, the circuit may be switched to another connection for use in another call. Having a dedicated circuit available for a voice call is justified on the grounds that speech conversation tends to have very few silent gaps between the periods of conversation, thereby providing good utilization of the connection.

Packet switching was developed in the 1960s in response to the growing use of computers and the need for communications between remote terminals and computers, and between computers themselves. In such computer communications, the occurrence of data tends to be very bursty in nature with variable gaps between bursts. Packet switching therefore switches each burst, or packet, of data according to a route through the network determined at the time the call was established. No dedicated connection therefore exists between the users, which allows the physical circuits to be more efficiently utilized by handling packets from more than one call.

In the past, networks have been established to support a particular service by the conveyance of a certain class of data. In some cases, several services can be offered by the same network, for example telephony and fax over the PSTN. For other services, a different network is required, and more often than not, the access to the network requires a different subscriber loop.

The existence of separate subscriber connections and networks for different services impedes the ability of the network operators to efficiently accommodate new services. While many networks are likely to remain separate for quite some time due to the massive capital expenditure already made in them, the access to these and other networks can be integrated such that from the user's perspective there exists a single standardized interface over which the user can access a variety of services. This is the basis on which today's ISDN is founded.

1.2 APPLICATIONS AND SERVICES

A telecommunications service is provided by a network, and is the product that is sold by the network operator to its customers. A service may be capable of supporting one or more applications used by the customer. In many cases, the service is known by the same name as the application, although it is important to note that they are different, the distinction being that a service is embodied within the network, while the application lies outside it and is typically found embedded within a piece of terminal equipment at the user. The following is a list of the main applications and services available today, and some comments concerning the subscriber loop connection used between the user terminal equipment and the network for the particular service.

1.2.1 Telephony

Telephony is the primary application for which the PSTN was designed. The telephony service provided by the PSTN supports a dedicated bandwidth channel between users in the range 300 Hz to 3400 Hz for interactive voice communications.

Although the switching and transmission parts of modern PSTNs are mostly digital, communication and signalling across the subscriber loop connection employs mostly analogue techniques. The subscriber loop is a two-wire subscriber cable which is brought to the subscriber's premises from the local PSTN exchange, and is normally terminated in a simple wall socket in the user's premises to allow connection of the user's telephone. Signalling over the subscriber loop takes the form of simple direct voltage or current levels, and sometimes pairs of tones to represent dialled numbers, while speech conversation across the subscriber connection involves the transmission and reception of analogue voltage signals. The basic principles on which the telephone and the analogue subscriber loop work have not changed significantly since the turn of the century, although the application of modern integrated circuit electronics has changed them technologically, and has helped to make them more reliable and less expensive.

1.2.2 Data communications

Data communications are possible across the PSTN by adapting the digital data from a computer or data terminal to be communicated within the 300 Hz to 3400 Hz bandwidth voice channel of an end-to-end connection. The adaptation is performed by equipment called a modem. The modem (short for modulator/demodulator) modulates a serial digital data stream onto an analogue carrier signal that can be transported within the voice channel. At the far-end, the received signal is demodulated and the serial data stream is passed to the receiving device. The modem is the interface for most computer-based equipment to the public networks for data transmission and reception.

The transmission quality and small bandwidth of the PSTN subscriber loop connection is adequate for telephony, but limits the practical modem data rate for many subscriber loops to speeds of 28.8 kbps or lower. An alternative which has been adopted by a number of network operating companies, as for example in Germany and in the Nordic countries of Europe, is to install and run a separate circuit-switched data network with higher quality subscriber loop connections that can handle higher data rates (for example up to 64 kbps).

1.2.3 Facsimile

Facsimile has recently emerged as an indispensable means of communication for business documents consisting of text and pictures, and is an application that has been designed for use across the PSTN. A paper document is scanned and digitized into a stream of picture elements which are then compressed and sent as digital data across the PSTN using a modem. Any document that can be scanned, which therefore includes hand-written or hand-drawn documents, may be communicated as a facsimile, which also makes the service particularly useful for the transmission of documents written in alphabets whose standard computer representation is not commonly found. The importance of facsimile is reflected in the fact that it is used the world over to complete transactions in all aspects of business, commerce, finance and service-related industries.

Applications such as facsimile and data communications use the same switched 'telephony' service of the PSTN. Hence, the standard voice service dial-up signalling is used to perform the call setup and tear-down of the connection, while further in-band end-to-end signalling associated specifically with the application may then take place to configure the mode of communications once the connection is established.

1.2.4 Telegraphy

As far back as 1847, telegraphy was used to transmit Morse code signals on networks run by the early American railroad companies, and may be considered as one of the first digital transmission methods. This type of telegraphy used a manually operated make-and-break switch to cause short and long DC current pulses to flow in a connection to a remote receiving actuator which in turn caused either an equivalent acoustic signal to be heard by the receiving operator, or an indication on a moving paper tape. This low-speed text transfer technology is now almost extinct, and has been replaced by telex and facsimile (fax) services.

1.2.5 Telex

Telex is essentially the combination of telegraphy and the typewriter, the name being derived from the name TELegraph EXchange. The first telex networks were introduced in the 1930s. A benefit of telex is that, unlike telegraphy, the telex equipment can be unattended as it provides automatic answering of incoming calls, and a hard copy of the information received. Text is transferred as digital codes where each code word represents either a text character or a control function such as a carriage-return or line-feed. These codes are standardized to allow equipment inter-operability, one example of which is the ASCII (American Standard Code for Information Interchange) digital alphabet.

Telex provides a low-speed data transfer, typically around 50 bps, that uses DC currents through subscriber connections to signal the digital '1's and '0's of the character codes between the network and the telex terminal. In older systems voltages as high as 80 V may be used, particularly on long subscriber connections with a high loop resistance, that cause transients to be induced in adjacent cables and resulting in noise and interference. More modern telex networks employ either lower voltage signals, or data modems. Telex networks can be implemented as part of the PSTN or as a separate circuit-switched network, and are still used today to provide basic business communications, in particular for hotel reservations and news gathering services.

1.2.6 Teletex

Teletex is an extension of telex, and provides an extended character repertoire for, amongst other things, the definition of page formats, and provides error detection and correction. To a large extent, this service has today been replaced by the facsimile service available across the PSTN.

1.2.7 Videotex

Videotex services provide access to database and information services such as on-line telephone directories, public transport timetables, weather reports, news summaries, and so on. These services are typically available across the PSTN using a videotex terminal that includes a keyboard and some kind of low resolution graphic screen. The videotex terminal is typically connected via a data modem to a PSTN subscriber loop and allows the user to dial a particular service on demand. Once connected, the user is then presented with a series of menus which allow the required information to be retrieved and displayed. Examples of videotex services are Prestel in the UK, Minitel in France, and BTX in Germany.

1.2.8 Packet–switched data

An alternative to the circuit-switched data service provided by the PSTN is the packet-switched data service provided by the packet-switched public data network (PSPDN). In a packet-switched network, data is transmitted in variable length packets. Due to the bursty nature in which the data is generated, the packets from several 'connections' can be multiplexed in the same physical connection. Unlike the PSTN in which a connection becomes dedicated for the duration of a call once it is established, the same physical connection within the packet network can support multiple logical connections, thus making more efficient use of the connection bandwidth. Consequently, the cost of a call over a packet-switching network is based on the amount of information sent rather than the time duration of the call, and is ideally suited to the bursty nature of computer host-to-terminal communications.

Access to the PSPDN can either be direct with a special subscriber loop connection, or indirect through the PSTN using a data modem where a gateway exists between the PSTN and PSPDN. Direct connection usually allows a higher data rate capability than is possible through the PSTN.

1.2.9 Leased lines

Leased lines are specially conditioned secure connections which are reserved for the exclusive use of a leasing customer between pre-defined locations. These are typically rented by a business from a network operator to link their separate offices and business locations to create a private corporate network.

The leased line is a dedicated connection where the destination is fixed and not switched as in the PSTN. No signalling is therefore required, and the connection is available to the user all the time. To justify this, the user traffic requirements are sufficiently high as to warrant such a permanent connection.

Digital leased line connections serve a variety of purposes, the most common being for data communications ranging from low speeds of 64 kbps up to speeds as high as 34 Mbps in Europe and 45 Mbps in North America. Consequently, leased line services are provisioned with special subscriber loop cables, usually a high quality four-wire cable with a separate cable pair for transmission and reception.

1.2.10 Wireless paging

By replacing the wired subscriber loop with a wireless radio connection, a user with a suitable terminal or handset has greater freedom of movement within the coverage area of the system. Two basic services are provided by such wireless extensions to the PSTN, these being paging and mobile telephony.

Paging services have been available since the mid 1970s and provide a one-way messaging service to remote users. Two basic types of paging service exist today, one which simply emits a tone to indicate that the user carrying the paging device must make contact with a pre-arranged person, and the second that delivers a message from the person wishing to make contact, and would typically be used to give the telephone number to be called. More sophisticated systems are being developed that will allow a voice message to be forwarded and replayed by a paging receiver, and a two-way paging device that will allow a message to be transmitted as well as received. A pan-European paging network called ERMES (European Radio Messaging Service) is currently in operation.

1.2.11 Mobile telephony

Mobile telephones allow freedom of the user such that he or she is no longer tethered to the wired subscriber loop. Two generic services are available, one called **cellular mobile radio** and the other **cordless** telephony.

In a cellular mobile radio network, a geographical area is divided into cells, each served by a base station connected to the PSTN through mobile switching centres. The network is sufficiently intelligent as to allow subscribers to roam freely from one cell to another, and is therefore suited to subscribers using vehicular transport. This has two important implications. Firstly, for incoming calls, the network must know of the location of the subscriber who is being called, as it is possible that he or she could be in any of the cells at the time the call is made. The base stations are therefore capable of periodically interrogating the mobile telephones present in their cell without any intervention from the subscriber, in order to keep an up-to-date record of the location of every subscriber. Secondly, should a subscriber cross a cell boundary during a call, the handover from one cell base station to the other must appear transparent to the subscriber, and is achieved with additional control signalling between the mobile telephones and the switching centres.

The first mobile radio networks, such as AMPS (Advanced Mobile Phone System) in America and TACS (Total Access Communications System) used in the UK, appeared in the early 1980s and used analogue radio technology with a low speech quality and large spectrum requirements which limited the subscriber density. Unfortunately, at this time little standardization existed for such networks with the result that many different incompatible systems evolved. However, second generation systems based on digital radio with higher speech quality and more efficient spectrum usage are currently being deployed. While these systems are standardized, there exists several competing standards based mainly on two technologies known as time division multiple access (TDMA) and code division multiple access (CDMA), a form of spread-spectrum technology. Systems deployed in Europe, such as GSM, use TDMA, while others being developed in America, such as IS-54 and IS-95, are based on CDMA.

Cordless telephone systems provide coverage over smaller areas, called micro cells, of typically 100 to 200 metres in radius and may not allow roaming from one area to another. Consequently the cordless handsets are less complex, are not required to radiate as much power, and are therefore smaller and more lightweight than their mobile counterparts. Like mobile technology, the first cordless telephone systems, such as CT-1 used in Europe, were analogue but have now migrated towards digital systems such as CT-2 and DECT (Digital European Cordless Telephone system).

The benefits of both cellular and cordless technologies are currently being evolved into a new generation of future personal communication networks (PCNs) based on existing technologies that will provide the subscriber with a small, lightweight, hand-held portable telephone, a range of both voice and data services, and the ability to roam in a network built up from micro cells. An example is the DCS1800 system based on GSM which is currently being deployed in several European countries.

1.2.12 Value-added network services

In addition to the standard services offered by network operators, a new generation of services has emerged in recent years which use the network to deliver processed information to the user across the network. These services essentially add value to the basic transport service provided by the network operator, by delivering information to the user which is already processed and presented according to its use in some service or application. The services may be provided either by the network operator, or by private organizations which may also require a leased line connection to their private network. Examples of value-added services are:

- information-related services such as the World Wide Web, news, financial or timetable information;
- processing services such as credit card processing, inter-bank money transfers and airline reservation systems;
- messaging services such as email and EDI (electronic data interchange used for direct order entry).

Access to these services is dependent on the targeted customer. For example, the SWIFT system used by banks for customer transfers, bank transfers, statements and confirmations, uses encryption techniques to provide secure communications over leased line connections. On the other hand, services such as public transport timetables, news and weather services, are targeted at a much wider audience and therefore available via videotex or teletex across the PSTN.

1.3 THE INTEGRATED SERVICES DIGITAL NETWORK (ISDN)

This brings us to ISDN, whose purpose it is to support a wide range of both new and existing voice and data services using a limited, but well defined, set of connection types and interfaces between the user and the network. The main feature of ISDN is reflected in its name, that is to provide an end-to-end **Digital Network** capable of supporting **Services** which, from the user's perspective, appear as though they are provided by an **Integrated** network by virtue of the fact that the services are accessed from a single connection to the network using a common set of well defined protocols.

Figure 1.2 shows the arrangement of networks prior to ISDN and how access to them becomes integrated with the implementation of an ISDN. Access to the services of the separate networks is achieved through an ISDN exchange with a single interface to the user such that the user perceives the network to be a single entity. The ISDN subscriber loop and the protocols it employs are defined such that the user can readily access the wide range of services provided by the individual networks that lie behind the user–network interface, and are sufficiently flexible that new services can be added in the future.

The definition of the ISDN subscriber loop is key to the successful implementation of ISDN, because it must:

- provide the user with a flexible digital interface with access to a wide variety of present and future services;

- allow the evolution of the individual network services to progress towards achieving an ISDN according to the different strategies of the networks operators.

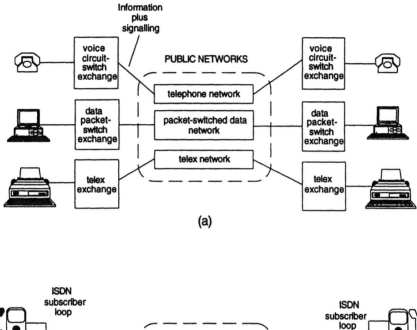

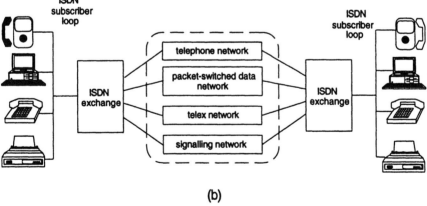

Fig. 1.2 (a) Separate access to networks prior to ISDN; (b) access integration provided by ISDN.

With many of the world's public networks evolving in different directions and at different paces, it is necessary to define a common starting point for ISDN. This is a minimum set of capabilities of the network infrastructure that would allow the implementation of ISDN to be achieved as an upgrade to the existing network that could take place over a period of time and without loss of the existing services. The ISDN requires the existence of a digital network infrastructure that can provide:

- digital circuit-switching exchanges and the digital transmission of circuits at a 64 kbps rate;
- a digital signalling network to provide circuit-switching control information within the network between switching exchanges.

The first feature is widespread in today's telecommunication networks, most of which are now coming to the end of digitalization programmes to convert from older mechanical exchanges and analogue transmission systems to fully digital switching and transmission systems. The second feature is one with which many networks are currently being upgraded in order to provide support for both ISDN as well as provide a more efficient network signalling standard for analogue telephony.

1.3.1 The digital subscriber loop

Today, we find the subscriber loop is very much the focal point of ISDN because it is the user's focal point for the provision of integrated services, and is also the missing link in achieving end-to-end digital connectivity across the public network. An objective of this book is to describe to the reader the requirements of the ISDN subscriber loop and the technology required to implement it. The transition from a POTS analogue subscriber loop to an ISDN digital subscriber loop involves the application of digital signal-processing technology for the transmission and processing of digital signals across the subscriber loop, and the application of digital communication protocols to provide a powerful, yet flexible interface for the request and control of an expandable range of services to the user across the interface.

One requirement which has been key to the definition of schemes for the digital transmission of data across the subscriber loop, is that the ISDN subscriber loop must be capable of being implemented with the existing copper cables currently used in the analogue telephony subscriber loop of the PSTN. To an extent, this has also determined the bandwidth available to the user across the subscriber loop given the performance of suitable digital transmission techniques on such cables. The narrowband ISDN, or N-ISDN, consists of two bi-directional 64 kbps channels, known as B-channels, for user data such as voice, data or video, and a single bi-directional 16 kbps channel known as the D-channel which is primarily used for call control signalling.

The alternative to the use of existing installed cable would be to re-wire the subscriber loop with a better quality of cable, thereby allowing less complex and hence less expensive digital transmission schemes and techniques to be employed that also provide higher bandwidths to the user, but this would render the implementation of the ISDN much too expensive for the owners of the networks due to the significantly higher installation cost of the cable. By using the existing cable, the subscriber loop re-wiring costs are not incurred, but more complex transmission techniques must be used to compensate for the poorer characteristics of the existing cable. These transmission techniques would be embodied in an electronic integrated circuit (IC) with the benefit that the incremental cost in upgrading existing subscriber loops to ISDN, particularly in the longer term, is considerably less than the alternative of re-wiring the subscriber loop. The reasons for this lie in the economics of the mass production of electronic ICs with which the circuit designs for transmission schemes and data processing functions for the ISDN subscriber loop are implemented.

1.3.2 Digital terminals

In order to use the ISDN, a subscriber must have an ISDN compatible terminal. The name 'terminal' is used as a generic name for any piece of equipment that can be connected to the subscriber loop and from which the user can access the services provided by the network. A simple example is a telephone. As well as providing a physical connection to the ISDN subscriber loop, an ISDN terminal must be capable of the necessary signalling capabilities that will allow it to request and acknowledge the provision of services to and from the network. For the case of an ISDN telephone, the action of the user lifting the handset off-hook to make a call will cause a message to be sent the network to indicate the off-hook status of the telephone and the request of a telephony service. On receipt of this message, the network will determine if it is capable of delivering the requested service, and if it has a free outward bound line to route the call. If so, the network acknowledges this to the telephone with an acknowledgment message. In a similar way, the dialling information is sent to the network, the call is routed to its destination, several status messages are exchanged between the terminal and the network, and finally, when the call is answered at the destination, a connection is established between the two users.

In an analogue subscriber loop, this signalling is achieved with defined voltage and current levels, and the transfer of dialling information is achieved by current pulses or by the transmission of tones, as in DTMF (dial-tone multiple frequency) dialling. The accuracy and speed with which these analogue signals can be generated and detected over the subscriber loop limits the flexibility and scope of their use. However, the subscriber loop in ISDN is digital, and signalling information now takes the form of digital messages which are exchanged between the terminal and the network.

The rules by which the subscriber loop messages are exchanged and processed are defined by communication protocols. The protocols which have been defined for the ISDN subscriber loop are sufficiently flexible to allow a wide range of services to be accessed. However, the necessary flexibility of the protocols required, together with the consequent complexity of their implementation, typically requires the application of microprocessor technology in order to process the information. The result is that, in upgrading the analogue subscriber loop to ISDN, the basic telephone is, through necessity, transformed from a low-technology apparatus based on simple analogue circuitry to a comparatively high-technology microprocessor-based terminal.

An aim of this book is to give the reader an insight into what the circuitry of various ISDN subscriber loop equipment looks like, its major building block functions, and how they operate, both individually, and together to perform the required functions of the equipment. As well as treating this area with a generalized approach, examples of some specific ICs are also described to illustrate the material and provide a more practical insight to the ISDN.

1.3.3 The capabilities of ISDN

The PSTN subscriber loop provides the user with a single analogue voice channel which is circuit switched across the PSTN. If the network is digital, a single voice channel will correspond to a 64 kbps digital channel within the network. The equivalent ISDN subscriber loop offers the user not one, but two digital Bearer channels, or B-channels, each having a data rate of 64 kbps equivalent to an analogue voice channel, plus a 16 kbps D-channel used for signalling. This is known as the 2B+D, or basic rate access (BRA) to the ISDN.

A high-speed interface containing more B-channels, called primary rate access (PRA) ISDN, has also been defined. The PRA interface consists of 30B+D channels in Europe, and 23B+D channels in North America, the difference being due to the different digital multiplexing hierarchies used in Europe and North America. The higher data rate required to transmit and receive these channels means that PRA ISDN interface must use specially conditioned subscriber loop cables and is therefore not as easy as BRA ISDN in terms of its deployment. Today, PRA ISDN is used primarily to connect equipment which makes use of a large number of bearer channels, such as multiplexing equipment and PBXs. In this book, we concentrate our attention mostly on the more widely applicable BRA ISDN interface, although many of the services and concepts are also valid for PRA ISDN. (BRA and PRA are also known as BRI (basic rate interface) and PRI (primary rate interface).)

The ISDN has been designed such that those services available to subscribers of existing PSTN and packet data networks are also available to subscribers of the ISDN, with the advantage that they can be accessed from a single digital subscriber loop. In addition, the availability to the user of two B-channels in the BRA ISDN interface allows the possibility of two independent simultaneous calls. For example, a telephony call could be established on one B-channel while the other was used to transfer a computer file or receive a fax. Alternatively, for some services, the two B-channels could be combined to effectively create a single channel at 128 kbps on a single circuit-switched connection for those applications requiring higher speed data transfer, such as videotelephony.

The ISDN also provides the benefit of digital transmission at speeds above the capability of analogue modems. Assuming that usage costs of the ISDN and PSTN are the same (which is true in most cases), the increased speed reduces the cost of a transaction such a fax or file transfer. At 14.4 kbps, a 1 Mbyte data file would theoretically take around 9 minutes and 15 seconds to transfer, although in reality this time can be expected to be 20% higher due to setup times and the transmission of additional protocol bits. Transferring the same data file in a 64 kbps ISDN B-channel would only require between 2 and 3 minutes and therefore the cost would be less than a third of the cost using the analogue modem[1]. However, the increased data transfer speed that ISDN provides does not make it suitable for all applications. The real issue which has to be addressed is how the characteristics a service provided by ISDN match the communication requirements of the application that may use the service. Clearly, some applications have requirements beyond the capabilities of BRA ISDN. Figure 1.3 shows some typical applications and their suitability to a range of communications services and technology.

From Fig. 1.3 it is clear that a range of telecommunication and data communication services are needed to economically cater for the range of business and private applications both now and in the future. BRA ISDN will satisfy a particular range of the application spectrum, and will coexist with other services and technologies intended to cater for those applications with communications requirements which fall outside the scope of ISDN.

1.3.4 The market acceptance of ISDN

As well as addressing existing applications, it is expected that ISDN will stimulate the growth of new and improved services and applications that improve our way of life, and contribute to the productivity of businesses and organizations through the people that work in them that make use of these services. The ISDN can be considered as an enabling technology for the provision of these services.

---

[1] It should be noted that this comparison does not take into account hidden usage costs such as line rental.

For example, the recent interest in the Internet and the World Wide Web (WWW) has created an unprecedented demand for ISDN connections because of its speed advantage over the PSTN that allows faster downloading of files and pages with a high graphic content. Not only does this make use of the WWW cheaper where connection tariffs are incurred, but also easier to use.

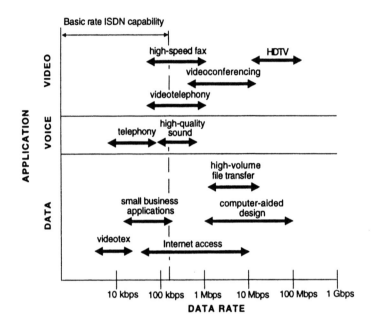

Fig. 1.3 A selection of applications and their typical data rates.

The success of ISDN has depended on numerous factors that have determined firstly the implementation and deployment of ISDN by the network operators, followed by the uptake of ISDN by subscribers.

Implementation of ISDN

Technology. ISDN depends on the availability of semiconductor integrated circuits that provide economical implementations of the necessary protocol processing functions both within the local exchange and user's terminal equipment.

Cost. The overall cost of upgrading the existing PSTN to ISDN must not be prohibitive to its wide scale implementation. In this respect, the ability of ISDN to use existing subscriber loop cabling is a key factor.

Standards. The existence of complete and unambiguous standards is essential for the inter-operability of ISDN equipment and networks. Standards will also help to create a competitive market for ISDN terminals and services.

International cooperation. Realization of international ISDN can only be achieved by the cooperation of the network operators and the telecommunications administrations. This has been particularly relevant in the establishment of ISDN within the European Community.

Uptake of ISDN

Availability. Widespread availability of ISDN is essential for its acceptance. Many of the unique services available with ISDN are only possible if both users are connected to it. In many countries where deployment of ISDN has been a focus for the last five years, existing networks have been upgraded such that more than 90% of subscribers could be connected to the ISDN.

Services and cost. The services offered by ISDN must provide real benefits to the user. In many cases, in particular during early deployment of ISDN services, this has meant continued use of many of the same applications as previously, but at a lower overall cost, improved functionality and ease of use.

The cost of having access to and using ISDN has three components: a one-time installation fee, a monthly or quarterly rental fee for the subscriber connection, and a usage tariff. Today, the installation fee for a BRA subscriber connection to the ISDN can be between two to ten times more expensive than an analogue PSTN, while monthly rental charges for a BRA ISDN connection are of the order of two to three times more expensive. It is expected that the price of monthly rental will settle to around a two times difference in the future[2]. BRA ISDN voice call rates using a single 64 kbps channel are typically the same as the PSTN, although calls that connect to other networks, such as the packet data network, and those that use the additional 'supplementary' services provided by ISDN typically incur additional charges.

The costs for PRA ISDN follow a similar pattern, although the monthly rental is expected to become more attractive compared to the equivalent number of PSTN lines. For example, the monthly rental for a PRA ISDN connection, which is equivalent to 30 PSTN connections, is between 8 and 15 times that of a BRA connection, and is expected to stabilize at around 10 times the BRA price.

As well as the cost of renting lines and using ISDN, the cost of ISDN terminal equipment is also a major factor in the general acceptance of ISDN, particularly with residential users. While much of the terminal equipment available five to ten years ago was prohibitively expensive, prices have dropped within the reach of

---

[2]A BRA ISDN connection is equivalent to having two analogue subscriber connections. The two times difference between BRA ISDN and the PSTN subscriber connections therefore means they are also equivalent in terms of pricing.

'home office' workers and even residential users. One of the biggest terminal equipment markets today is for ISDN PC plug-in adapter cards and ISDN 'modems' which allow a PC to be connected to the ISDN for Internet connection and remote connection to corporate local area networks (LANs). Simple low-end cards which rely on the PC to perform most of the protocol processing required are priced at around $130 today, with more sophisticated cards with on-board protocol processing costing anything up to $1000 or more. ISDN telephones can typically be purchased for a price of around $200.

With many commercial ISDNs now being operational for five years or more, many consider that the early introductory phase of ISDN, particularly in Europe, is now coming to an end. What has begun to happen, and is expected to continue over the next few years, is that ISDN equipment and service costs will continue to decrease, and as a consequence the uptake of ISDN is expected to grow rapidly, particularly in markets such as the SOHO (small office home office) market where ISDN is an ideal match for the services and applications required.

1.3.5 Applications for ISDN

A key factor in the uptake of ISDN is the applications that will attract a user to make use the service. In reality, there is no so called killer application for which ISDN services are a prerequisite. Instead, ISDN services allow many of the same applications to become more widespread, and provides them with a better quality of service thereby allowing them to deliver improved performance and capabilities to the user. In many cases, this improved performance yields cost savings and productivity improvements. When ISDN has become established as a mainstream service, new applications that can take advantage of its digital services will start to appear.

A selection of applications representative of those for which ISDN is used today, are described below.

Voice and speech applications. Although basic telephony is primarily a service, the ISDN has enabled higher quality speech communications through the ability to incorporate digital compression into terminals that can provide effective bandwidths of 7 kHz, or double that of the PSTN. This quality improvement has led to its use in conferencing and broadcasting applications. The standard voice compression schemes are discussed in Chapter 7.

As a service, telephony also benefits from ISDN through significantly faster digital signalling in the network that has reduced connection times to fractions of a second, and through digital signalling in the subscriber loop which has enabled a range of supplementary voice services that improve that flexibility of call handling.

Videotelephony. Videotelephony provides communication of a real-time image as well as voice between two or more users. Much of today's videotelephony is used by large organizations for videoconferencing. The equipment is typically large and cumbersome and is used to connect groups of people together rather than individuals. More recently, desktop videoconferencing stations based around PCs have become available, as too have small stand-alone videotelephones aimed at commercial and business users. The cost of this equipment is typically in the range of $5000 to $10 000, and so is still prohibitively expensive for the residential market. Much of this cost today lies in the implementation of standard videocodecs which compress and decompress the video information such that it may be transported across a narrow bandwidth digital communications channel, such as that provided by ISDN. However, it is expected that in time the cost of videocodecs will decrease sufficiently that videotelephony will become a major application for the residential market. Videotelephony over ISDN is discussed in more detail in Chapter 7.

LAN-to-LAN interconnectivity. Local area networks (LANs) are now common in the workplace and allow users to share resources and communicate with one another within a single site. In large organizations, there is a need to interconnect people at different sites such that data can be shared. Equipment called routers provide LANs with the access they need to public network connections, and ISDN has quickly become the preferred public network service to provide the interconnection between routers due to cost savings gained in the use of an ISDN compared to digital leased line connections. Whereas a user pays a fixed rate for a leased line connection regardless of how much or how little it is used, a circuit–switched data service provided by an ISDN is tariffed according to the connection time. Consequently, if the cost of use of the two services is compared, there will be a period of time up to which use of the ISDN is more cost effective, and beyond which a leased line is more effective. Of course, this period of time depends on the respective tariffs of the two networks, but typically lies between four to five hours of usage per day at today's tariffs. LAN-to-LAN interconnectivity using the ISDN is discussed in more detail in Chapter 8.

Leased line backup. Many corporate private networks make use of leased lines because they were established before ISDN was available, or because the volume of traffic they carry exceeds the limit at which the switched data service provided by ISDN would be more cost effective. ISDN is used to supplement existing leased line connections for the following two main reasons.

- In cases where the application that makes use of the service is considered critical to the operation of a business or organization, the availability of the service is of paramount importance. Should the leased line service become unavailable for any reason, then the multiplexing equipment that provides the customer interface to the public network would provide an automatic switch-over to a dial-up ISDN service. Due to its digital signalling, the speed of connection is fast enough that the switch-over to ISDN can be imperceptible to the user.

- In cases where the data traffic generated by a user occasionally peaks beyond the leased line capacity, then additional ISDN connections can be made for the duration of the peak in order to accommodate the excess traffic. Each call made will add 64 kbps of additional capacity until the prevailing bandwidth requirement is met. This ability to dynamically adjust the aggregate connection bandwidth according to requirements is also known as **bandwidth on demand**.

ISDN PBXs. Most organizations rely on optimizing their use of public telephone services, and hence cost, through the use of a private branch exchange, or PBX. For an organization with a large number of employees, each requiring the use of a telephone, the idea of the PBX works on the assumption that not all employees will need to use the telephone at the same time. This means that in practice only a smaller number of public subscriber lines need be installed, and the PBX then takes care of switching calls from employees to the available public subscriber lines. The PBX is in effect a switch which is owned or rented by the company or organization and connected to the PSTN by a number of public subscriber lines. The advantages gained are that fewer public subscriber lines are required with a consequent reduction in line rental cost, while calls made internally to the PBX system do not incur any PSTN charges. The disadvantage is that if all the public lines are in use, then new incoming or outgoing calls will have to wait until a line becomes free. However, the probability of this happening can usually be set at an acceptable level by the installation of a sufficient number of subscriber lines above the expected call rate and duration.

While PBXs have for many years provided advanced voice call features to their users, many of today's PBXs extend both the voice and data services of the public ISDN to these users. In particular, the data communications capabilities of ISDN allow ISDN-based PBXs to be used as a private circuit-switched data communications infrastructure that interconnects PCs, workstations, and terminals to multi-user computers, both within the customer premises as well as to the public ISDN. In addition, ISDN type signalling between PBXs, for example at different sites of the same organization, allow them to appear to the user as a single 'virtual' PBX with services uniformly available to users regardless of whether they are connected to the same PBX or different PBXs interconnected through the public network.

Computer telephony integration. Computer telephony integration (CTI) is the functional control of a telephone, and hence the services to which it connects, by a computer. In its most simplest form, a computer such as a PC is connected to a telephone through an interface that allows software running on the PC to control the call functions of the telephone. As the user interface of the telephone is notoriously bad, particularly in controlling new service features, the PC is used to provide an improved user interface. More recently, interfaces between the PC and the telephone, such as the universal serial bus (USB), have been defined to allow both control and voice information to be exchanged. This will pave the way for a range of applications that use the PC to process the voice signals from the telephone, and to playback processed signals over the ISDN.

While such applications can be implemented equally by a telephone connected to the PSTN as by one connected to the ISDN, there are a range of CTI applications that rely on the advanced signalling features that are inherent in ISDN. Most of these applications use the calling party number that can be embedded in the call signalling information to the called party as a means of identification to retrieve details concerning the calling party. For example, in a customer support centre the information may be used to automatically retrieve from a database and display the service record of a calling customer.

Low-bandwidth multimedia communications. Today's work-practice in large organizations is highly oriented towards inter–disciplinary teams, and with many organizations having global presence, the members of these teams will often be dispersed between various locations. The team members will have a need to frequently collaborate on their assignment which will involve the sharing of information such as design data, spreadsheets, documents and so on. The process of collaboration involves communication between team members, and is enhanced by the ability to communicate both verbally and visually while at the same time manipulating the work information. The process is sometimes referred to as collaborative computing, and is today embodied in numerous commercially available software products.

Although such products may typically be used with either the PSTN or ISDN, the ISDN has the advantage that its increased bandwidth capabilities allow it to support simultaneous voice, data and video communications with a single call.

Point of sale (POS) transactions. The speed with which automatic calling can be achieved and the speed with which data can be transferred makes the ISDN ideal for point of sale transactions such as credit card verification, the benefit being in the decreased time taken to process a customer purchase, and consequently lower cost of the call. ISDN is also being used to allow fast and efficient updating of stock records in retail chains.

1.4 SUMMARY

The ISDN gives the user access to a range of services and networks through a single standardized digital subscriber connection. With a BRA ISDN connection and suitable terminal equipment, users can communicate with voice, audio, video and data, at rates of up to 128 kbps, and access a range of services which allow more efficient use of the network and facilitate more efficient and productive communication practices. In addition, BRA ISDN has the ability of becoming as widespread as the existing analogue PSTN through its use of already installed subscriber loop cables.

The benefit to the user of ISDN can be summarized as follows.

- The **cost** of communications is **reduced** as multiple services are available from a single interface to the network rather than from numerous different interfaces. Furthermore, applications that traditionally use low-speed modems can use the 64 kbps data capability of ISDN that result in shorter transmission times and lower costs.
- The ISDN provides the user with a **standardized flexible interface** to the network through which he can select and configure the services required.
- The **quality of communications** is improved because the ISDN end-to-end connection is completely digital, and the digital signalling of ISDN provides the user with faster and more reliable call control, with more information available to the user concerning the call.
- The ISDN services permit more **effective** and **productive communications**.

2

ISDN standards and architecture

ISDN standards and the ISDN architectural model are the two main themes introduced in this chapter, and provide us with a framework with which to describe the basic terminology and technical attributes of the ISDN subscriber loop in subsequent chapters. Throughout this book we refer to the standards that define the interfaces for ISDN, and it is therefore deemed important to briefly discuss not only the core standards themselves, which takes place through the book, but also the organizations that were responsible for generating them and their relationship to one another in the worldwide standards arena.

2.1 ISDN AND THE CCITT/ITU-T

Prior to 1993, the organization which governed the international standardization in public telecommunications networks was the International Telegraph and Telephone Consultative Committee, more popularly known by the initials CCITT. The CCITT was, together with the CCIR (International Radio Consultative Committee) which dealt with standards recommendations relating to radio and spectrum allocation, part of the International Telecommunication Union (ITU) based in Geneva, Switzerland, and whose members are representatives from the governments of countries belonging to the United Nations. The objective of the CCITT was to achieve compatibility, functionally and operationally, between the public telecommunication networks of the different nations through the process of providing recommendations on the standards to be adopted by individual nations. The documents produced by the CCITT are therefore known as **Recommendations** instead of standards.

Until 1991, the process of updating the Recommendations by the CCITT operated in four-year cycles during which Study Groups within the CCITT were active on a programme of work decided upon during a Plenary Assembly which marked the end of one four-year period and the beginning of another. At the Plenary Assembly, Recommendations made by the Study Groups as the result of work during the previous four years were approved and published. The

Recommendations were then adopted either completely or in some modified form by the national standards organizations of individual member nations.

In 1993, the ITU was restructured to meet the requirements of a changing telecommunications industry. Under this restructuring, the activities of the ITU are divided into three sectors: development, standardization and radio communication. The standardization activities of both the CCITT and CCIR are now consolidated in a single **Telecommunication Standardization Sector**, known as the ITU-T. The CCITT ceased to exist on 28th February 1993, and the new ITU-T came into existence on 1st March 1993.

All major groups of ITU-T Recommendations are designated an arbitrary but unique letter as a name, those belonging to the ISDN Recommendations being known as the I-Series Recommendations. Up to 1988, the year of publication of a CCITT Recommendation could be determined from the colour in which it was bound, so that the ISDN Recommendations from 1984 are often referred to as the Red Book I-Series Recommendations, or those from 1988 as the Blue Book I-Series. 1992 was to have seen the publication of the White Books. However, due to the growing delay experienced in publishing an increasing volume of revised and new standards after each Plenary Assembly, the ITU-T have now adopted a continuous procedure of approval and updating their Recommendations. From now on we shall refer to both CCITT documents, that is those documents that were generated prior to the 1993 reorganization, and the newer ITU-T documents simply as ITU-T documents.

Much of the work on ISDN within ITU-T is the responsibility of Study Group XVIII. In addition, Study Group XI deals with aspects of switching and signalling within the ISDN, Study Group VII deals with the data communications aspects of ISDN as part of its overall study area of data communication networks, while Study Group II covers the international ISDN and ISDN operations.

2.2 THE EARLY DAYS OF ISDN

The concept of an ISDN has been actively researched since the early 1970s, and the first set of ITU-T ISDN Recommendations was published in 1984, although reference to some of the general concepts that are now embodied in ISDN first appeared in earlier ITU-T Recommendations. The 1984 Recommendations were by no means complete, particularly those parts defining the higher level functions of the ISDN signalling protocol control, and contained many inconsistencies, but were sufficiently well defined at the lower levels to allow semiconductor manufacturers to start designing integrated circuits to perform these functions, thereby making it economically feasible for equipment manufacturers to embark on the design and manufacture of the first generation of ISDN equipment.

However, four years until the publication of the following, more complete set of Recommendations, was considered by many nations as too long a delay in their existing plans for the deployment of ISDN in their public networks, and so they embarked on the definition of their own national ISDN standards, for example the VNx series in France and 1TR6 in Germany. Although these national standards, in general, adopted ITU-T Recommendations where possible, those parts which were not complete had no guarantee of being compatible with those of other nations, thereby impeding the progress towards an international ISDN service. However, the individual national programmes for the deployment of ISDN could now be accelerated, with the result that numerous national ISDN field trials took place during the period from 1983 to 1988 with many countries following shortly afterwards, with pilot ISDN services offered to business and commercial organizations where a demand for such services was perceived to exist.

In 1988, the ITU-T published the Blue Book Recommendations which contained a more complete standardization of ISDN protocols, but still lacked important detail in the definition and implementation of ISDN services which have since been published. However, the basic core functions deemed necessary for basic ISDN services were in place, and the operators and equipment manufacturers have since embarked on migration paths towards more universal implementations of ISDN.

The role of the ITU, however, is in the area of global standards, which naturally leads to the inclusion of many options to satisfy the requirements of such a broad scope. It has therefore been the role of regional standards organizations to decide which options are relevant and turn the Blue Book Recommendations into implementable standards. The implementation of a single common worldwide standard for ISDN is not feasible due to the fact that ISDN is an upgrade to existing telecommunications networks, each of which have evolved with different influences for many years. Instead, it is likely that several variations will exist based on geographic areas within which there exists a strong economic community, such as that which exists within Europe, within North America, and within South East Asia.

2.3 ISDN STANDARDIZATION IN EUROPE

The standardization of European telecommunications is undertaken by the European Telecommunications Standards Institute (ETSI) which is based in Sophia Antipolis near Nice in southern France. ETSI is a comparatively new organization, having been established in 1988 by CEPT, the European Conference of Post and Telecommunications Administrations which, among its other responsibilities, provides a European representation within the ITU-T.

Many countries had already embarked on the development of early ISDN services using proprietary definitions where standards were lacking. However, these developments were only a short term measure while standardization activities were completed, and in parallel, a unified approach to a pan-European ISDN was coordinated by CEPT. This was achieved through the definition of a Memorandum of Understanding (MoU) which required the provision of a first stage of international ISDN capable of supporting a limited range of services by the introduction of a standardized common channel signalling protocol to be used between exchanges in different countries at the international interface, and the use of standards to establish compatibility of certified terminal equipment in any CEPT country. This initiative was endorsed by 26 public network operators from 20 European countries of the CEPT, and have resulted in the pan-European ISDN known as **Euro-ISDN**. Figure 2.1 shows the deployment of some of the national ISDNs within Europe and the migration toward Euro-ISDN. Euro-ISDN was officially launched in December 1993 with the BRA service capable of being accessed from approximately 50% of all European lines[3] rising to over 90% by the beginning of 1996.

Prior to the establishment of ETSI, CEPT was itself responsible for technical standardization work with the aim of standardizing telecommunications equipment among its member countries. One of the achievements of CEPT during this time was the introduction of a special series of standards called NETs (an acronym of the French translation for European Telecommunications Standard) which were generated by a group within CEPT called the Technical Recommendations Applications Committee, or TRAC. The NET standards are aimed at type approval for telecommunications equipment that can be connected to the public network, such as telephones, fax machines, modems, packet-switching terminals, ISDN terminal equipment and terminal adapters. The relevant NETs are as follows.

> **NET3** - Connection to ISDN Basic Rate Access.
> **NET5** - Connection to ISDN Primary Rate Access.
> **NET7** - ISDN Terminal Adapters.
> **NET33** - Digital Telephony.

The NET specifications are in the process of being superseded by Common Technical Regulations (CTRs) which are based on the newer standards generated by ETSI known as Technical Basis for Regulations (TBRs). TBRs are likely to be self-contained test standards which are passed to TRAC for approval. TRAC then add the necessary regulatory text to the document to create the CTR. The intention behind the generation of both the NETs and CTRs is to allow approval of terminal equipment gained in one European country to be valid for all countries

---

[3]Actual ISDN usage is significantly less than this, being of the order of 1-2% of subscriber access lines across Europe.

within the EEC (European Economic Community). The CTRs will provide the specification with which authorized testing laboratories test manufacturers equipment for conformity to the approved ISDN standards prior to the equipment being marketed and sold as certified for connection to the public telecommunications network. A programme called the Certification Testing Services (CTS) is sponsored by the European Commission to provide certification services to manufacturers. CTR3 will define certification procedures for BRA terminals while CTR4 will define those for PRA. The scope of CTR3 and CTR4 will cover only the basic ISDN service and will not include supplementary services.

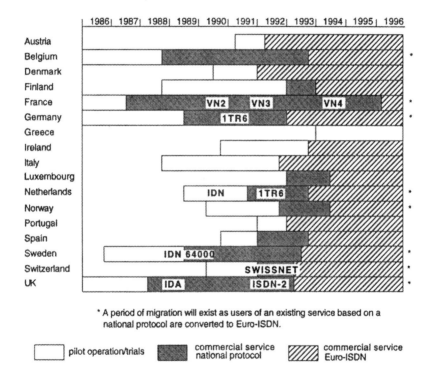

* A period of migration will exist as users of an existing service based on a national protocol are converted to Euro-ISDN.

pilot operation/trials   commercial service national protocol   commercial service Euro-ISDN

Fig. 2.1 The deployment of European BRA ISDNs.

ETSI has now become the focal point of all European telecommunication standardization, including ISDN. Unlike the ITU-T which consisted only of government bodies, and CEPT whose members are national administrations, ETSI consists of the national administrations, public network operating companies, manufacturers, users, and research bodies. These organizations all contribute toward the definition of standards which, as a result of broader participation and cooperation with other standards organizations both within Europe and worldwide, are more representative of the requirements of those concerned and therefore more widely accepted within the European arena.

2.4 ISDN STANDARDIZATION IN NORTH AMERICA

During the years up to 1984, the operation of the US telecommunications network, referred to as the Bell System, was dominated by AT&T and as a consequence many of the standards in operation came about as a result of de-facto AT&T practices. However, because of its monopolistic powers, the Bell System was split up into seven Regional Bell Holding Companies (RBHCs), each of which control a number of subsidiary Regional Bell Operating Companies (RBOCs) which provide communication services over a portion of the territory covered by the RBHC. In addition, an association called the Exchange Carrier Standards Association (ECSA) was formed to provide standards for the operation of the telephone system in North America. ECSA subsequently sponsored the formation of a public standards committee, Committee T-1, for the development of interconnection standards for the national telecommunication system.

Committee T-1 is now one of the principal national standards organizations in North America for ISDN, and compiles candidate standards for submission to the American National Standards Institute (ANSI). The membership of T-1 includes operators, manufacturers and users, and has six technical committees that address standards in particular areas. For example, committee T1S1 deals with signalling systems, protocols, ISDN architecture and services, while T1E1 deals with network interfaces, which includes work on the ISDN physical layer protocol. As well as submitting standards to ANSI, the work programme of Committee T-1 includes liaison with the international bodies such as the ITU-T.

As part of the divestiture of AT&T into the RBHCs, a provision was also made such that the RBOCs could be supported by a centralized organization whose responsibility would be to provide engineering and technology support to the RBOCs. This centralized organization is Bell Communications Research, Inc., known more popularly as Bellcore, and until recently was owned by the seven RBHCs. Bellcore is also active in the national and international standards arena, and at the national level generates Technical References, or TRs, which are used for specification and conformance testing of equipment within the national network.

The main telecommunications regulatory organization within North America is the Federal Communications Commission, or FCC, which was created by the American Congress in 1934 to regulate and promote telecommunications in North America. Since the divestiture of the Bell System in 1984, the telecommunications operators are currently made up of companies which fall into one of two categories; the local exchange carriers (LECs) who deal with the local telephone operations, and the inter-exchange carriers (IECs) who handle the long distance communications between the local exchange areas. Each LEC company, which for example may be one of the RBOCs or an independent telephone company, operates as a monopoly regulated by the FCC within its local area. The IEC market however is open to competition, despite its domination by AT&T, who, as a consequence remain regulated by the FCC, while competing companies can currently operate free from such regulation.

ISDN provides an opportunity to both IEC and LEC companies to offer new and improved services to the customer, but requires their cooperation in order that an end-to-end connection may be made between different regions. In a similar approach to the coordinated development of ISDN in Europe by the network operators, cooperation toward national ISDN in North America was initiated in 1991 under a program called **National ISDN-1**. There are four key aspects to this initiative required for its success. They are:

- agreement by the switch manufacturers to implement common interfaces, both hardware and software, between the user's terminal equipment and the central office switch;
- agreement by the switch manufacturers on the implementation and deployment of a common channel signalling protocol for use between exchange switches;
- agreement by the LECs on an accelerated deployment of ISDN.

The National ISDN-1 (NI-1) programme was launched in November 1992 and provides a basic ISDN service without many of the more advanced ISDN services and features. However, the National NI-1 is only the first step toward a full implementation of ISDN in North America, and is currently being followed by the deployment of National ISDN-2 (NI-2) which, amongst the addition of other features, expands the supplementary service and packet-switched data capabilities of NI-1. Beyond NI-2, further implementation steps are being planned in National ISDN-3 (NI-3).

2.5 NARROWBAND AND BROADBAND ISDN

Two types of ISDN have been identified for use in the public networks; **narrowband ISDN** and **broadband ISDN**. Figure 2.2 shows the relationship between them.

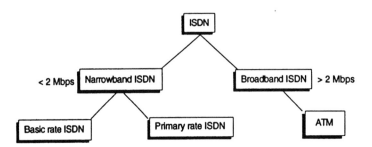

Fig. 2.2 The relationship between different classes of ISDN.

From a historical perspective, narrowband ISDN was conceived first, and consequently its state of development and standardization is comparatively advanced to that of broadband ISDN. Initially, narrowband ISDN was known simply as ISDN, and was developed around the technology of the time to provide widespread availability of digital services to both business and private users. The original precept was that the ISDN should make use of existing subscriber loop cable and the switched 64 kbps digital network. The interface to the user provides two 64 kbps channels for user information, and a 16 kbps signalling channel. This form of ISDN is known as basic rate access (BRA) ISDN.

At that time, the popularity of private networks using connections at 2 Mbps or 1.5 Mbps was growing[4], and so a higher speed ISDN at these same data rates, but using the equivalent of leased line connections to the network instead of the standard telephone subscriber cable, was also introduced. This higher speed ISDN is known as primary rate access (PRA) ISDN which offers the user either 30 or 23 64 kbps channels together with a 64 kbps signalling channel. Both primary rate ISDN and basic rate ISDN come under the category of narrowband ISDN.

However, technology has progressed over the last ten years much faster than the rate at which it could be applied within the public networks. Consequently, it soon became apparent that many services, particularly those which involve video, would require significantly higher data rates than can be provided by narrowband ISDN. To meet this demand for new high data rate services, a broadband ISDN was proposed to provide the subscriber with flexible data rates up to hundreds of megabits per second. The technology required for the switching, transmission and

---

[4]The American digital network hierarchy is based on a multiplexing structure that is different from that used in Europe. Although both use the 64 kbps channel as the basic unit, transmission of multiple channels takes place in the form of a multiplex where the next level in the hierarchy is made up of 24 channels in the American system giving a data rate of 1.5 Mbps, and 32 channels in the European system giving a data rate of 2 Mbps. In the European system, one of the channels is used for framing and synchronization of the multiplex, while an additional bit is used for the same purposes in the American system.

subscriber access to such a network is radically different to that which exists at present, and requires the construction of completely new networks created from fast switching exchanges, and optical fibre transmission and subscriber access technology to accommodate the multi-megabit data rates required. The border-line between narrowband and broadband ISDN is ill-defined, but is generally considered to be that any service providing a user data rate above 2 Mbps falls within the broadband category, while lower data rates are considered to be narrowband. Broadband ISDN is a technology which is expected to emerge early in the next century as widely available telecommunications service. Aspects of broadband ISDN are discussed further in Chapter 10.

This book deals with the narrowband ISDN subscriber loop and its operation and implementation, as it is very much a reality of today and growing in its importance for the provision of digital services to users of the public networks.

2.6 THE ITU-T I-SERIES RECOMMENDATIONS

The ITU-T I-Series Recommendations deal primarily with the standardization of

- the ISDN user–network interfaces,
- services offered across the user–network interface,
- the ISDN inter-network interfaces,
- the network capabilities which allow user-to-network and network-to-network interworking so as to achieve the first two points above.

The structure of the I-Series Recommendations is illustrated in Fig. 2.3. The I.100 Series Recommendations provide essential basic information such as an explanation of the general concepts of an ISDN, an introductory description of ISDNs, and a glossary of terms. The remaining Recommendations of the I-Series are split up into six groups which, making use of the general ISDN concepts, provide specific standards for a particular area of the ISDN. A list of the individual Recommendations is given in Appendix B.

Where a particular topic of a Recommendation falls more appropriately into an area covered by another Study Group within the ITU-T, then it is allocated an I-number whose Recommendation simply refers in whole to the corresponding Recommendation of the relevant Study Group. An example is Recommendation I.463 (see Appendix B) which deals with the support of data terminal equipment with V-Series type interfaces by an ISDN, and which refers the reader to the Recommendation V.110.

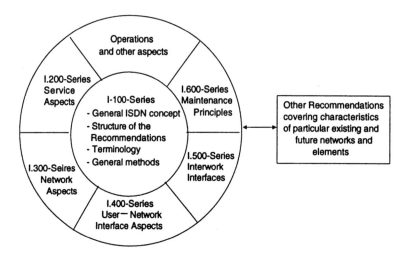

Fig. 2.3 The structure of the I-Series Recommendations.

2.7 THE ISDN SUBSCRIBER LOOP REFERENCE MODEL

The ITU-T Recommendation I.441 defines reference configurations for the ISDN user–network interface which are used to classify, in general terms, the types of equipment used, their interfaces to the cabling which exists within the customer premises, and the possible ways in which the equipment and the interfaces can be interconnected or configured.

An objective of standards such as those produced by the ITU-T, is that they should describe how things work without dictating the way in which they are implemented. To this end, the reference configurations of I.411 are described in terms of **functional groups** and **reference points**. Functional groups are groups of functions which are required to gain access to the ISDN network, and are typically implemented in one or more pieces of equipment. It may be that more than one function may exist in a single piece of equipment, and, that not all functions may be necessary to provide access to the required ISDN service. The points which divide the functional groups are referred to as the reference points, and will typically correspond to a particular physical interface connecting two pieces of equipment together.

Another important objective of the I-Series recommendations is that a wide range of user applications, equipment and configurations, be supported by a set of standardized user–network interfaces which can be universally implemented.

2.8 REFERENCE CONFIGURATIONS AND REFERENCE POINTS

Figure 2.4 shows an example reference configuration. The boxes, marked TE1, TE2, TA, NT2, NT1 and LT are each a functional group which represents different categories of equipment. The TE1, TE2 and TA are all types of terminal equipment (TE), while the NT1 and NT2 are network terminations (NT) that mark the boundary between the public network and the customer premises. The LT is the line termination that terminates the subscriber connection at the local exchange. The reference points between the functional groups are designated the letters T, S and U. In practice, these are all interfaces that allow the different pieces of equipment in the subscriber loop to be interconnected. The S interface is also known as the **user–network** interface as it represents the common interface at which TEs can be connected to the ISDN.

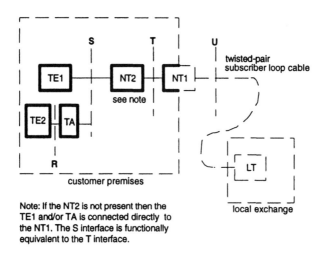

Note: If the NT2 is not present then the TE1 and/or TA is connected directly to the NT1. The S interface is functionally equivalent to the T interface.

Fig. 2.4 The ISDN subscriber loop reference model.

2.8.1 The TE1

The TE1 is a terminal equipment of type 1. Examples of this type of TE are telephones, fax machines and videophones which can be directly connected to the S interface and have the ability to establish and terminate a call. During the call, the corespondents will make use of one or more of the available ISDN services which may involve the communication of voice, data, or video information. The functions of a TE1 include interface and protocol processing for call setup and termination, maintenance functions, processing of user information, and connection functions to other equipment.

2.8.2 The TE2

A terminal equipment which does not have an ISDN user–network interface, but instead some other interface over which information may be carried, is designated a TE2, or terminal equipment of type 2. There already exists a large amount of computer and communications terminal equipment designed for connection to other types of networks, and the ISDN Recommendations make provision for an interface to such equipment to allow them to be connected to the ISDN, thereby obviating the need to replace them.

A TE2 cannot directly interface to the ISDN because by definition it contains no compatible interface. Instead, a terminal adapter (TA) is used to adapt the TE2 such that together they have the same characteristics as the TE1 functional group, as indicated in Fig. 2.4, and can therefore connect directly to the ISDN user–network interface at the S reference point. The interface between the TE2 and the TA is designated the R reference point.

2.8.3 The terminal adapter (TA)

A terminal adapter (TA) adapts a non-ISDN terminal to the ISDN. It will contain functions which include layer 1 and higher layers of the OSI reference model. This means that as well as providing a physical layer interface between the ISDN and the terminal, the TA will also perform call processing functions in order to establish and terminate calls, and possibly also perform processing of user information as it is adapted for communication across the ISDN.

Two other important functions of the TA are **rate adaptation** and **multiplexing**. The interfaces of TE2 type equipment that transmit and receive data at rates lower than the 64 kbps of an ISDN bearer channel must be adapted to the bit rate of the ISDN through insertion of additional bits into the user bit stream. However, the addition and removal of inserted bits is a waste of available bandwidth, which under certain circumstances may be avoided by multiplexing several user bit streams into a single ISDN bearer channel.

Stand-alone TAs are typically used to adapt data equipment having a standard synchronous and asynchronous R interfaces, such as those defined in the ITU-T V and X-Series Recommendations, to the ISDN. In this role, the TA can be thought of as analogous to an analogue fax or data modem, but capable of comparatively higher data transfer rates with fewer errors due to its end-to-end digital capability. In the case of a system such as a PC or workstation, the TA is the plug-in card, and the R interface is effectively the machine's system bus.

2.8.4 The NT1

On the network side of the S interface is the network termination that marks the point at which the public network ends and the customer premises begins. The network termination of type 1 (NT1) provides a conversion at the physical layer between the S interface cable that runs inside a customer premises and the subscriber loop cable at the U interface that connects to the local exchange. Some typical functions of the NT1 are the termination of the physical interfaces, maintenance functions, and power feeding to the terminal either from local mains supply or from the network as a backup when mains power fails.

2.8.5 The NT2

Unlike the NT1 that provides only a physical translation between S and U interfaces, an NT2 may also incorporate more complex functions such as switching and multiplexing. For example, in some customer premises, most notably those of companies and organizations, calls are internally routed by means of a private branch exchange (PBX). The PBX operates like a small exchange switch that provides its users with internal voice services, as well as access to external lines connected to the PSTN or ISDN. An ISDN-based PBX belongs to the NT2 functional group, as would other switching and multiplexing equipment such as local area network (LAN) routers.

Theoretically, the NT2 is connected on the network side to an NT1 which provides it with a connection to the ISDN, but in some cases such as the PBX, the NT1 may be integrated within the NT2. The reference point between the NT1 and NT2 is designated as the T interface and is a subset of the S interface.

2.8.6 The LT

At the far side of the U interface reference point exists the line termination, or LT, which marks the end of the ISDN subscriber loop. The LT is typically located within the local exchange switch equipment as a line card containing the terminations for a number subscriber lines. The primary functions of the local exchange are to provide control over the local subscriber loops during calls, implement local services requested or subscribed to by the users, provide signalling information to other exchanges in order to coordinate the call and services between the local and remote users, and provide interfaces to other public networks such as the packet-switching data network.

The distance between the local exchange and NT may be greater than that supported by the U interface transmission system. Such cases exist most frequently in rural areas, and may be overcome either by use of a repeater, or by placing the LT in a remote unit within range of the customer premises and then connecting the remote unit to the local exchange with a high-speed digital connection. These systems are frequently used in America where they are known as digital loop carriers.

Strictly speaking, the U interface and the LT are not considered to be part of the user–network reference configuration as defined in Recommendation I.411. No user interface to the network is provided beyond the NT1 in the direction of the network, and Fig. 2.4 shows this by indicating with a dotted line those parts which lie outside the I.411 user–network configuration.

2.8.7 The reference configurations

Not all reference groups and reference points must be present in a configuration in order to provide an ISDN user–network interface. For example, Fig. 2.5a shows a valid configuration in which the T interface is not present due to a combination of the NT1 and an NT2 in a single functional entity. This would represent the configuration for a PBX system.

Figure 2.5b shows a valid configuration in which both the S and the T reference point coincide at the same physical interface, and where the NT2 does not exist. This simple configuration would be typical of that used to replace an analogue telephone connection in a private residence where the TE1 is the ISDN telephone and the NT1 is located close to the point where the twisted-pair cable from the local exchange enters the building.

The S interface makes provision for the connection of up to eight TE1s to the same S interface cable as shown in Fig. 2.5c, and can therefore simultaneously support a range of TE equipment such as a telephone, PC and fax machine with a single connection to the network. This configuration is likely to be taken advantage of by small offices that require services beyond basic telephony.

Figures 2.4 and 2.5 indicate the physical location of the various types of equipment and interfaces. A distinction is made between the customer premises, such as a house or a building, and the area that lies outside it which is considered to be in the domain of the public network operator. Equipment and cabling within the customer premises is owned by, and is the responsibility of the user, whereas equipment and cabling that lie outside it are the responsibility of the network operating company. At the boundary between these two areas lies the NT1 which provides the interface between the S/T interface cabling running throughout the customer premises, and the twisted–pair cable brought from the local exchange to the customer premises. Both Europe and America place the customer premises boundary on different sides of the NT1. In Europe the NT1 is considered the property of the network operating company, while in America it is considered to be the property of the user. It is therefore permitted to have a TE which combines the TE1 and NT1 functional groups and has a U interface connection directly to

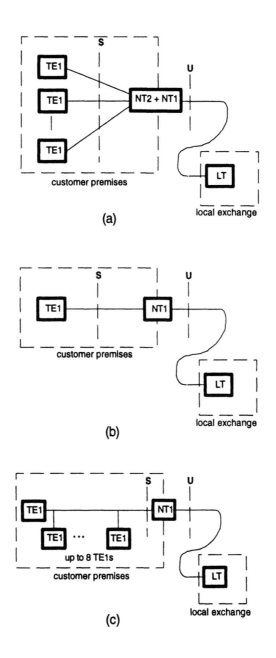

Fig. 2.5 Example reference configurations. (a) Combined NT1 and NT2 as for example in a PBX; (b) a simple residential configuration; (c) a small office configuration supporting multiple Tes.

the public subscriber loop cable. Such a configuration in Europe is so far not allowed, and would be difficult to achieve because of the need to make available various maintenance functions within the NT1 that are independent of the user and can be controlled from the local exchange.

2.9 CHANNELS AND INTERFACE STRUCTURES

Information is transmitted and received between the user's terminal equipment and the network across the R, S/T and U interfaces. The information-carrying capacity of the interface is defined in terms of channels, and a combination of channels at a particular interface is referred to as its interface structure. The interfaces at the S/T and U reference points are synchronous serial digital interfaces, and their associated channels represent specific portions of the overall bandwidth of the interface.

2.9.1 Channel types

The ISDN provides three different channel types designated B, D, and H. These channels are defined in the ITU-T Recommendation I.412.

2.9.1.1 The B-channel

The fundamental unit of bandwidth that is switched within today's digital telecommunication networks is a 64 kbps channel. This is the equivalent bit rate required to transport a single 3 kHz bandwidth PCM encoded voice channel, and is derived by sampling the band-limited analogue voice signal at 8 kHz, and for each sample, assigning an eight-bit PCM code that results in a channel bit rate of 64 kbps. Other higher bandwidth channels are based on multiples of 64 kbps. ISDN extends the use of such digital communications into the subscriber loop, and it is therefore not surprising that we find that the basic channel used in the ISDN subscriber loop to convey user information, known as an ISDN **Bearer Channel**, or **B-channel**, is also a 64 kbps digital channel. Hence, when a single channel connection is established across the ISDN, an end-to-end digital 64 kbps channel exists between the two corresponding TEs.

2.9.1.2 The D-channel

The D-channel is either a 16 kbps or 64 kbps channel depending on the interface structure. Its main purpose is to carry call control signalling information between the user terminal and the network exchange that are used to establish, terminate and supervise calls. For example, whenever a user picks up the receiver of an ISDN telephone and starts to dial, the information such as the setup request and dialled digits are encoded and embedded in call control messages sent in the D-channel to the network exchange.

The call control protocol is packet based, and allows the D-channel to be used by other applications that employ packet based communications. In addition, the generation of messages by the call control occurs only when a call is established, terminated or altered in some way. Consequently, there is adequate residual bandwidth in the D-channel to support additional low-speed packet communications. The ISDN allows the user to make use of the D-channel to access a public packet-switched data network, provided that the local exchange has a suitable connection to it.

2.9.1.3 The H-channel

An H-channel is a higher bandwidth channel for user applications that require high-speed information transfer capabilities, such as videoconferencing or Local Area Network interconnection. A range of data rates are defined for H channels which are organized in a hierarchy based on both the European and North American standard rates. The first two levels in the H channel hierarchy are the H0 and H1 channels. The H0 channel has a data rate of 384 kbps which is six times that of the B-channel. Two data rates are defined for the H1 channel, the first at 1536 kbps and designated H11, and which is aligned to the North American transmission rates, while the second, H12, is defined as 1920 kbps and is aligned to European transmission rates. Higher data rate H2 and H4 channels are also defined for future applications in broadband ISDN. Today, these high bit rate channels are only available in the public network through leased line services or early broadband networks, and are very expensive to use in comparison with the telephony network. In the future, as the demand increases for applications that require this high bandwidth, particularly in the area of video and multimedia, broadband ISDN implemented with ATM (asynchronous transfer mode) networks will provide the public network infrastructure to allow these types bandwidths to be switched between subscribers. More is said concerning these broadband networks and their impact on the subscriber loop in the last chapter.

The B-, D- and H-channels are combined in different ways to create the interface structures that are used at the reference interfaces. Two categories of access interface structure are defined for narrowband ISDN; basic rate access (BRA) and primary rate access (PRA).

2.9.2 The basic rate access interface

The simplest and the most common interface structure is the basic interface structure, which is composed of two independent B-channels, each at a rate of 64 kbps, and a D channel at a rate of 16 kbps giving a combined bit rate of 144 kbps. The 2B+D structure is referred to as a basic rate access (BRA) interface.

The two B-channels can either be used independently, so that the BRA interface can support two simultaneous calls, or together as one single channel of 128 kbps to support services that require higher than 64 kbps data rates, as for example in videotelephony. In both cases, call control signalling messages in the D-channel are used to control the establishment of the call during which the terminal indicates to the network its intended use of the B-channels.

However, combining two B-channels into a single coherent 128 kbps stream that remains coherent on an end-to-end basis across the network, has its difficulties. The problem arises in that most telephony networks were designed to switch only a single 64 kbps channel. A 128 kbps channel is therefore established as two separate calls each of 64 kbps. However, because the calls can be routed independently and therefore take different routes through the network, their end-to-end transit delays across the network would differ. This means that data in the two B-channels that starts off being transmitted together, do not arrive together at their destination, thus causing data incoherence. The normal way to resolve this problem is to have some function within the terminals capable of detecting the difference in transit delays and equalize them through buffering. This process, known as **bonding**, is discussed in more detail in Chapter 7.

2.9.3 The primary rate access interface

The PRA interface structure is composed of either B-channels or H-channels, and again, different structures are used according to the North American and European hierarchies. One purpose of PRA subscriber loop is to provide the user with more B-channels, as for example would be useful for the interconnection of a PBX to the ISDN. Alternatively, the PRA interface can deliver higher bit rate channels for connection to networks capable of switching them. In either case, because of the higher data rates involved, a PRA connection cannot use a standard analogue subscriber cable between the customer premises and the local exchange, but must instead use specially conditioned four-wire cables.

2.9.3.1 B-channel PRA structures

A PRA interface composed of B-channels can have one of two structures depending on the bit rate of the interface:

- 23B+D at 1.544 Mbps (North America and Japan);
- 30B+D at 2.048 Mbps (Europe).

The difference between the combined bit rate of the channels associated with a particular structure and the bit rate of the corresponding PRA interface (for example, 30B+D gives an aggregate bit rate of 1.984 Mbps compared to the bit rate of 2.048 Mbps) is used as timing and framing information for the data stream. In some cases, it may also be used for supervisory channels to provide status monitoring and control of the interface from the network, while remaining transparent to the user.

It is possible that the access to the ISDN involves more than one PRA interface. Only a single D-channel is required at an access point, as the same D-channel can be used to control a large number of calls. If the user–network access contains more that one PRA interface, then the D-channel from one of the interfaces may be used as the D-channel for the access point, while the remaining D-channels can be used as additional 64 kbps B-channels.

2.9.3.2 H-channel PRA interface structures

H-channels have higher bit rates than B-channels and are more suited to specialized high-bandwidth services and applications. The H-channel PRA structures optionally include D-channels, and if not present, can be provided by the D-channel of another interface where these interfaces are combined at the user–network access.

The basic H channel is the H0 channel having a bit rate of 384 kbps. Possible interface structures are as follows:

- 4H0 or 3H0+D at 1.544 Mbps (North America and Japan);
- 5H0+D at 2.048 Mbps (Europe).

In the case of the 3H0+D structure, if the D-channel were provided by another interface, then its structure could be altered to 4H0. However, for a similar situation with the 5H0+D structure, it is not possible to add a further H0 channel within the 2.048 Mbps rate, and the D-channel capacity would remain unused.

The remaining H channels (H11 at a channel rate of 1.536 Mbps and H12 at a channel rate of 1.920 Mbps) interface structures are as follows: 1H11 at the interface rate of 1.544 Mbps with the D-channel provided by another interface within the same access, and 1H12+D at the 2.048 Mbps interface rate where the D-channel may also be optionally carried by another interface within the same access if so desired.

BRA and the B-channel PRA interface structures are the principal types implemented in narrowband ISDN networks today as a B-channel interfaces easily with the 64 kbps channels found in today's public switched networks. The main emphasis in the following chapters of this book is on the BRA ISDN subscriber loop as it will be the most widely implemented interface to the public ISDN.

2.10 ISDN SERVICES

One of the key concepts of ISDN is the provision of services to the subscriber over a single standardized user–network interface. The network is equipped with a set of standardized protocols and functions which enable services to be offered to users connected to the ISDN by means of terminal equipment capable of accessing these services. The I-200 series of Recommendations covers the definition, descriptions, and details concerning the provision of ISDN services.

2.10.1 Bearer services, teleservices, and supplementary services

There are two families of services as viewed by the user connected to the ISDN network; **bearer services**, and **teleservices**. The difference between these two types of service are reflected in the capabilities of the terminal and the network required to implement and access the service. Bearer services and teleservices each comprise a basic service and extended services known as **supplementary** services as shown below in Fig. 2.6.

Telecommunication services			
Bearer service		Teleservice	
Basic bearer service	Basic bearer service + supplementary service	Basic teleservice	Basic teleservice + supplementary services

Fig. 2.6 The classification of telecommunication services.

Network and terminal capabilities can be generally partitioned into two levels which group together the higher and lower layers of the open systems integration (OSI) seven-layer reference model. Referring to Fig. 2.7a which shows the ISDN user–network reference model, bearer services involve the transfer of information between users connected to the points marked 1 or 2 using functions which support the necessary OSI layers one to three, that is the physical, data-link and network layers shown in Fig. 2.7b. The functions supported by these layers are referred to collectively as **low-layer functions** and are used simply to transport the application information across the network. The user may of course choose to provide further levels of protocol processing for the communication in the terminal, but where only a bearer service is specified, the network is not responsible for checking the compatibility of the higher layers between the communicating users.

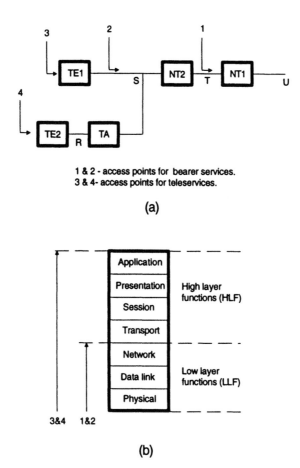

1 & 2 - access points for bearer services.
3 & 4- access points for teleservices.

(a)

(b)

Fig. 2.7 (a) Access points for bearer services and teleservices; (b) high-layer and low-layer functions.

Teleservices involve the transfer of information between users connected to the points 3 or 4 in Fig. 2.7a and use functions in both the terminal and the network which may require support from all levels of the OSI layers. This includes not only the low-layer functions, but also **high-layer functions** performed by protocols in OSI layers 4 to 7 as seen in Fig. 2.7b. In this way, the teleservices may be thought of as an extension to the bearer services, adding information processing features to the more basic bearer service which it uses to transport the information across the network. This additional information processing capability may either be required by both the terminal and the network, as for example in an electronic mail service within the ISDN, or only by the communicating user terminals as in the case of a facsimile transmission between two fax machines.

As can be seen from Fig. 2.7a, access of a teleservice is through functions provided by the terminal as well as the network, and so the service is characterized in terms of bearer capability, which defines the technical features of the bearer service, and the terminal features. Recommendations I.240 and I.241 identify six teleservices to be supported by an ISDN, these being telephony, teletex, telefax 4 (Group 4 facsimile), mixed mode (combined text and facsimile), videotex and telex. It is expected that further service definitions will be added later.

Each of the bearer services or teleservices can be further added to by a number of supplementary services which build additional functionality onto the basic service. The supplementary service cannot exist on its own, but supplements the existing bearer service or teleservice. A simple example is the Call Forwarding supplementary service which may be added to the basic telephony service, and allows calls to an unattended telephone to be automatically forwarded to another number where the call may be answered. The feature would be activated by the user pressing a special button on the terminal followed by the number to which calls are to be forwarded. The local exchange would interpret and register these commands and from then on forward calls to the given number until the feature was deactivated by the user.

ISDN services must be supported by both the terminal and the network in order for the service to be available to the user. During the early phases of introducing ISDN into the national telephone networks, it may be the case that although a terminal is equipped with certain service capabilities, the network is not yet capable of providing such a service.

2.10.2 Service description

An ISDN service is defined in standards documentation according to both its static and dynamic characteristics. The static description of a service refers to those features which remain constant throughout the duration of the call. These are described by a set of **attributes**, each of which can have one of a range of values dependent on the service. A bearer service will be described by a set of low-layer attributes, while a teleservice is described in terms of both low-layer and high-layer attributes. These values are embedded as part of the information elements within call control messages between the terminal and exchange switch. For example, in order to establish a call, a 64 kbps, 8 kHz structured unrestricted digital information bearer service may be requested by the terminal during the setup phase of the call. During this setup phase, the terminal will send a setup message in the D-channel to the ISDN containing values of the relevant attributes of the service requested. '64 kbps', '8 kHz structured' and 'unrestricted digital information' are values given to the main attributes of this bearer service, and must be supported by both the terminal and the network for the service to be operational. If the ISDN and the called terminal are capable of supporting the requested service, the network will respond to the calling terminal with a confirmation message. In such cases that either the network or the called terminal cannot support the service being requested, then the call is rejected.

The dynamic description of a service indicates the flow of information involved between the user and the network from the time the service is invoked to the time it is completed. Only the information transactions across the user–network interface are indicated. The description takes the form of a flow chart supported by a text description of the service. The benefit of the flow chart is that it provides the reader with a visual form of the transactions, as well as the flow of events and states, which is quicker to assimilate and understand.

The use of flow charts to describe the operation of services and protocols is common throughout the ITU-T Recommendations, and is a result of the definition by the ITU-T of a **Specification and Description Language** (SDL) that is used for the specification and description of the behaviour of telecommunication systems. The graphical representation of SDL is in the form of flow charts, although the scope of SDL itself is much wider.

2.11 ISDN PROTOCOL STACKS

With the increasing reliance on computers for signalling and management within today's telecommunications networks, the ITU-T have adopted the International Standards Organization (ISO) open systems integration reference model, known as OSI-RM, in order to define the communications between elements within a telecommunications network. The OSI-RM is a seven-layer model protocol stack which is used to define the operation of a computer communications system in such a way that the generic functions of a layer, and the interfaces between layers, are standardized, with the aim of allowing computers and software from different manufacturers to communicate with one another. Unfortunately, many of the principles on which telecommunications systems are founded were defined long before the OSI-RM came into being, the main consequence of which is that although ISDN standards are modelled on OSI-RM principles, they do not entirely conform to it. A short description of the OSI-RM is given in Appendix A.

The most basic function of any terminal connected to the ISDN is the call control signalling which allows the user to establish a connection to another user or service connected to the network. Call control signalling is implemented with the low layer functions (physical layer, data-link layer and network layer) across the user–network interface, and allows the user to access both bearer services and teleservices. Extensions to the basic call control signalling protocol provide control of supplementary services.

The protocol layers for an ISDN BRA subscriber loop equivalent to the diagram in Fig. 2.5b are shown in Fig. 2.8. The two B-channels, B1 and B2, and the D-channel are shown as individual connections although in reality the physical layer provides a horizontal integration of all channels in the user–network interface. However, this allows the diagram to indicate the concept of a **control plane** (C-plane) and **user plane** (U-plane) which are useful to distinguish between information flows and protocols associated with call control signalling and those associated with processing user information. The protocols associated with call control processing, and which are vertically aligned with the D-channel in Fig. 2.8, are said to belong to the C-plane. Similarly, the vertical protocol layers which process user information in the B-channels are said to belong to the U-plane.

Figure 2.8a shows an example of the protocol layers used by a simple telephony service. In the C-plane, the D-channel is used to transfer call control signalling messages between the terminal and the network exchange. Its operation is defined by the equivalent of the first three layers in the OSI seven-layer reference model, these being the physical layer, the data-link layer and the network layer. Protocols in the U-plane are responsible for processing the speech information which is conveyed in a B-channel to and from the network. In the case of the telephony service, a single processing function is defined in layer 6 which implements the A- or μ-law companding algorithm as part of pulse code modulation. Use of a particular B-channel, B1 or B2, is controlled by the network which will allocate the B-channel during the setup phase of the call, although a calling terminal can state its preference for use of a particular B-channel which is then confirmed by an acknowledgment message from the network.

In contrast to the ISDN telephone which contains few U-plane protocols, an ISDN facsimile terminal is shown in Fig. 2.8b. This contains a complete suite of terminal teleservice protocols for processing the B-channel user information. Two types of transport across the network are possible for this teleservice. The one indicated by the protocols shown makes use of a digital circuit–switched network such as the PSTN, while the alternative method requests a connection through a packet–switched network. The protocols used at layers 2 and 3 in the B-channel of the U-plane are used to provide a packet-oriented low-layer end-to-end communication with the called facsimile machine, while the higher layer protocols of layers 4 to 7 handle the establishment and negotiation of the document transfer procedures using in-band signalling in the B-channel. In addition, these protocols handle the document processing functions associated with the transmission and reception of the facsimile image itself.

Call control signalling is a key requirement of all ISDN terminals regardless of the service requested and how the user application makes use of that service. The

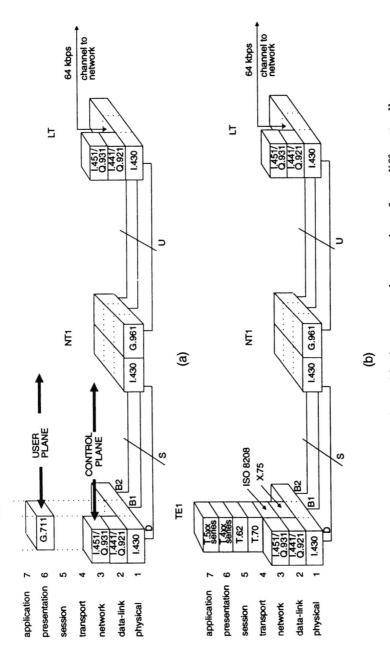
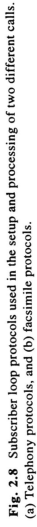

Fig. 2.8 Subscriber loop protocols used in the setup and processing of two different calls. (a) Telephony protocols, and (b) facsimile protocols.

C-plane protocols and physical layer transceivers for the S and U interfaces of the ISDN subscriber loop may therefore be considered as foundation elements of ISDN subscriber loop equipment. These foundation elements are described in the following three chapters.

3

The S/T interface – the ISDN basic rate access physical layer user–network interface

The physical layer of a communications system contains the functions responsible for the transmission and reception of a bit stream across the physical media which connect the communicating entities. The functions associated with the physical layer, such as line coding, timing recovery and electrical interfacing to the line, are typically implemented using integrated circuit technology. In this chapter we describe the functions and operation of the physical layer ISDN BRA S/T interface which forms the user–network interface to the ISDN for customer premises equipment. As well as discussing the requirements of the ITU-T Recommendations, a specific example of how these requirements are implemented in a commercially available semiconductor IC is also discussed.

Figure 3.1 shows a summary of user–network configurations. The NT2, such as a PBX or some other kind of multiplexing equipment, is an optional item, and when not present the S reference point replaces the T reference point. However, functionally, the S and the T reference points are identical and are defined in a single interface standard. ITU-T Recommendation I.430 [1] provides the details of the S/T interface specifications. The corresponding ANSI T1.605-1991 [2] for use in America, and the ETSI ETS 300 012 [3] standard for use in Europe are both equivalent to I.430, but contain minor differences to accommodate regional implementations. These specifications cover the electrical and functional characteristics of the user–network interface which relate to the transmission system used to connect users' terminals to the ISDN NT equipment. We shall refer to this transmission system as the **S transmission system** which consists of the S interface transceivers within the TE and NT equipment at both ends of the cable, the cables and the connectors. The S transmission system may be used in a point-to-point configuration to connect a single terminal to the NT, or in a point-to-multipoint configuration where more than one terminal is connected to the NT.

The S transmission system is a four-wire system that uses a pair of conductors to transmit information in the TE-to-NT direction, and another pair in the NT-to-TE direction. The benefit of the four-wire digital transmission system is that it requires no hybrid function within the transceivers for full duplex transmission over a single pair, and results in less complex circuitry in the transceiver[5]. The disadvantage is that four wires are needed instead of two. However, the adoption of a four-wire system reflects a requirement to keep the design of the S transceiver simple in order to contribute toward a low terminal cost that is acceptable to the market place, despite the expense of more costly cable installations[6].

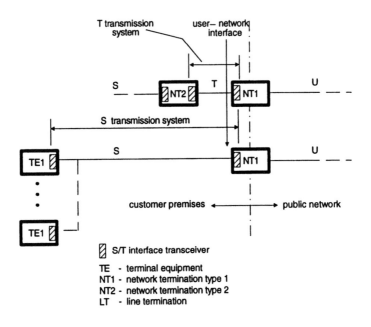

Fig. 3.1 The user–network S interface and the S transmission system.

The S interface specifications cover all aspects of the physical layer, which is layer 1 of the OSI-RM. While the majority of the layer 1 functions may be implemented in hardware, typically as a single integrated-circuit transceiver, the control of the transceiver for activation/deactivation and management such as loopback operation are more usually implemented in firmware running on a

---

[5]One can compare the complexity of an S interface transceiver with that of the U interface transceiver, described in the next chapter, that operates over a single two-wire twisted-pair cable. However, it should be noted that there are less complex proprietary alternatives for full-duplex digital transmission over a single twisted-pair cable, most notable being time-compression multiplexing which is discussed in more detail in Chapter 8.

[6]Four-wire voice-grade unshielded twisted-pair cabling, also known as UTP3, is also extensively used in 10BaseT Ethernet local area networks, and is extensively deployed in existing buildings and enterprises.

microprocessor within the equipment. The following sections up to and including 3.6 describe functions that are typically implemented in hardware, while section 3.7 describes the state machine operation of the layer 1 at which lies the boundary between hardware and software implementation. A specific example of an S interface transceiver is described in section 3.9 and used to illustrate typical hardware and software partitioning schemes within the ISDN BRA layer 1.

3.1 FRAMING

The S transmission system, when fully activated, operates by the continuous transfer of frames in both directions between the NT and the TE. The data rate used is 192 kbps and is made up of the two B-channels at 64 kbps, the D-channel at 16 kbps, and 48 kbps of framing and maintenance channels which are terminated within the layer 1 and are transparent to the user. These channels are time-division multiplexed and framed according to the diagram shown in Fig. 3.2.

Frames in the NT-to-TE direction and the TE-to-NT direction are defined over a 250 μs period and contain two successive instants of the 2B+D information in order to accommodate the necessary framing and maintenance channel bits. The line signal is encoded in order to improve transmission performance and combine clocking information with the data. An alternate mark inversion (AMI) coding is used in which a binary 1 condition at the transmitter output causes it to go into a high impedance state, resulting in no line signal, and alternate binary 0s result in alternative polarity line signals having a nominal level of plus or minus 750 mV. Therefore, although each information bit has a binary value, its line signal can take one of three levels as indicated in Fig. 3.2. A benefit of AMI is that the wiring polarity (which way round each pair is connected) is unimportant as both positive and negative pulses represent a logic 0 value. A positive level binary 0 will be indicated in the text as 0+, while a negative level binary 0 as 0−.

The beginning of a frame is indicated by a violation of the line code described above. The first pair of bits in each frame always consists of a framing bit, F, and its associated DC balance bit, L. The L-bit of the FL pair is always at a 0− level, and the next binary 0 which occurs will also have a 0− level to create the violation. However, if the B- and D-channels are inactive and are set at binary 1 there will be no opportunity in the remainder of the frame to introduce the line code violation, and a mechanism must be built into the frame in order to guarantee the occurrence of a framing violation. This is the purpose of the auxiliary framing bit, F_A, which ensures that a framing violation exists by the time 13 bits of the frame have been transmitted after the opening F-bit.

It is important that the AMI line code achieve a zero DC balance in the line signal (in other words the average DC line level of all bits in the frame is zero) to eliminate charge accumulation on the cable which would otherwise result in a degraded transmission performance. However, the use of line code violations for

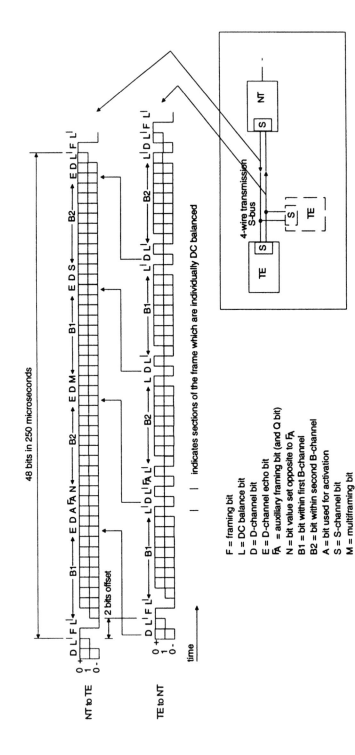

48 bits in 250 microseconds

NT to TE

2 bits offset

TE to NT

time

| | indicates sections of the frame which are individually DC balanced

F = framing bit
L = DC balance bit
D = D-channel bit
E = D-channel echo bit
FA = auxiliary framing bit (and Q bit)
N = bit value set opposite to FA
B1 = bit within first B-channel
B2 = bit within second B-channel
A = bit used for activation
S = S-channel bit
M = multiframing bit

4-wire transmission
S-bus

Fig. 3.2 I.430 frame structure.

framing introduces the possibility of imbalance. To overcome this problem, the frames are split into sections, and each section is individually balanced by the inclusion of a DC balance bit, L, whose level is determined such that the average level within the section is zero. The NT-to-TE frame contains only two such sections, the first consisting of the first two bits, F and L, whose levels are fixed at 0+ and 0– respectively, and the remainder of the frame, which is balanced by the last L-bit of the frame. The TE-to-NT frame contains ten sections, each with its own balance bit, where the additional eight balance bits are placed in the corresponding position of the E-, A-, N-, M- and S-bits which are only used in the NT-to-TE direction.

3.2 POINT-TO-MULTIPOINT D-CHANNEL ACCESS

In a point-to-multipoint configuration, up to eight TEs can be connected to an NT. As there are insufficient B-channels on the BRA interface to allow all terminals to use a B-channel at the same time, they are allocated on a call-by-call basis to the TEs. If both B-channels are already in use when another call arrives at the interface, or another terminal requests to make a call, then the call attempt is rejected by the local exchange.

While B-channels are thus allocated on a call-by-call basis, the D-channel is shared by all TEs connected to the S bus cable. Therefore, although there is a single transmitter in the NT-to-TE direction, there can be several transmitters in the TE-to-NT direction, with the consequence that transmissions in the D-channel from one terminal may collide with those from another. A collision detection mechanism is therefore provided by means of the D-channel echo bit, E, in the NT-to-TE frame, which reflects the binary value of the previously received D-channel bit in the TE-to-NT frame. The D-channel bits sent by a TE to the NT are therefore echoed back to the TE so that it can compare the binary value it sends with that of its echo. If they are the same then no collision has occurred. However, when another TE attempts to access the D-channel at the same time, their transmissions will interfere and cause the echo bit from the NT to differ from the value of the D-channel bit sent by one of the TEs, indicating that a collision has occurred. If a TE thus detects a collision, it immediately stops its transmission in the D-channel.

The occurrence of collisions is significantly reduced by a priority access arrangement which is executed by each TE. Priority is based on the type of message to be sent, and determines the success of each TE that attempts to simultaneously access the D-channel. The priority mechanism is implemented with a counter, C, which counts the number of successive binary 1s which appear in the D-channel echo bits. In the idle condition, the D-channel, and hence the echo bits, contains continuous binary 1 data. When the D-channel is occupied by a terminal, it will transmit messages that are framed using HDLC (high-level data-link control) flags, where the opening and closing flags consist of two binary 0s separated by six binary 1s. Furthermore, to ensure that these flags are unique and cannot occur in the data contained in the message itself, HDLC uses a bit-stuffing process to ensure that no more than five consecutive binary 1s can occur between opening and closing flags. Hence, under normal operating conditions a binary 0 will occur at least once within six consecutive bits between the opening and closing flags of the message. The counter is reset each time a binary 0 appears in the D-channel echo bits and the S transceiver may attempt to transmit into the D-channel only when the counter value has reached a value of between 8 and 11 inclusively, as shown in Table 3.1 below. A TE will not attempt a new transmission for values of C less than 8 because during the transmission of an HDLC frame, C may have a value between 0 and 6, and if C equals 7 it could be due to the transmission of an HDLC abort indicator in the HDLC frame (a 0 followed by seven 1s).

Table 3.1 shows that TEs with signalling messages to send in the D-channel have priority over TEs which use the D-channel to send data packets to a packet-switching network. Each type has either a normal or low priority. Consider what happens when two TEs connected to an NT each have more than one signalling message ready to send over the D-channel. Both TEs have a normal priority and one of the TEs has already gained access to the D-channel and is in the process of transmitting a message. When it has finished transmitting, its priority is changed to low before it starts to monitor the D-channel echo bits and attempt to transmit

Table 3.1 D-channel access priority

Counter value	Type of message	Priority
8	Signalling	Normal
9	Signalling	Low
10	Data	Normal
11	Data	Low

its second message. However, because the second TE remains on a normal priority it is allowed to start its transmission one D-bit before the low priority TE. As the transmissions always start with the first binary 0 of the opening flag, the counter in the low priority TE is reset before it reaches 9. When the low priority TE does eventually gain access to the D-channel, its priority is reset to normal. This automatic toggling of priority between high and low provides a fair access scheme for two terminals both making use of the D-channel for call control at the same time. A similar mechanism also applies to the transmission of data packets.

For TEs with the same priority and which simultaneously compete for access to the D-channel, only one will win and continue to transmit. If each TE transmits a different message, the winning TE is determined by the bit pattern of the message. After transmission of the opening flag, there will come a time when one TE transmits a binary 0 while another transmits a binary 1. The binary 1 line condition is high impedance and so the actual line state is forced to 0 level which is then reflected in the D-channel echo bit. The TE which transmitted the binary 1 will therefore detect a collision, cease its transmission, and again start to monitor the D-channel in preparation to retransmit its message. This process continues for all remaining active TEs until only one is left transmitting.

3.3 MULTIFRAMING

The F_A-bit in the TE-to-NT direction has an alternative use as a low-speed maintenance channel. There is also a similar channel created by the S bits in the NT-to-TE direction. Both the ITU-T I.430 Recommendation and the American standard ANSI T1.605-1991 define how these channels may be optionally used as maintenance channels to indicate, for example, various status conditions such as loss of power at the NT, and to allow TEs to request that the B-channels may be looped back in the NT for testing purposes. The ETSI document [3] does not define how these channels may be used. The channels, known as the S- and Q-channels, can be used on both point-to-point links and point-to-multipoint links.

When in use, the Q-channel operates using a multiframe technique as follows, where a multiframe consists of 20 consecutive normal frames. In the NT-to-TE direction the F_A-bit is set to a binary 1 every fifth frame to indicate the presence of a Q-bit in the F_A position in the TE-to-NT direction. In a point-to-multipoint configuration, all the attached TEs will detect this change in the NT-to-TE F_A-bit and therefore synchronize their use of the Q-bit in the same frame in the return direction to transfer Q-channel information to the NT, thus avoiding the interference that otherwise would be caused if a TE was using the F_A-bit as a Q-bit while another TE used it for framing. Although collisions of Q-channel messages may be caused by the simultaneous transmission of a message by more than one TE, no collision detection or recovery mechanism is defined, but is left for implementation by some other means, typically by a microprocessor within the TE or NT which could verify the validity of the message and perhaps require that the same message be received more than once before being executed as a simple means of collision detection. As the F_A-bit is now set to a binary 1 every fifth frame when multiframing is used, an additional N-bit is used in the NT-to-TE frame that has an opposite binary value to the F_A-bit and is used to create a framing violation if there is an absence of binary 0s before the F_A-bit position. A TE may use the F_A-bit in its transmitted frame as a Q-bit whenever it detects that

Table 3.2 Q- and S-channel bits within the multiframe

Frame	NT to TE F_A-bit pos.	NT to TE M-bit pos.	TE to NT F_A-bit pos.	NT to TE S-bit pos.
1	1	1	Q1	SC11
2	0	0	0	SC21
3	0	0	0	SC31
4	0	0	0	SC41
5	0	0	0	SC51
6	1	0	Q2	SC12
7	0	0	0	SC22
8	0	0	0	SC32
9	0	0	0	SC42
10	0	0	0	SC52
11	1	0	Q3	SC13
12	0	0	0	SC23
13	0	0	0	SC33
14	0	0	0	SC43
15	0	0	0	SC53
16	1	0	Q4	SC14
17	0	0	0	SC24
18	0	0	0	SC34
19	0	0	0	SC44
20	0	0	0	SC54
1	1	1	Q1	SC11
2	0	0	0	SC12

SCxy = sub-channel x, bit y

the binary value of the F_A-N bit pair in the received frame changes from 01 (normal condition) to 10.

Q-channel messages are four bits in length, and therefore require the transmission of 20 frames (one bit every five frames) to be completely transferred between TE and NT. The beginning of the Q-channel message is indicated by the presence of a binary 1 in the M-bit position which indicates the first frame in the 20-frame multiframe.

Whereas the Q-channel transfers information in the TE-to-NT direction, a second channel, the S-channel, transfers information in the NT-to-TE direction. The 20 S-channel bits in the multiframe are split into five sub-channels of four bits each. Messages for only the first two sub-channels are defined, the others being reserved for future definition. Identification of S sub-channel bits is similar to the Q-channel bits within the same 20-frame multiframe, although now one

S-bit occurs in every frame. The structure's of both the Q- and S-channel bits are shown in Table 3.2.

A **persistence** message transfer mechanism is employed in the S- and Q-channels. Each message is repeated at least six consecutive times, or as many times as necessary to achieve the desired effect. For example, a loopback request sent from the terminal to the NT in the Q-channel will be acknowledged by the NT with a corresponding code in the S-channel back to the terminal. A message is not considered to have been properly communicated unless the receiver, which can be either the TE or NT, has received the same message three times in a row.

3.4 WIRING CONFIGURATIONS

The ISDN standards provide examples of the point-to-point and point-to-multipoint wiring configurations that may be used to provide connection of terminal equipment to the NT. Four common configurations are shown in Fig. 3.3, and while they fulfil a large number of interconnection requirements, the standards do not preclude the use of other configurations provided they meet basic cable loss and signal delay criteria for the S transmission system.

3.4.1 S cable

The cable itself will contain between four and eight conductors. One pair of conductors is used to transmit information in the NT-to-TE direction, while another pair is used in the TE-to-NT direction.

The remaining two pairs of conductors may be used to supply power between the terminals and the NT where the power source is generated locally within the customer premises. Alternatively, power may be delivered to the terminals by means of a **phantom** power connection that uses the data transmission pairs of the S cable for power feeding as well as transmission.

In a customer premises building, the S cable will typically be terminated in wall-mounted sockets. Connecting cords consisting of a similar type of cable of up to 10 m in length with plugs at either end connect the terminal to the S cable. Similarly, a connecting cord of up to 3 m in length may be used between the NT1 and the S cable. The cable must be terminated at either end with a 100 Ω resistor located either within the NT or TE equipment, or in the wall socket. In the case of the point-to-multipoint bus configurations, the terminal-end terminating resistor is contained within the wall socket at the farthest end of the bus cable such that the cable is properly terminated, even if there is no terminal connected to this socket.

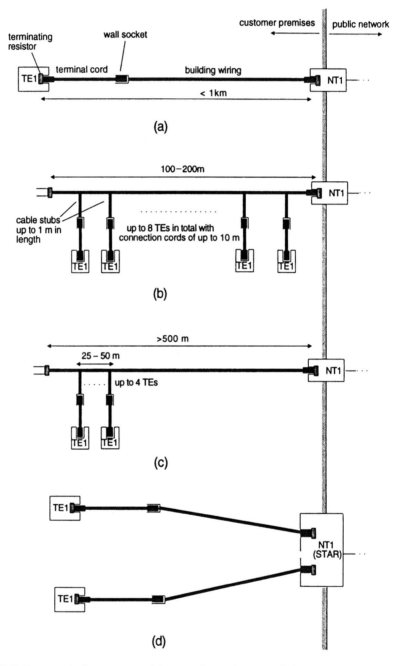

Fig. 3.3 S transmission system wiring configurations. (a) Point-to-point; (b) short passive bus; (c) extended passive bus; (d) NT1 star.

3.4.2 The S connector

Eight-pole plugs and sockets are used with the S cable and the connecting cords. The connectors and their pin assignments conform to the Draft International Standard DIS 8877. The type of connector is more popularly known according to its American designation RJ45 and is shown in Fig. 3.4 together with its pin assignments.

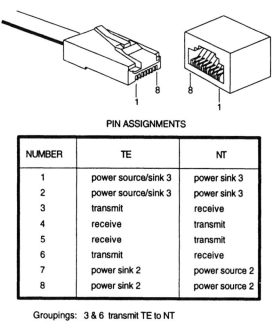

PIN ASSIGNMENTS

NUMBER	TE	NT
1	power source/sink 3	power sink 3
2	power source/sink 3	power sink 3
3	transmit	receive
4	receive	transmit
5	receive	transmit
6	transmit	receive
7	power sink 2	power source 2
8	power sink 2	power source 2

Groupings: 3 & 6 transmit TE to NT
4 & 5 transmit NT to TE
1 & 2 power supply no.3
7 & 8 power supply no.2

Fig. 3.4 The S interface cable, plug and socket.

3.4.3 Point-to-point configuration

The simplest wiring configuration, shown in Fig. 3.3a, provides a point-to-point connection of a single terminal to the NT, while those shown in Fig. 3.3b–d illustrate point-to-multipoint connections where more than one terminal, typically up to eight, may be connected by the same cable to the NT in a **multidrop** bus, or a star configuration.

The length of the point-to-point connection is determined by two parameters: the cable attenuation and the round-trip delay of the signal. The attenuation is determined from the transmitter output level and receiver input sensitivity and is defined by the standards as 6 dB at 96 kHz (half the bitrate) using high-capacitance cable (120 nF/km at 1 kHz). The maximum tolerable round-trip delay is defined by the relative bit positions of the D-channel bits and their echo bits. Although D-channel collisions cannot occur on a point-to-point configuration, echo-bit verification is still performed by the TE. Figure 3.2 showed the timing offset between transmit and receive frames on the S cable, assuming it has zero length. As the length of the cable increases, the position of the TE-to-NT frame is shifted more to the right where the limiting case for this shift is when a D-bit is one bit position before its corresponding E-bit. This represents a worst-case shift of seven bits. If we assume that a one-bit timing margin is required to process the E-bits within the NT, then a safe value to use for the limiting shift would be the time equivalent of six bits.

A final factor which enters the delay calculation is the timing jitter between the input and output of the terminal S transceiver which is caused by the introduction of timing noise in process of extracting timing information from the NT-to-TE frame in order to provide a clock signal within the transceiver with which the TE-to-NT frame may be sent. The permitted value is set by the standards as between −7% and +15% of a bit period. The maximum permitted overall round-trip delay therefore comprises three components: the two-bit offset between the two frames as shown in Fig. 3.2, the six bit delay from the limiting case of correct D-channel echo-bit operation and the worst case of 0.15-bit delay due to timing jitter, resulting in a value of $8.15 \times 5.2\,\mu s = 42\,\mu s$. (One bit period is given by $1/192\,000 = 5.2\,\mu s$.) Similarly, a minimum delay is calculated by considering a zero length cable corresponding to the case where the TE is located directly adjacent to the NT, and consists of the two-bit offset delay less the 7% timing jitter, resulting in a delay of $1.93 \times 5.2\,\mu s = 10\,\mu s$. The permissable round-trip delay of the signal across the S cable is therefore defined as between 10 and 42 μs. Typically, the round-trip delay on a single segment of cable will be between 10 and 26 μs, but may be extended to the maximum of 42 μs when repeaters are used.

The attenuation and delay specifications translate to a nominal maximum length of 1 km of high-capacitance cable for the point-to-point configuration. The timing recovery mechanism used in both the NT and TE in this configuration is made adaptive to be able to cope with the wide range of delays encountered on different cable lengths. The use of a terminal extension cable of up to 25 m in length is permitted in the point-to-point configuration, provided the overall cable meets the above attenuation and delay specifications.

3.4.4 Short passive bus

The short passive bus configuration shown in Fig. 3.3b has up to eight terminals connected at random points along the full length of the cable. The NT must therefore be capable of receiving signals from the TEs arriving with each other but with different delays, depending on the TEs' position on the cable. The limiting factor which determines the cable length is in this case the timing, or round-trip delay, rather than the attenuation.

The round-trip delay for the short passive bus is set by the standards at between 10 and 14 µs (13 µs when the same NT S transceiver may be operated in both point-to-point and point-to-multipoint configurations). The lower limit is calculated as for the point-to-point configuration when the TE is adjacent to the NT, while the upper limit is set by considering the timing conditions of a signal received from a TE located at the end of the cable, and from which a length of between 100 and 200 m in length is specified depending on the impedance characteristics of the cable.

The presence of more than one TE transmitting its signal toward the NT creates a wide variation in the timing of the signal transitions received by the NT, and therefore make it difficult for the S transceiver in the NT to implement an adaptive timing recovery technique. Instead, a fixed timing recovery method is used whereby the time at which the NT samples the received signal is fixed relative to its transmit clock, and therefore no longer dependent on the transitions of the received signal.

3.4.5 Extended passive bus

This configuration represents a combination of the previous two, where up to four TEs can be grouped together at the end of a cable of at least 500 m in length, as shown in Fig. 3.3c. The limiting factor is now the differential round-trip delay between the connected TEs which is set to 2 µs, and from which the maximum distance of 50 m between TE connection points is derived, and is sufficient to allow the NT to use an adaptive timing recovery technique. The overall round-trip delays for the extended passive bus are also specified at between 10 and 42 µs.

3.4.6 NT1 star

An alternative point-to-multipoint configuration may be achieved with only point-to-point wiring in an NT1 star configuration, as shown in Fig. 3.3d. Although each TE in the figure is connected to the NT by a dedicated link, the NT1 still provides only a 2B+D BRA toward the network exchange and must process the D-channel echo bits from each S interface to provide contention resolution as if the

TEs were connected to a single passive bus. Each port of the NT1 star may also support passive bus connections.

3.5 ACTIVATION AND DEACTIVATION

The S transmission system goes through an activation phase before it is ready to transmit the frames shown in Fig. 3.2. The purpose of the activation phase is to ensure that the receivers at one end of the cable are correctly synchronized to the transmitters at the other end of the cable, and is achieved by the exchange of signals, called INFO signals, whose content enables the receivers to extract timing information in order to generate internal receiver clock signals, and to synchronize the reception of frames before information is retrieved from them.

Figure 3.5 shows INFO signals that may be measured by test equipment connected to a short S transmission cable between a single TE and NT in a point-to-point configuration. Five different INFO signals are used. The first, INFO 0 is used to indicate the absence of any active signal from the S transceivers, and is considered present when all transceivers in the S transmission system are fully deactivated. This is shown in the upper trace of Fig. 3.5a in the NT-to-TE direction, and would also be present in the TE-to-NT direction under idle conditions. When the TE needs to establish a call to the network, it must initiate the activation of the S transmission system by transmission of an INFO 1 signal, shown in the lower trace of Fig. 3.5a, to the NT. The signal consists of a 0+0–pair followed by six 1s (shown in AMI line code). In response to the INFO 1 signal, the NT sends the INFO 2 signal shown in the upper trace of Fig. 3.5b back toward the TE which allows the clock recovery mechanism in its S transceiver to lock onto the recovered clock signal, and for its framing logic to synchronize itself with the framing information contained in the signal. The INFO 2 signal conforms to the frame shown in Fig. 3.2 with all B- and D-channel bits set to binary 0. The F_A-, M- and S- bits are also shown as binary 0, although INFO 2 frames are allowed to carry information in the multiframe channels which result in slightly different forms of INFO 2 signal. The E- and L-bits are calculated according to their normal rules. To indicate that the link is not fully activated yet, a bit called the activation bit, A, is also set to a binary 0, and later set to a binary 1 when activation is achieved. Each INFO 2 frame also contains the requisite framing violations, the first, shown in the diagram as V1, created by the last D-channel bit of the previous frame and the F framing bit of the current frame, and the second, V2, caused by the L-bit following the F-bit and the first B1-channel data bit. When the TE has achieved frame synchronization, it responds by transmission of the INFO 3 signal shown in the lower trace of Fig. 3.5b to the NT. The INFO 3 signal also corresponds to the standard framing and may contain data in the B- and D-channels as required, although Fig. 3.5b shows the INFO 3 signal

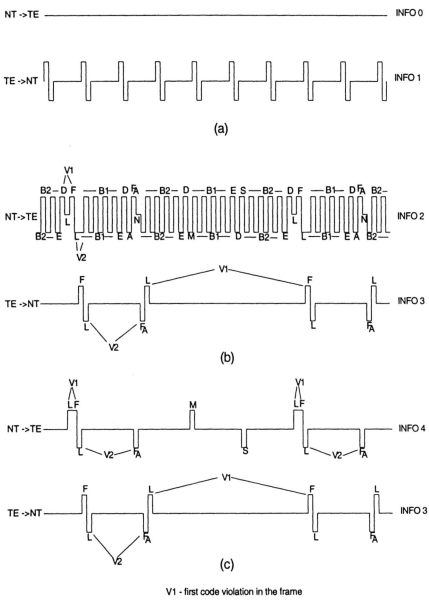

Fig. 3.5 INFO signals during activation of the S interface. (a) TE-to-NT activation initiated with INFO 1; (b) NT-to-TE activation initiated with INFO 2; (c) the fully activated S interface.

with inactive B- and D-channels having their contents set to binary 1. The framing violations V1 and V2 are also shown in the diagram, although the absence of B- and D-channel binary 0s means that they are created using the available F-, F_A- and L-bits. This frame indicates to the NT that the TE has achieved synchronization and the activation sequence is then completed by the NT sending an INFO 4 signal to the TE. The INFO 4 signal, which is shown in the upper trace of Fig. 3.5c, also contains B- and D-channel data, and multiframing channel data which is shown here in an inactive state. The link is now fully activated, with INFO 3 frames in the TE-to-NT direction and INFO 4 frames in the NT-to-TE direction, and the INFO 4 frame A bit set to a binary 1.

The activation sequence in the NT-to-TE direction, when the network wishes to establish a call to the TE, is the same except for the fact that the NT goes from its idle state, in which it sends INFO 0, to sending INFO 2. The INFO 1 signal, which is only sent by a TE, is not used in the NT-to-TE activation. The two activation sequences described above are illustrated further in the first two diagrams of Fig. 3.6, which also indicates the corresponding I.430 defined states of the TE and NT.

Two timers are shown in Fig. 3.6: T3 in the TE and T1 in the NT. T1 and T3 are used to recover from hang-up situations where, for example, one side of the link is caused to wait indefinitely for a change in signal from the other side, which never occurs due to some failure. If the timer reaches its defined limit, then it will cause the corresponding interface to reset to a known state from which a recovery can be performed. T1 and T3 are given values which represent the maximum time allowed for the subscriber loop to activate, which includes the activation of the U transmission system between the NT and LT. The values of both T1 and T3 are dependent on specific network implementations and are typically defined by the network operator. A typical maximum value is around 30 s.

Figure 3.6c shows the deactivation sequence which is always initiated from the exchange side of the S transmission system, even if a call is terminated from the TE and no more active calls remain. Timer T2 is used within the NT to ensure that it is brought to a fully deactivated state before the TE can attempt to reactivate it again, and requires that the TE S transceiver recognize the received INF0 signal and respond with INFO 0 within 25 ms. Finally, Fig. 3.6d illustrates what happens when the TE momentarily loses frame synchronization, which is deemed to occur if, while in the active state, the TE receives three consecutive frames without the correct framing violations when the FA bit is set to a binary 1, or for two consecutive frames when the FA bit is a binary 0.

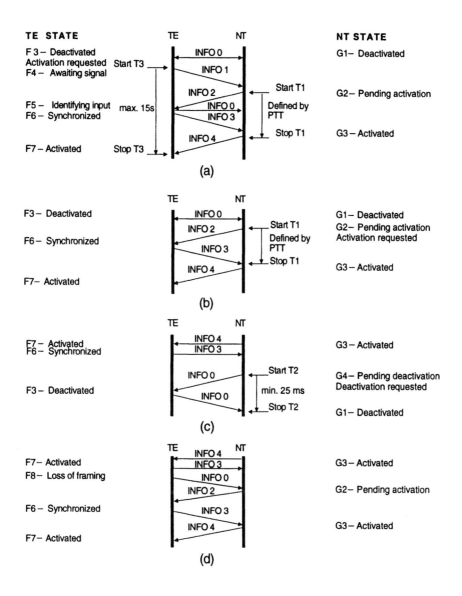

Fig. 3.6 S interface activation and deactivation sequences. (a) Activation from the TE; (b) activation from the NT; (c) deactivation from the NT; (d) loss of synchronization by the TE.

3.6 POWER FEEDING ACROSS THE S INTERFACE

Most analogue telephones connected to the PSTN are powered from the network itself, and many network operators are required to provide this power even under mains supply failure conditions in order to make available a continuous basic telephony service. The importance of this is that emergency services may still be contacted in a situation in which the mains supply has failed. The same provision is also defined for the ISDN subscriber loop, although the increased complexity of an ISDN subscriber loop results in a significantly higher power consumption than its analogue counterpart. In addition, power feeding across the ISDN subscriber loop involves the delivery of power to both NT1 and TE, so a number of power feeding schemes are defined to provide the flexibility of implementation to network operator and the necessary power requirements of the TE. In this section we deal with the options available for power feeding across the S interface.

Power feeding is used across the S interface to provide power from the NT to the TE. The NT itself may receive its power either from the network and/or locally from a mains supply or a battery.

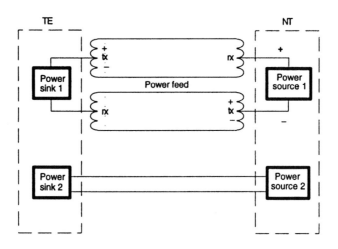

Fig. 3.7 Power feed configurations for the S interface.

Two power feeding configurations are shown in Fig. 3.7. The first involves what is known as a phantom mode power feed which uses the four-wire S transmission system to superimpose a DC supply voltage on the transmitted signals by means of centre-tapped transformers used to couple the S transceiver in the NT and TE(s) to the cable. The power source in this configuration, known as power source 1, is injected on the cable side of the transformer at the NT and

Table 3.3 Power source specifications for the S interface

Condition	Power source 1		Power source 2	
	ITU-T/ETSI* (nominal 40 v)	ANSI* (nominal 48 v)	ITU-T/ETSI* (nominal 40 v)	ANSI* (nominal 48 v)
voltage/power at NT in normal mode	34 to 42 v at up to max. power (at least 1 W)	34 to 56.5 v at up to max. power	42 v max., min. defined by TE requirements. Max. power 8 W (ETSI-7 W)	56.5 v max., min. defined by TE requirements. Min.power 8 W
voltage/power at NT in restricted mode	34 to 42 v at up to 420mW	34 to 56.5 v at up to 420mW	42 v max., min. defined by TE requirements. Min. power 2 W	56.5 v max., min. defined by TE requirements. Min. power 2 W
voltage /power at TE in normal mode	24 to 42 v at up to 1 W	24 to 56.5 v at up to 8 W (non-portable TEs) 1 W (portable TEs)	32 to 42 v at min. power of 7 W	32 to 56.5 v at up to 8 W max. power
voltage/power at TE in restricted mode	32 to 42 v at up to 380 mW for a designated TE plus 20 mW for all other TEs combined.	32 to 56.5 v at up to 380 mW for a designated TE plus 20 mW for all other TEs combined.	32 to 42 v at min. power of 2 W	32 to 56.5 v at min. power of 2 W

* ITU-T specification, I.430 (1993)
ETSI specification, prETS 300 012
ANSI specification, T1.605-1991

recovered from the cable side at the TE where it is used to feed a DC-to-DC converter within the TE to power its circuitry. The advantage of the phantom method is that the power can be supplied using the same cable pairs as the signal. The disadvantage is that the amount of phantom power that can be delivered to a terminal is limited, and requires larger transformers to couple the S transceivers to the cable.

The second configuration is to provide power from a power source in the NT, known as power source 2, to the TEs through the separate pair of conductors within the S cable that are independent of the transmit and receive signal conductors, and therefore capable of delivering higher levels of power to the TEs. Power may be fed to TEs from either power source 1 or 2, or from a combination of both. Provision of power feeding and how it is achieved is dependent on the network operator that governs the ISDN service provision, and may vary between different countries and regions.

A third configuration is also possible where the power is supplied by a TE to power other TEs on the same multidrop bus. However, this configuration is not often used.

Two states exist in which power is supplied to TEs, these being the **normal** state and the **restricted** state. The state determines the maximum power levels supplied to TEs by the power sources. Under normal conditions, power source 1 will deliver up to 1 W of power to an activated terminal on the S bus. The restricted condition may be enforced under emergency conditions when the network or NT is forced to supply power in the event of a mains supply failure. When phantom power feeding is employed from power source 1, this condition is indicated to the TEs by the NT reversing the polarity of the feed voltage causing the attached terminals to enter a restricted mode of operation in which the terminal designated to remain operable can consume only 380 mW of power. In order to limit the power supplied in restricted mode to a minimum so as to prolong active battery life, a single TE is chosen from those terminals in a passive bus configuration as the **designated** terminal which must have at least voice call capability. All other TEs will deactivate and enter a low-power state on the presence of a restricted condition, while the designated terminal will continue to provide a basic voice service. The power capabilities of power sources 1 and 2 in normal and restricted modes are given in Table 3.3.

3.7 LAYER 1 STATES AND PRIMITIVES

Figure 3.6 has already indicated some of the states the physical layer is considered to be in during the activation and deactivation phases. The concept of states and state machines is one that can be generally applied to many forms of communication system and is used extensively to describe the behaviour of the processes which operate within the layers of the OSI-RM.

In general, a layer entity associated with a particular communications link will be in one of a defined number of states, which determines how it should process the information it receives from the layer above or below. In the case of the S interface, which resides in the lowest layer of the OSI-RM, the 'layer below' is the physical medium to which it is attached, and from which it receives signals and information from the S cable. Layers at the same level of the OSI-RM which are associated through a communication process are referred to as **peer** layers. The entity may change its state in response to external stimuli which it receives in the form of software or hardware messages called **primitives**, and additionally in the physical layer due to the INFO signals received from the line. As a result of this change of state, the entity may send out other primitives to other layers, or other INFO signals to the peer S transceiver.

3.7.1 Primitives

Primitives represent the exchange of information and control between a layer entity and adjacent layers and consist of commands and their respective responses associated with the services requested of a lower layer. Primitives are communicated between entities in adjacent layers across a service access point (SAP). These statements apply to primitives communicated between any adjacent layers. The control between peer entities is exercised by four types of primitive: REQUEST, INDICATION, RESPONSE and CONFIRM. Figure 3.8 illustrates the relationship between these primitives.

- **REQUEST.** A REQUEST primitive is used when a layer requests a service from the layer beneath it.
- **INDICATION.** Any activity in the lower layer as a result of providing a service to the upper layer is indicated using an INDICATION primitive, and may be caused as a consequence of the REQUEST primitive at the peer entity.
- **RESPONSE.** The RESPONSE primitive is used by a layer to acknowledge the receipt of an INDICATION primitive from a lower layer.
- **CONFIRM.** Receipt of the CONFIRM primitive by a layer which requested a service indicates that the activity has been completed.

Primitives are named according to a convention which is applied to all primitives. A primitive name has three fields arranged in the following order:

[layer interface]-[service type]-[primitive type]

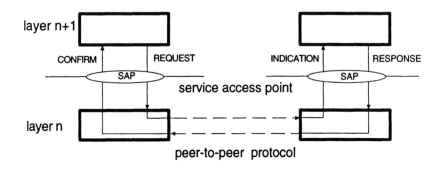

Fig. 3.8 Primitives used by layers to access services in adjacent layers.

The layer interface defines the boundary between two layers across which the communication of the primitive takes place. For example, primitives which are communicated across the interface between the physical layer and data-link layer above have the prefix PH, while primitives communicated across the intra-layer interface between the management entity and the physical layer have the prefix MPH.

The **service type** indicates the service or action which should be applied by an entity, while the **primitive type** is one of the four peer service types described above.

3.7.2 Layer 1 primitives

When describing the behaviour of the S interface, or any other layer within the OSI-RM, it is beneficial to consider a peer-to-peer communication instead of the behaviour of a layer in isolation. It is therefore more convenient to consider the peer-to-peer physical layer behaviour between a TE and NT2 because the NT2 will typically provide equivalent peer layers above the physical layer which interact with it during its operation, whereas an NT1 simply provides a physical layer bridge between the S and U interfaces of the BRA subscriber loop and does not contain any higher layers. The physical layer operation of the NT1 can therefore be considered as a subset of that of the NT2.

The primitives relevant to the physical layer are shown in Fig. 3.9 where the TE is shown connected to an NT2 (PBX). Both the TE and NT2 contain a vertical function called a management entity whose purpose it is to provide services to all layers, and in particular the provision of activation control, TE data-link address establishment and maintenance, and error statistics gathering. More will be said about the management entity later when aspects of the data-link layer are discussed in Chapter 5.

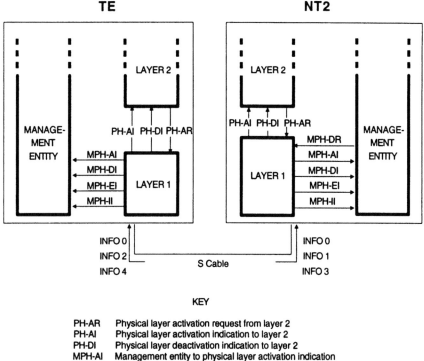

Fig. 3.9 Layer 1 primitives.

In Fig. 3.9 we can see that the layer 1 within the TE accepts the PH-ACTIVATION REQUEST (PH-AR) primitive from the layer 2 that requests activation to commence, which initiates the complete activation sequence shown previously in Fig. 3.6a. Six INDICATION type primitives, two to the layer 2 and four to the management entity, are sent by the layer 1 as a result of the S Interface changing state. For example, the PH-ACTIVATION INDICATION (PH-AI) is sent to the layer 2 whenever the S interface reaches a fully activated state and informs the layer 2 that it can start to send messages across the S interface towards the network. At the same time the management entity is also informed that the layer 1 is in the activated state by the MPH-ACTIVATION INDICATION (MPH-AI) primitive. The PH-DEACTIVATION INDICATION (PH-DI) primitive is used to inform the layer 2 that the physical layer has been deactivated and should suspend its use of the S interface to send information to the NT. The MPH-INFORMATION INDICATION (MPH-II) primitive is used to inform the management entity of the status of the power feed (connected or disconnected) while the MPH-ERROR INDICATION (MPH-EI) primitive informs it of the occurrence and recovery from errors such as loss of frame synchronization. Deactivation of the physical layer under normal operating conditions can only be achieved from the network side of the S interface with a MPH-DEACTIVATION REQUEST (MPH-DR) primitive from the management entity in the NT.

3.7.3 Layer 1 states

The physical layer states which exist for the TE side of the S interface are described in more detail in the following text, which is illustrated by the state transition table in Table 3.4. Based on the current state of the physical layer and the event which occurs while in this state, the table shows the primitives which are issued by the physical layer and the new state which it is then considered to be in. Table 3.4 is valid for TEs which are powered from power source 1 or 2.

F1 **Inactive.** In this state the TE is inactive and not transmitting, which means that if it is connected to the S cable, an INFO 0 signal should be present. Additionally, if the TE is powered from either power source 1 or 2 then this state is entered when the TE detects a loss of power. For TEs which have a local source of power and which cannot detect the presence or absence of power source 1 or 2, then this is the state in which the layer 1 is considered to be in when the local power is removed.

F2 **Sensing.** When the TE is powered on it enters this state when it has not yet determined the type of signal it is receiving. Should its power be removed while still in this state, the TE returns to the F1 state. However, if the NT is active at the time of power-up and the TE detects an INFO 2 or INFO 4 signal, then it immediately enters either state F6 or F7 respectively. Should the NT be inactive such that INFO 0 is present, then the TE transitions to state F3.

F3 **Deactivated.** This is the idle state in which the TE has power applied to it and there is no call based activity across the S Interface and therefore INFO 0 is present in both transmit and receive directions. From this state, the interface may be activated either locally by receipt of a PH-AR primitive from

Table 3.4 TE state transition table for the BRA S interface

	STATE INFO Sent							
	F1 – Inactive	F2 – Sensing	F3 – Deactivated	F4 – Awaiting Signal	F5 – Identifying Input	F6 – Synchronized	F7 – Activated	F8 – Lost Framing
Event	INFO 0	INFO 0	INFO 0	INFO 1	INFO 0	INFO 3	INFO 3	INFO 0
Power ON and detection of power feed	F2	no change	no change	no change	no change	no change	no change	no change
Loss of power	no change	F1	MPH-II(d) F1	MPH-II(d) MPH-DI PH-DI F1	MPH-II(d) MPH-DI PH-DI F1	MPH-II(d) MPH-DI PH-DI F1	MPH-II(d) MPH-DI PH-DI F1	MPH-II(d) MPH-DI PH-DI F1
PH-AR	not allowed	not allowed	Start T3 F4	not allowed	not allowed	no change	not allowed	no change
Expiry T3	not allowed	not allowed	no change	MPH-DI PH-DI F3	MPH-DI PH-DI F3	MPH-DI PH-DI F3	no change	MPH-DI PH-DI F3
Receive INFO 0	not allowed	MPH-II(c) F3	no change	no change	no change	MPH-DI PH-DI F3	MPH-DI PH-DI F3	MPH-DI PH-DI MPH-EI2 F3
Receive any signal	not allowed	no change	no change	F5	no change	not allowed	not allowed	no change
Receive INFO 2	not allowed	MPH-II(c) F6	F6	F6	F6	not allowed	MPH-EI1 F6	MPH-EI2 F6
Receive INFO 4	not allowed	MPH-II(c) PH-AI MPH-AI F7	PH-AI MPH-AI Stop/reset T3 F7	PH-AI MPH-AI Stop/reset T3 F7	PH-AI MPH-AI Stop/reset T3 F7	PH-AI MPH-AI MPH-EI2 Stop/reset T3 F7	no change	PH-AI MPH-AI MPH-EI2 Stop/reset T3 F7
Lost Framing	not allowed	not allowed	not allowed	not allowed	not allowed	MPH-EI1 F8	MPH-EI1 F8	no change

the data-link layer, or remotely by it detecting the presence of an INFO 2 signal.

F4   Awaiting signal. Should the S interface be activated locally from state F3 with PH-AR, the physical layer starts the timer T3 (see Fig. 3.7), enters state F4 and sends an INFO 1 signal, and then waits for a response from the NT. When a signal is received from the NT, the TE enters state F5. The T3 activation timer has a duration of up to 30 seconds, and if it expires before the layer 1 reaches the activated state it causes the interface to deactivate.

F5   Identifying input. Having received a signal from the NT, the TE ceases transmission of INFO 1 and sends INFO 0 while it determines the type of signal it has received. If the received signal is an INFO 2 then the layer 1 progresses to state F6, or if an INFO 4 is detected it proceeds to state F7 and issues a PH-AI and MPH-AI primitives to the data-link layer and management entity respectively.

F6   Synchronized. When entering state F6 the TE sends INFO 3 to indicate to the NT that it has become synchronized to its INFO 2 signal and is ready to be fully activated. Receipt of INFO 4 from the NT causes the physical layer to enter state F7 and send a PH-AI to the data-link layer, as well as MPH-AI and MPH-EI2 primitives to the management entity. The MPH-EI2 is an error primitive and in this case indicates to the management entity that the TE has gained synchronization. Its use is intended for situations where synchronization is lost and the NT reverts back to sending INFO 2 in an attempt to resynchronize the interface.

F7   Activated. In this state the interface is activated and the TE sends INFO 3 toward the NT while the NT sends INFO 4 toward the TE. If the activation timer T3 is still running, it is stopped and reset. D-channel data transfer can now take place across the S interface to further establish a call on one of the available B-channels. Deactivation of the TE is caused by the NT ceasing its INFO 4 transmission and the TE detecting INFO 0, which then enters the deactivated state F3 and sends the PH-DI and MPH-DI primitives. Should frame synchronization be lost while in state F7, the TE ceases its transmission of INFO 3 resulting in an INFO 0 line signal, sends an MPH-EI1 primitive to the management entity informing it that the TE has lost framing synchronization, and enters state F8 awaiting resynchronization or deactivation by the NT.

F8   Lost framing. The fact that the TE has gone from sending INFO 3 to INFO 0 as a result of losing frame synchronization will take a finite amount of time to be recognized by the NT, after which it sends an INFO 2 signal in an attempt to resynchronize the interface. During this time the NT will continue to send INFO 4 and may be sufficient to resychronize the TE which sends PH-AI, MPH-AI and MPH-EI2 primitives and re-enters state F7. However, should the TE detect an INFO 2 signal, it sends only the MPH-EI2 primitive informing the management entity that framing has been recovered and enters the F6 state.

The NT2 side of the S interface is less complicated than the TE and has only four states. The smaller number of states and permissible state transitions allow them to be illustrated using the state transition diagram shown in Fig. 3.10 instead of in a tabular format.

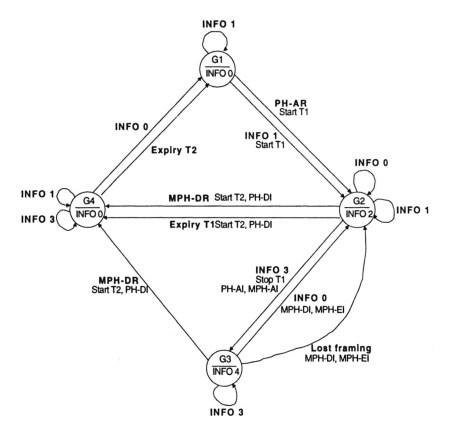

BOLD names are events received.
NORMAL names are actions or primitives sent as a result of the event.

Fig. 3.10 The NT state transition diagram.

G1 Deactivated. This is the idle state in which the NT is not transmitting and INFO 0 is present on the interface. An activation may be requested by the NT by issuing a PH-AR primitive to the physical layer, which then starts an activation timer T1 and enters state G2. Alternatively the interface may be activated by the TE in which case the NT is notified by the reception of an INFO 1 signal (see Fig. 3.6a), starts the T1 timer and moves to state G2.

G2 Pending activation. In this state the NT is sending INFO 2 to the TE to provide it with synchronization information. During the normal course of an activation sequence, the TE responds with an INFO 3 which, when detected by the NT layer 1, causes it to stop the T1 activation timer and enter the active state G3. The activation will be aborted in this state if an MPH-DR primitive is received from the management entity or the activation timer expires. In both cases a deactivation timer, T2, is started, a PH-DI is sent to the data-link layer and the NT layer 1 enters state G4.

G3 Activated. This is the normal activated state in which the NT sends INFO 4 toward the TE while the TE sends INFO 3 toward the NT. Deactivation is initiated at the NT by receipt of a MPH-DR causing it to send INFO 0 and enter state G4. Should the TE deactivate under abnormal operating conditions, for example persistent loss of framing, then the NT will detect an INFO 0 signal causing it to enter state G2 awaiting further re-activation. The NT itself may also detect a loss of framing due to noise corrupting the INFO 3 signal it receives from the TE, in which case it also goes back to state G2 to await re-activation.

G4 Pending deactivation. Once in this state, the NT layer 1 has signalled to the TE its intent to deactivate by ceasing transmission and setting INFO 0 on the line. The normal course of deactivation would be for the TE to respond likewise with INFO 0 at which point the interface is considered fully deactivated. However, it is possible for the NT layer 1 to receive a further PH-AR while in this state, which causes it to start the activation timer and re-enter state G2.

3.8 LOOPBACKS AND MAINTENANCE

A loopback allows the transmit signal path to be connected to the receive signal path at some point in the subscriber loop to allow maintenance activities to be performed. For example, should a fault be suspected somewhere within the subscriber loop, its approximate location may be identified by the exchange by successively enabling a loopback of the 2B+D channels at each equipment along the subscriber loop once it has been activated.

A total of seven loopback modes are defined in I.430, only one of which is identified as being recommended and one other as desirable, and are both shown in Fig. 3.11. All other loopbacks are optional. The ANSI and ETSI specifications also define required loopbacks from the seven ITU-T loopbacks and are also shown in Fig. 3.11 to illustrate their differences.

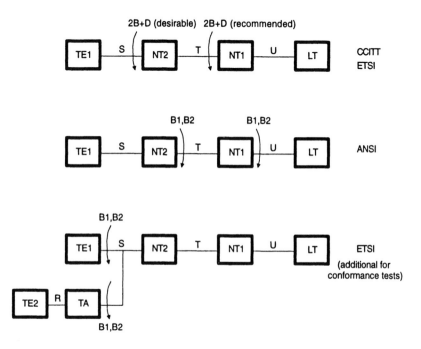

Fig. 3.11 Loopbacks required for ITU-T, ETSI and ANSI BRA S interface specifications.

3.9 S/T INTERFACE TRANSCEIVERS

The establishment of international standards and the emergence of a mass market for ISDN terminals and equipment has led several semiconductor manufacturers to develop chip-sets for ISDN applications, a key part of which are integrated circuit transceivers which implement the S interface functions. As the T interface is a subset of the S interface, an S transceiver can also be applied at the T interface between the NT2 and NT1. These transceivers are designed to conform to the ITU-T I.430 standard and consequently provide the standard functions such as line coding, framing, clock recovery and activation and deactivation control which were discussed previously in this chapter. Devices differ, however, in the way they interface to the other components of the equipment into which they are built. This interface, known as the **system interface**, must provide a data port through which access is gained to the 2B+D channels, and a control port through which the operation and functions within the transceiver can be controlled and monitored.

Table 3.5 gives a list of common S/T transceiver devices, and indicates the type of control and data ports found at their system interface. The data port interface is a serial bit-synchronous time division multiplexed (TDM) bus that consists of a number of 64 kbps time-slots or channels in which the B- and D-channel data is conveyed between the S transceiver and other components in the system. Some of these interfaces were originally defined by equipment manufacturers as a system or board level interface, while others were defined by the semiconductor manufacturers as a means of serial inter-chip communications for bit-synchronous data.

Two approaches are common to the way in which the control ports on such devices are implemented. The control port may exist either as a physically separate interface, or may be integrated as a time-slot channel within the TDM bus of the data port interface. The different configurations and formats adopted by manufacturers for the control and data ports on their devices make interfacing components from different manufacturers difficult. Here, there is clearly a need for some kind of standardization to provide the benefit of interworking between different manufacturers' devices. Such a standard was defined during the latter part of the 1980s by a group of four manufacturers, Alcatel, Italtel, GPT and Siemens, and is known as the general circuit interface (GCI). The GCI is also a serial TDM bus that contains an integral control channel and a set of defined commands, status values, together with a protocol for the transfer of control information in the control channel of the GCI. The GCI is dealt with in more detail in Chapter 6 in which other hardware aspects of ISDN equipment are discussed.

We now examine a specific example of an S interface transceiver and look at the general features of the device which relate to the implementation of the standard features discussed above.

3.9.1 The National Semiconductor TP3420A

Like many of the available S/T interface transceivers, the National Semiconductor TP3420A S/T interface device [4], or SID as it is more commonly known, is designed to operate at both sides of the S and T interfaces within terminal and NT equipment. Figure 3.12 shows a block schematic of the device and how it may be

Table 3.5 A selection of Basic Rate ISDN S Interface chips

Manufacturer	Device	Control port	Data port *serial data time-slot bus*
Advanced Micro Devices	AM79C30A (with HDLC and codec)	microprocessor bus	GCI
AT&T	T7250A (with HDLC)	microprocessor bus	proprietary format
Intel	29C53	microprocessor bus	SLD^{TM}
Level One	LTX200 (with HDLC)	microprocessor bus/integrated with data port	ST-Bus^{TM}
Matra -Harris Semiconductor	29C91	integrated with data port - DSB	DSB
Mietec	MTC2072	integrated with data port - V*	V*
Mitel	MT8930	microprocessor bus/integrated with data port	ST-Bus^{TM}
Motorola	MC145474/5	SCP^{TM}	IDL^{TM}
NEC	uPD98201/2	microprocessor bus	TDM
National Semiconductor	TP3420A	MICROWIRE^{TM}	TDM, IDL, GCI (framing only)
Siemens Semiconductor	PEB2080/1, PEB2085 (with HDLC)	integrated with data port - IOM$^{®}{}_2$	IOM$^{®}{}_2$
SGS Thomson Microelectronics (STM)	ST5421	integrated with data port - GCI	GCI

Note: GCI and IOM are compatible with one another

Key: IOM - ISDN Oriented Modular interface (Siemens)
     GCI - General Circuit Interface
     IDL - Inter-chip Data Link (Northern Telecom)
     SCP - Serial Control Port (Motorola)
     SLD - Subscriber Line Datalink (Intel Corp.)
     DSB - (Matra-Harris Semiconductor)
     V* - (ITT/Alcatel)
     TDM - a generic 64kbps multiplex

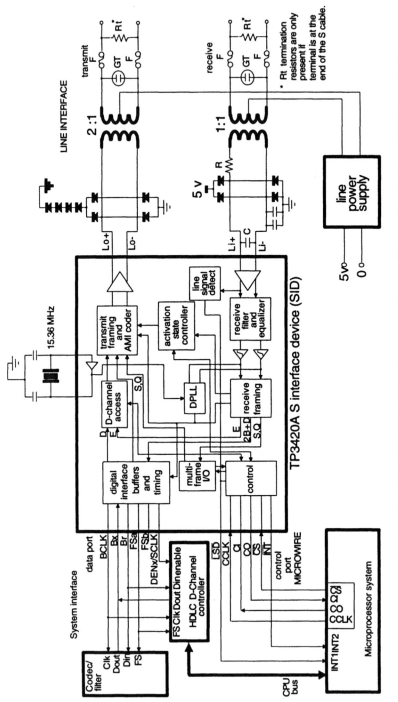

Fig. 3.12 Block schematic of the TP3420A S interface transceiver and its application in an ISDN terminal.

applied in an ISDN telephone where it is connected to an HDLC controller to provide framing for the D-channel signalling messages, and a codec/filter to provide speech data in one of the two available B-channels.

3.9.1.1 Line interface circuit

The line interface circuit couples the line-side input and output of the SID to the S cable through transformers which provide electrical isolation between the cable and the equipment, as well as the ability to create a phantom power circuit to deliver power from the NT to the terminal. In Fig. 3.12 the primary windings of the transformers are connected to the S cable and the secondaries to the SID. The design of the input and output stages of the transceiver and the line interface circuit, including the transformer, must meet the signal and line impedance characteristics as specified in I.430, and are typically provided as part of the manufacturers' application information. In particular, the following characteristics are important:

- pulse amplitude shape and timing must fall within the specified mask limits;
- receiver input impedance and transmitter output impedance with frequency;
- imbalance of the transmission system.

The **imbalance** of the transmission system refers to differences in the impedance between the conducting paths in each line pair which will cause a common-mode signal injected into the system at one point to be translated into a differential signal at another point. For example, a common-mode signal injected at the transmit line of the NT will cause the same voltage to be applied to both conductors in the transmit pair such that no potential difference exists between them. However, differences in the resistance of the conductors and a slight offset in the position of the transformer centre-tap may cause there to be a difference in voltage between the conductors at the TE receiver and subsequently cause problems with the transmission performance of the system. The degree to which the impedances of the transmission paths are matched is referred to as longitudinal balance, and an imbalance may cause both the injection of noise into the transmission system as well as generate DC offsets in the case of remote powering across the S bus.

The resistor marked R and capacitor marked C in the line receive interface act as a filter to reject high-frequency noise from the line, while the diodes provide a low-impedance path to electrical surges that could otherwise damage the device. The diodes in the line transmit interface also prevent clamping of signals from other terminals connected in passive bus configuration. As additional protection, gas discharge tubes (GT) can be used on the line side of the interface to absorb the energy in high-voltage surges, and fuses can be added to isolate the terminal should a possibility exist that wiring faults could cause high voltages to be applied to the S cable. However, the capacitance and resistance of these devices will affect the signal and impedance characteristics of the line interface and must be chosen carefully.

A line signal detect (LSD) circuit is present in the receive signal path in order to detect activity on the S cable, and thus alert a terminal to incoming call activity. In an idle state, with INFO 0 on the line, the SID is typically placed into a powered-down state in order to conserve power. The LSD circuit will detect the presence of a signal and proceed to activate the LSD output of the SID. This can then be used as an interrupt to a microprocessor causing it to command the SID into a powered-up state in which the state machine within the SID can respond to the INFO signals appearing at its input.

3.9.1.2 Line equalization

In the receive path of the transceiver, the received signal passes through a filter which removes noise above 200 kHz, and an equalizer whose purpose it is to compensate for distortions introduced into the signal by its transmission along the S cable. The frequency characteristics of a real cable depart from the ideal of constant amplitude and linear phase (or constant delay) by attenuating and delaying higher frequency components of a signal more than their lower frequency components. For the transmission of a positive pulse down the cable, the result is the dispersion of a received pulse which appears smeared out. This may occur to such an extent that it interferes with the following pulse, and is known as **intersymbol interference** (ISI). The ISI may be sufficient on the transmission of high bit rate pulse streams over long cables as to cause a transmitted logic 1 pulse to be received and sampled as a logic 0 pulse. The equalizer is in effect a filter which attempts to compensate for the characteristics of the cable by adapting its characteristics to be the inverse of those of the cable, and thus boost the higher frequency components of the signal such that the ISI is minimized at the point at which the received pulse is sampled. The equalizer is made adaptive so as to be able to compensate for different cable characteristics and configurations.

In the SID, the adaptive equalizer is implemented as a switched-capacitor filter whose gain and bandwidth can be altered to generate a set of filter characteristics which progressively boost the high-frequency components of the signal. Adaption is achieved with a circuit which measures the rising and falling times of the pulse edges and adjusts the filter characteristics until the difference between the rise and fall times is as small as possible. In addition to the adaptive equalizer, an adaptive threshold circuit is used to continuously adjust the sampling threshold of the receiver to approximately 50% of the average pulse peak amplitude in order to compensate for different cable attenuations.

3.9.1.3 Timing and synchronization

A key feature of the ISDN subscriber loop is that it provides digital connectivity from the user to the network. Inside the network, information is transmitted and received synchronously, based on an 8 kHz master network clock that allows bit-synchronous delivery of data across switched network connections. Extending the synchronous transmission across the ISDN subscriber loop from the network to the user requires synchronization of the data streams at both U and S interfaces to that of the network clock.

Synchronization of signals to the network clock creates a synchronization chain that follows a path around the subscriber loop starting from where signals are transmitted at the local exchange (LE), to the terminal, and then back again to the LE. Achieving synchronization can be thought of as setting up a chain of synchronizers in the receiver of each transceiver in the subscriber loop. Propagation of the 8 kHz master clock down the subscriber loop chain is achieved by successively embedding 8 kHz frame synchronization information within a transmitted frame, and then recovering it within the receiver at the other end of the line. For example, in the S interface at the NT1, the F frame synchronization bits in the S interface frame (see Fig. 3.2) are derived directly from the FS signal at the system interface of the S transceiver, while at the system interface of the transceiver in the TE, the F-bits are used to generate an FS signal, thus passing on the synchronization to other components within the terminal. Within a transceiver itself, the frame synchronization is used to synchronize other clock signals such as the line transmit clock and the system interface clock.

Although the clocks in the subscriber loop are synchronized to the master network clock, the round-trip transmission delay between transceivers, which varies with the length of cable, causes a timing offset between the line transmit and receive frames of the S transceiver closest to the network, in other words at the NT1. As a consequence, the receiver at the NT1 cannot use its transmit clock in order to sample the received signal. Instead, the receiver must recover clock information from the received signal and use this to generate sampling instants. However, the delay now means that there will be a phase difference between the network synchronization and that recovered from the received frame. This is accommodated in the receiver with a buffer, known as an elasticity buffer, which re-synchronizes the 2B+D channel data in the received frame with the frame synchronization input to the NT1 transceiver. This is illustrated in Fig. 3.16.

The TE S transceiver however, relies on the recovery of both frame synchronization and clock signals from the received signal to generate FS and clock signals at the system interface, and frame synchronization and clock for the transmitted frames.

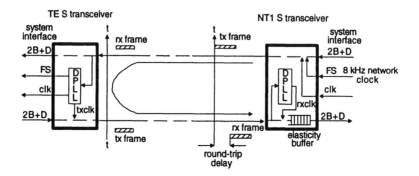

Fig. 3.13 Network synchronization across the S interface.

3.9.1.4 Adaptive timing

The extraction of timing information from the received signal to generate a receive clock is achieved by a digital phase-locked loop (DPLL) whose free-running frequency is determined by the 15.36 MHz crystal oscillator. The received line data rate of 192 kbps is an exact sub-multiple of this frequency. The DPLL locks onto the transitions in the received signal and thus generates a clock signal of the same frequency and phase as that used to generate the transmitted signal at the far end of the S interface. Any changes in the far-end transmit clock, such as frequency drift over time, are tracked by the receiver DPLL to ensure correct sampling of the received signal. This type of timing recovery is referred to as **adaptive** timing and is used in the TE SID for all cable configurations. The line transmit clock and system interface BCLK in a terminal application are also derived from the recovered receive clock.

3.9.1.5 Fixed timing

Features have been integrated into the SID which allow it to function both within the terminal and an NT. With respect to clock recovery, the operation of an NT SID in a passive bus configuration prohibits the use of timing extraction from the received signal because of the different distances between the individual terminals connected to the S bus and the NT. Each terminal transmitter will transmit a signal which is synchronized to the signal they receive from the NT, but because of the different distances between the terminals and the NT, the received signals will be slightly out of phase with each other, as shown in Fig. 3.14. Similarly in the TE to NT direction, the path differences lead to additional phase differences that cause the pulses from different terminal transmitters to overlap. This makes it difficult for the DPLL in the NT to lock onto and maintain synchronization with the received signal from the terminals. As a result of this, a fixed timing method is used within the NT based on sampling the received signal at a fixed time after the transition of the transmit clock.

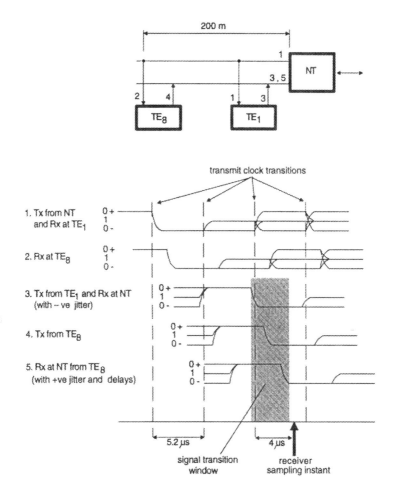

Fig. 3.14 NT fixed timing mode on a short passive bus configuration.

The fixed-time sample instant is derived from the difference in round-trip delay between a terminal adjacent to the NT and a terminal which is the furthest permissible distance from the NT in a passive bus arrangement, which is 200 m. The terminal closest to the NT will experience a total delay of around 10 µs, while the terminal at 200 m will experience an additional 4 μs delay due to the cable length. When referenced to the transmit signal clock which occurs every 5.2 μs, a received signal transition may occur at any time between −0.4 μs and +3.6 μs from the edge of the transmit clock. To ensure that the receive signal is not sampled during this time when a signal transition may occur, the received signal sample time is fixed at 4 μs after the transmit clock.

The fixed-timing method is used in the NT SID for the short passive bus configuration. Additionally, when fixed timing is used the adaptive equalizer is bypassed as it is not required for the short cable lengths found in passive bus configurations. The differential round-trip delay of 2 μs for the extended passive bus configuration is not sufficient to cause problems for the DPLL, which therefore makes use of adaptive timing for the extended passive bus as well as point-to-point configurations.

After the equalizer and filtering stages, the received data stream passes to a block which retrieves information from the various fields within the frame. The B- and D-channel information is sent to the data port, the S- and Q-bits are sent to the multiframe I/O block which is accessed by the control port, and the echo bits, E, are sent to the D-channel access controller.

3.9.1.6 The system interface data port

The data port of the SID is where the B- and D-channel information is transferred between the SID and those devices within the equipment which process the information, for example the codec/filter which transmits and receives PCM speech information into one of the B-channels, and the HDLC controller which transmits and receives signalling messages for call control. The data port consists of a clock (BCLK), serial data input (Bx) and output (Br), two frame synchronization signals (FSa and FSb) and a D-channel transmit enable signal (DENx). The data port can either act as a master in TE applications where the BCLK and FS lines are outputs from the SID, or as a slave in NT applications where BCLK and FS are inputs derived from other devices acting as the data port master.

Figure 3.15 shows the timing of one of four data port formats which can be selected by commands issued to the SID through the control port. The different formats allow the SID to be interfaced to a range of products from different manufacturers, including GCI (although without the control functions), and IDL. When in master mode, the FSa and FSb outputs indicate the presence of the B1- and B2-channel eight-bit time-slots and can be used to enable a device to receive or output its data into a B-channel. In Fig. 3.12, FSa is taken directly to the frame synchronization input of the codec/filter such that its data will always be transferred in the B1-channel on the system interface. The SID has the capability of exchanging the B1- and B2-channels between the system and line interfaces such that if the bearer channel selected for the call is B2 and not B1, then the SID may be instructed to internally cross-over the B1- and B2-channels so that on the line side interface the speech data from the codec appears in the B2-channel.

The two D-channel bits are transferred to the Bx line from the HDLC controller by the DENx signal which is enabled according to the D-channel access mechanism described earlier (section 3.2). In the receive direction, the HDLC controller will use the FS and BCLK signals to identify the D-channel time-slot and read the corresponding bits on the Br line into the HDLC receiver. (In some cases, the HDLC controller does not have this facility, and an external logic circuit must be applied to identify the D-channel time-slot and feed the data bits and clock for the duration of the time-slot to the HDLC controller.)

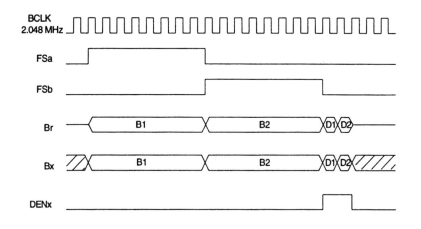

Fig. 3.15 Example of the data timing at the system interface of the TP3420A.

3.9.1.7 The system interface control port

The control port is used to send control information to the SID and retrieve status information from it. The MICROWIRE™ control port is a proprietary interface defined by National Semiconductor for inter-chip control of a range of peripheral devices, and is a simple serial interface which allows a byte of information to be transferred across the interface in both directions between a master and slave device. In Fig. 3.15, the microprocessor system is the master MICROWIRE device and provides the clock (CCLK) and chip-select (CS) signals to the SID which is the slave MICROWIRE device. When the microprocessor has a command byte to send to the SID, it activates the chip-select line to the SID (typically an output port bit of the microprocessor system) and generates eight clock pulses on CCLK. On the falling edge of each pulse, a new data bit appears on the output CO line of each MICROWIRE port which is then read into the corresponding CI input of each interface on the rising edge of the clock. Thus, as the byte is transferred from the microprocessor to the control register of the SID, a byte is also transferred from the status register of the SID to the microprocessor. The microprocessor controls the SID with single byte commands such as activation request and deactivation request, which directly correspond to the control primitives described in section 3.7, while the SID informs the microprocessor whenever a change of state takes place at the line interface with status bytes that correspond to the indication primitives. The SID provides an interrupt line to the microprocessor to prevent the microprocessor from continually polling the SID to determine when it has new information to be read. Additional commands control features such as loopbacks, multiframing data transfer, and control of the SID's power-down mode.

The SID contains an internal activation and deactivation state machine which is initiated by the activation request or deactivation request command, and takes the SID through the relevant states indicated in Fig. 3.8 to achieve activation or deactivation of the line interface. When activation is complete, the SID will interrupt the microprocessor with an activation indication status byte. This information is then used by the microprocessor to generate software MPH-AI and PH-AI primitives shown in Fig. 3.9 and Table 3.4 from the layer 1 driver to the management entity and layer 2 software. Deactivation of the line interface will cause the interrupt service routine to read a deactivation indication status from the SID which similarly causes software to generate MPH-DI and PH-DI primitives. This state machine significantly simplifies the driver software which interfaces the SID to the higher layer software protocols. When the SID is deactivated, it may be placed in a power-down mode for minimum power consumption.

The MICROWIRE port also provides the microprocessor with access to multiframe registers within the SID that enable communication of maintenance information in the Q-bits in the TE-to-NT direction, and the S-bits in the NT-to-TE direction. As described in section 3.3, no error checking is defined in the standards for multiframe data transfer. The SID therefore implements an optional three-times checking facility which only informs the microprocessor of receipt of multiframing information if the SID has received three consecutive multiframes with the same information, thus reducing the probability that it has been corrupted due to passive bus collisions. Since only the ANSI standard has defined the meaning of multiframing information while other standards have left this for future definition, the SID does not automatically execute the actions defined by ANSI for the multiframing information it receives. Instead, the specific actions to be taken, such as enabling loopbacks, are read from the SID and interpreted by software on the microprocessor which then issues the relevant command to the SID. This provides the system with the flexibility to be easily adapted for the different standards.

REFERENCES

1 ITU-T (1993) Basic User–Network Interface – Layer 1 Specification. Recommendation I.430.

2 ANSI (1991) Integrated Services Digital Network (ISDN) – Basic Access Interface for S and T Reference Points (Layer 1 Specification). T1.605-1991.

3 ETSI (1991) Integrated Services Digital Network (ISDN); Basic user-network interface Layer 1 specification and test principles. ETS300012.

4 National Semiconductor (1994) TP3420A ISDN S/T interface device, in Telecom Databook, pp. 2.73–2.102.

4

The U interface and the basic rate access digital transmission system

---

The connection between the network termination (NT) at the user premises and the local exchange (LE) of an ISDN subscriber loop is provided by a digital transmission system whose reference point is designated as the U interface, as shown in Fig. 4.1. A key feature of the BRA transmission system at this interface is that it is designed to operate over the same cables installed by the network operator used previously as analogue subscriber loops to connect analogue telephones or terminal equipment to the PSTN. Such re-use obviates the need to install, at considerable cost, new cables for ISDN subscribers. As the existing cable was originally designed and installed to provide a single 3 kHz analogue voice channel between the terminal equipment and the network, it presents significant impairments to the transmission of high-speed digital signals over the long distances served by subscriber loop cables. These impairments require the application of digital signal processing techniques in order to compensate and thus realise an acceptable transmission performance. Unlike the S interface which employs a four-wire transmission system, the U interface is a two-wire transmission system for bi-directional transmission across the interface. The signal processing functions are contained in transceivers at either end of the U transmission system and are implemented as VLSI integrated circuits.

Figure 4.1 shows the relationships between the functional groups and reference points for the entire ISDN subscriber loop. Whereas in previous diagrams the local exchange has been represented as a single functional group, it is shown in Fig. 4.1 as consisting of a line termination (LT) and an exchange termination (ET) to maintain equivalence with the configurations used in standards documents. The split into these functional groups allows the representation of other subscriber loop configurations besides the direct connection of the customer premises to the PSTN local exchange. For example, the LT may be a remote concentrator or multiplexer serving several subscriber loops and connected to a switching exchange, ET, with a high-speed digital link. This interface between the ET and LT is designated the V reference point. The material in this chapter deals primarily with aspects of the transmission system between the LT and NT1.

We start this chapter with a discussion of the causes of impairment that affect the transmission of a digital signal along the twisted-pair cable connecting the subscriber premises to the network, and the techniques which are used to overcome them. The general functions of a U interface transceiver are then described, followed by a discussion of the ITU-T, ANSI and ETSI standards, and the main differences between them. This is followed by a description of a commercially available U interface transceiver in order to illustrate how the principles discussed earlier in the chapter are implemented. Finally, the design and operation of an NT1 is discussed.

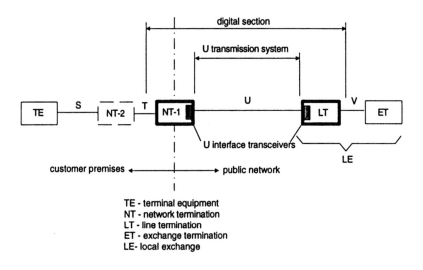

TE - terminal equipment
NT - network termination
LT - line termination
ET - exchange termination
LE- local exchange

Fig. 4.1 The U interface and U transmission system.

4.1 SUBSCRIBER LOOP CABLES

The U transmission system cable is the subscriber loop cable connecting a user premises to the local network exchange. In the distribution of cables from the network exchange, many individual cables are typically bundled together into larger cables, and taken in underground ducts to a junction box located within a convenient distance of the subscribers premises. Individual cables are then taken directly to the subscriber either underground as before or above ground on telegraph poles.

The subscriber loop cable is a twisted-pair copper cable which may either be a single continuous length, or consist of several sections of interconnected cable with an overall length of up to 8 km or more. Each section of cable may be of a different length, gauge and type, with different physical and electrical parameters. In addition, sections of unterminated cable called **bridged-taps** may be bridged across the main subscriber loop at certain points in order to provide access to the loop from another location. These factors combine together to give each subscriber loop cable a complex set of characteristics which will influence the transmission performance of a signal along it. The different cable configurations mean that the characteristic of a particular subscriber loop cable may be significantly different from another. Therefore, any compensating mechanism which forms a part the transceivers at the ends of the U transmission system must be capable of tuning its operational parameters to account for the cable differences.

Although the characteristics of a particular cable do not vary significantly from day to day, ageing will cause them to change very slowly over a much longer period of time. In addition, the need to maintain, repair and perhaps reconfigure the subscriber loop cables, may also cause changes to the subscriber loop transmission characteristics to which the transceivers must also adapt in order to maintain an acceptable transmission and reception across the cable.

4.1.1 Noise sources

The following sources of noise in the subscriber loop are impairments which the U interface transceivers must be designed to handle within the overall performance requirements of the U transmission system. Defined levels of simulated noise are injected onto the cable during transmission performance testing to ensure that transceivers meet this requirement.

4.1.1.1 Impulse noise

Impulse noise is mostly generated by network equipment transients and lightning that results in infrequent bursts of high-amplitude noise which can significantly corrupt the line signal. The characteristics of impulse noise vary widely from location to location, and with time. However, studies have shown [1] that in many cases the energy of impulse noise occurs at low frequencies below 50 kHz. A suitable compensation would therefore be the inclusion of a high-pass filter before the echo canceller in the receive path. Although a high-pass filter has the property of introducing more high-frequency noise into a signal, it is usually justified in this case as the effects of impulse noise can be more severe. An alternative method is the use of error correcting codes implemented in the data stream which, together with hardware at the receiver, allow the detection and correction of erred data.

Impulse noise can be expected to be much worse for subscriber loops that employ pulse-dialling, as used in many European PSTNs, as opposed to DTMF signalling, as used for example in the USA. In addition, the large AC voltages typically used in European networks for ringing can cause significant transients if they are switched to the subscriber loop with a non-zero amplitude.

4.1.1.2 Crosstalk

Crosstalk arises in telephone subscriber loop cables because of the capacitive coupling from one cable pair to another in a cable bundle. For example, a transmitter which drives a signal onto a cable at one end of the transmission system will cause an amount of signal energy to be coupled into adjacent cable pairs. For these adjacent cable pairs, the coupling is greatest close to the transmitter where the signal energy is strongest. This crosstalk signal can therefore be present at the receiver of an adjacent cable pair at the same end of the transmission system as the source of the crosstalk, as well as a receiver at the opposite end of the cable pair.

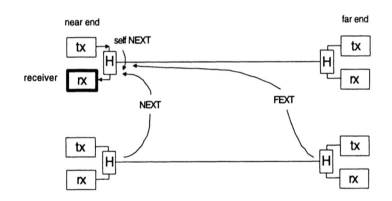

Fig. 4.2 Near-end crosstalk (NEXT) and far-end crosstalk (FEXT).

Two sources of crosstalk signal are illustrated in Fig. 4.2, where we consider the crosstalk noise experienced by the near-end receiver shown in bold. By the same process described above, a crosstalk signal will be present at a far-end receiver from an adjacent far-end transmitter. But at the near-end receiver, this far-end crosstalk (FEXT) noise signal will be attenuated by its transmission down the cable, and therefore does not pose too much of a problem for the near-end receiver. However, the near-end crosstalk (NEXT) signal does not undergo any similar attenuation and is directly superimposed onto the duplex signal on the cable close to the affected transceiver, and therefore appears in the receive path signal within it, with a power level significantly greater than that of the FEXT noise. Consequently, NEXT is the most dominant crosstalk impairment. A typical cable bundle may contain many transmitters, each of which contribute to the overall NEXT received by a particular receiver. Within the cable bundle, NEXT may be caused by the same signals from other U transmission systems, referred to as intra-system NEXT, or from other signals from different transmission systems, referred to as inter-system NEXT. Test arrangements for U transmission systems attempt to synthesize the intra-system NEXT, which is injected onto the subscriber cable close to the receiver during transmission performance tests.

As well as limiting the high-frequency components of the transmitted signal, NEXT can also be reduced by decreasing the signal power. However, a reduction in signal power will also reduce its range, so a compromise has to be met according to the expected levels of crosstalk and range requirements of the transmission system.

NEXT and ISI are the two major impairments which affect digital subscriber loop transmission systems. Both ISI and NEXT depend not only on the cable characteristics but also on that of the line code used by the transceivers. In the case of NEXT, the level of crosstalk at a receiver increases as a power-function of frequency, which indicates that it can be minimized through the design of a transmission system with a low-bandwidth line code.

4.2 DIGITAL SUBSCRIBER LOOP TECHNOLOGY

The ISDN requires the existing analogue subscriber loop of the PSTN to be converted into a digital subscriber loop suitable for BRA connections to the public network. The technology required is significantly higher in complexity than its analogue counterpart due to the higher bandwidth data rates employed by BRA ISDN and the impairments which hinder the transmission and reception of the digital signals across the subscriber loop cable connecting the user's premises to the public network.

4.2.1 The hybrid

Signals transmitted from either end of the U transmission system are combined onto a single cable pair by a **hybrid** circuit. Within each U transceiver, the transmit and receive signals are processed independently, and the hybrid performs the combination and separation of the transmit and receive signals at the interface of the U transceiver to the subscriber cable. The hybrid is known as a two-to-four wire converter because of its ability to combine and separate independent signal paths into a single path. The signal is digital and has an effective full duplex bit rate of 160 kbps made up of the 2B+D channels of the BRA, together with a 16 kbps operations and maintenance channel.

Operation of the hybrid is shown in general terms in Fig. 4.3a. The signal on
the two-wire system consists of the transmit signals from four-wire systems at
both ends superimposed on one another. If we were to observe the signal on the
two-wire cable close to the NT end, then the signal component due to the transmit
signal from the NT end would appear the strongest as the signal from the LT end
is significantly attenuated from its transmission down the entire length of the
cable. As we move away from the NT end and closer to the LT, the LT signal
component gets larger as the NT component diminishes, until at the LT end the
situation is reversed. While the hybrid injects the local four-wire transmit signal
into the two-wire transmission system, it must also recover the far-end transmit
signal from it as the local four-wire receive signal.

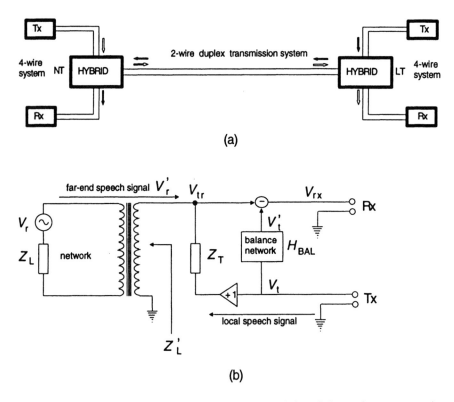

(a)

(b)

Fig. 4.3 Operation of the hybrid. (a) The principle of four-wire to two-wire
conversion; (b) analysis of the hybrid operation.

The local four-wire receive signal is recovered in the hybrid through a process
of cancellation. The principle is illustrated in Fig. 4.3b. If we consider either end
of the transmission system, the four-wire transmit signal, V_t, is used to create a

replica of its two-wire component, V_t', which is then subtracted from the composite two-wire signal V_{tr}. The residual signal, provided the replica is accurate enough, is the signal received from the far-end, V_r' which is equal to V_{rx}.

The replica of the local two-wire transmit signal, V_t', is generated by passing the four-wire transmit signal through a balance network. Exact cancellation of the two-wire transmit component is only achieved if the transfer function of the hybrid balance network, HBAL, is equal to $Z_T/(Z_L' + Z_T)$ where Z_T is a terminating impedance and Z_L' is the complex impedance of the transmission cable viewed from the four-wire side of the isolation transformer. In practice, it is found that the biggest variations in cable impedance, Z_L, across different cables occurs at lower frequencies, and that it is not possible for the HBAL transfer function to exactly match these variations, particularly at these lower frequencies. The filter network which implements the HBAL transfer function can therefore only provide an approximation to $Z_T/(Z_L' + Z_T)$ over the frequency range of interest (up to 0.5 MHz), and consequently an element of the local four-wire transmit signal, V_t, is always present in the received signal. The hybrid therefore provides a finite attenuation to a signal between its four-wire input (*Tx*) and four-wire output (*Rx*), which is referred to as its **trans-hybrid loss**.

4.2.2 Echo and echo cancellation

The residual element of the local transmit signal present in the local receive path is referred to as echo. The hybrid itself provides a first stage of echo cancellation, but, as explained above, due to the mismatch of the hybrid balance filter to the line impedance, it typically provides only as much as 10 to 20 dB, which in itself is not sufficient to meet the overall performance requirements of the U transmission system.

The performance of the transmission system is defined as a certain bit-error-rate (BER) when the transmission system is transmitting and receiving a pseudo-random bit sequence, and subjected to a noise signal that has a very specific level and frequency characteristic which is intended to simulate the electrical noise conditions typically found in the subscriber loop environment (see next section). Under such conditions, a typical BER performance is not more than one received bit in error for every 10^7 bits of data transmitted, more usually expressed as $1/10^7$ or 10^{-7}. To achieve this, the echo signal must be suppressed by as much as 60 dB, including that due to the hybrid. A dedicated echo canceller is therefore required.

4.2.2.1 Overview of the U transceiver

Figure 4.4 shows the functions contained in a typical U interface transceiver. The data to be transmitted is scrambled if necessary so that the resulting bit stream contains sufficient timing information. A suitable line code and pulse shaping is

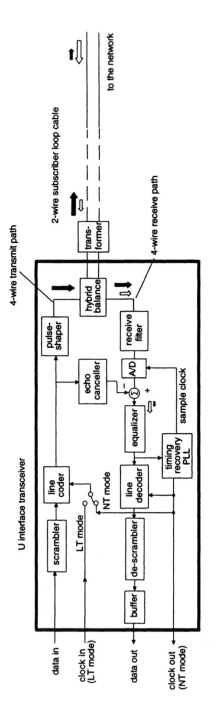

Fig. 4.4 The functions of a U interface transceiver at the NT side of an ISDN subscriber loop.

applied to improve transmission performance and the signal then transmitted onto the subscriber loop through the hybrid. The line coder will convert data that arrives at its input at a specific data rate into symbols that are transmitted at a corresponding baud rate. In the receive path, the received symbol pulses are filtered and sampled before the symbol level is converted into a digital value by the A/D converter. Although the symbols themselves represent digital data, their transmission across the cable will distort the amplitude and timing of the pulses such that suitable compensation must be applied at the far-end transceiver by its echo canceller and equalizer before a decision can be made by the line decoder as to what data value the received symbol represents. Instead of sampling the received signal at a rate that would allow complete reconstruction of the signal through interpolation, most transceivers take only one sample per baud symbol, where the timing recovery aims to generate a sample clock phase that samples the signal in the middle of the baud period. This sub-Nyquist sampling scheme is preferable in that a slower A/D converter may be used, and the resulting echo canceller and equalizer circuits are smaller and less complex. Furthermore, the decisions made at the output of the line decoder are based only on samples of the receive signal made at the baud rate, so information about the signal between these sample instants is of no consequence. Finally, the received data stream at the output of the line decoder is de-scrambled, buffered and sent to the system interface.

4.2.2.2 Operation of the echo canceller

The echo canceller operates on a similar principle to the hybrid, by making a replica of the received echo from the transmit signal and subtracting it from the receive signal. To do this, it must emulate the characteristics of the echo path response with a high degree of accuracy in order to achieve the required level of cancellation. For example, a signal transmitted down the subscriber loop cable can be expected to be attenuated by as much as 36 dB at 40 kHz, so at the two-wire interface to the hybrid the local transmit signal, represented by the large black arrow in Fig. 4.4, is 36 dB in level above that of the receive signal represented by the white arrow. If we assume that the four-wire to two-wire and two-wire to four-wire conversion losses of the hybrid are negligible while the trans-hybrid loss is 10 dB, then in the four-wire receive path just before the receive filter, the echo signal will be 26 dB above the receive signal. If the echo canceller then reduces the echo signal further by 50 dB, the received signal is raised to 24 dB above the echo, sufficient to achieve the required BER with enough margin to absorb other system effects and unwanted noise signals.

To achieve 50 dB or more of echo cancellation, DSP techniques must be used to accurately replicate the echo signal. In Fig. 4.4, the receive path signal is sampled at the baud rate after the receive filter and converted into a digital signal by an analogue-to-digital converter. At the same time, the digital transmit signal is processed by a digital echo canceller filter whose coefficients have been adapted to match as closely as possible the echo path response. The output of the filter is then subtracted from the digital form of the received signal to remove the echo, and the remaining signal is passed to an equalizer which compensates for intersymbol interference.

4.2.2.3 Realization of digital echo cancellers

There are two approaches taken in the design of the echo canceller for BRA ISDN U interface transceivers. The decision on which approach to use depends on the degree of nonlinearity present in the echo path which must be coped with by the echo canceller. With reference to Fig. 4.4, the echo path is that through the pulse shaper, the hybrid which includes a line interface transformer, receive filter and A/D converter. Sources of nonlinear distortion may be pulse asymmetry generated by the transmitter, and nonlinearities in the line interface transformer and A/D converter, each of which may add distortion components to the main signal. If a linear echo canceller is used, any nonlinear distortion introduced into the echo signal through the echo path must be well below the level of cancellation performed by the echo canceller, otherwise it becomes a significant impairment in the received signal after cancellation has occurred.

4.2.2.4 Linear echo cancellers

If the linearity criteria can be met, then a linear echo canceller can be implemented with a form of digital filter called a **transversal** filter. Figure 4.5a shows a representation of the transversal filter. Transmit baud symbols are input to the filter and successively shifted down the taps in the delay line as new symbols enter. For each shift, the value of the symbol at each tap is multiplied by a coefficient and the results summed to provide an output. This calculation is a convolution in the time domain of samples of the transmit signal and the impulse response of the filter represented by its coefficients, and is the equivalent of frequency domain filtering.

It can be noticed from the structure of the filter that only the present and past input samples are used in the calculation of the filter output, with the result that its response is inherently stable as there are no feedback paths from the output back into the filter structure. Consequently, the impulse response of the filter is finite in that an input sample will contribute to the output signal only as long as it remains in the delay line, after which it is discarded. This type of filter therefore belongs to a class of filter referred to as finite impulse response (FIR) filters.

The filter coefficients are adapted to match the echo path impulse response and require a period of training during which only the transceiver is active. This allows the coefficients of the filter to be determined such that the output of the echo canceller matches, and therefore cancels, the echo signal in the receive path.

The most widely used adaptation algorithm is the **least-mean squares** (LMS) algorithm, which iteratively chooses the value of coefficients that converge to a solution which generates the smallest mean-squared error signal present in the receive path after cancellation has occurred. Many practical adaptation algorithms require an initial training period during which a known signal is transmitted in order to initialize the convergence procedure. A more in-depth treatment of transversal filters and their adaptation algorithms can be found in [2].

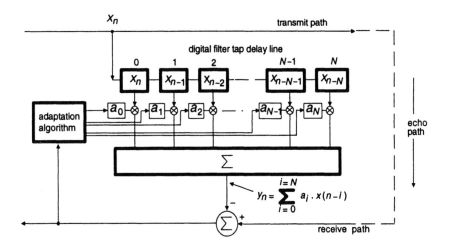

(a)

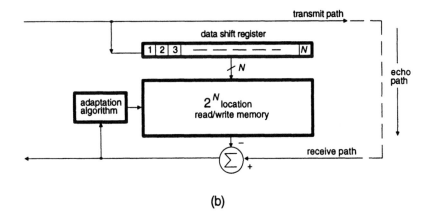

(b)

Fig. 4.5 The structure of digital data echo cancellers. (a) The transversal filter; (b) the memory-based echo canceller.

An important criterion in the design of an echo canceller transversal filter is the number of taps it requires, as this determines the size, and hence its contribution to the cost of the U transceiver. The number of taps is determined by the echo path impulse response, and must be sufficient to cancel the post-cursor[7] tails in the echo response which are generally long due to the low-frequency energy content of the echo signal. Typically, as many as 48 to 50 taps are required depending on the line code used, although the use of echo pre-cancellation in the hybrid and the judicious use of filtering of the received signal may reduce this number to below 40. An alternative solution to reducing the size of filter is to combine a smaller sized FIR filter with an infinite impulse response (IIR) filter which employs a feedback loop that causes the impulse response to be extended beyond the delay line length. With as few as two taps, the IIR filter can be made to have an exponential response to match the decay of the long echo tail. However, care has to be taken in the design of the IIR filter as the amount of feedback combined with certain types of output signal can cause it to become unstable and oscillate.

4.2.2.5 Nonlinear echo cancellers

Use of a nonlinear echo canceller relaxes the linearity requirement of the echo path, but at the expense of the size of the echo canceller as the nonlinear response of the filter is typically stored in a memory. Figure 4.5b shows how this works. The transmit data (before the line coder in Fig. 4.4) is taken to an N-bit shift register and used as an address to a memory which contains the nonlinear echo response for the transmitted bit pattern contained in the shift register. This is a type of look-up table operation where the transmit data is used as the look-up address. Any adaptation which takes place occurs only for the memory location which is being accessed.

The benefit of the memory-based echo canceller is that it is very simple and requires no multipliers as in the case of the transversal filter. However, its disadvantages are that its size grows with the length of shift register used according to 2^N, and similarly the convergence of the adaptation process also slows down as N increases.

4.2.3 Intersymbol interference (ISI) and equalization

Intersymbol interference (ISI) is one of the main contributors to the degradation of a pulse signal as it is transmitted down a cable, and is caused by the frequency dependent attenuation and delay characteristics of the subscriber loop cable. An amplitude-modulated pulse, such as may be used to transmit digital data down a cable, has a number of frequency components with different amplitudes. For a

---

[7]The cursor is the point in time at which the signal is sampled.

pulse to travel down the subscriber loop cable and be received at the far end with the same shape, although at a reduced amplitude level, the different frequency components of the pulse must all be subjected to the same attenuation and delay. Unfortunately this is not the case. A typical subscriber loop cable shows an increase in attenuation and delay with frequency, with the result that the pulse becomes dispersed as it travels down the cable, and thus interferes with adjacent pulses. Due to the cable delay, spreading of the pulse occurs both before and after the central peak of the pulse when viewed at the receiving end, and is referred to as pre-cursor and post-cursor interference, where the cursor is the centre of the baud period. With this interference comes the danger that the shape of the interfered pulses may be sufficiently changed such that at the instant a particular pulse is sampled by the receiver, its value corresponds to an incorrect representation of the data originally transmitted, resulting in receiver data bit errors.

ISI was discussed briefly in Chapter 3 as it is a phenomenon that also requires compensation in the S transmission system. However, the effects of ISI in the U transmission system are in general more severe due to the greater length of the subscriber loop cable and the presence of bridge-taps, and therefore require significantly more complex digital signal processing (DSP) techniques to overcome them. Two types of equalizer which are commonly found in U interface transceivers are adaptive linear equalizers (LEQ) and decision feedback equalizers (DFE). Either or both of these equalizers may be employed to reduce ISI to acceptable levels.

4.2.3.1 Linear equalization

The LEQ is a filter that is designed to have an impulse response which is as close as possible to the inverse of that of the cable, thus cancelling its dispersion effects. An LEQ is used for example in the S interface transceiver to cancel ISI. Immediately following the equalizer is a threshold-level detector circuit which decides what data value corresponds to the output level from the equalizer. The LEQ is made adaptable by being able to change its characteristics according to the difference between the output of the threshold-level detector and its input, as shown in the receiver of Fig. 4.6a. This difference represents the error signal between the equalized signal level and what the level detector thinks that it ideally ought to be, and directs the LEQ to change its characteristics in order to minimize the error. This process works provided that any noise in the system is not sufficient to corrupt the signal at the input to the detector such that it makes a wrong decision. In practice, such errors can be tolerated provided that the correct decision is made most of the time, as the adaptation process will typically involve an integration of the error signal that would require a high error rate in order to significantly de-tune the equalization process.

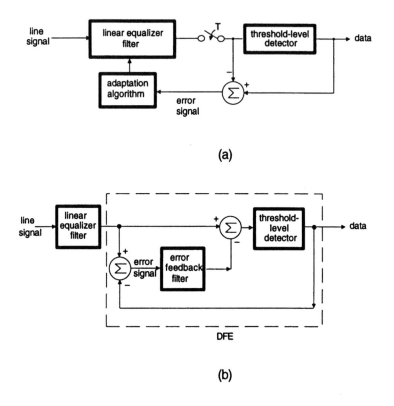

Fig. 4.6 Equalizer structures used in digital transceivers. (a) The adaptive linear equalizer; (b) the decision feedback equalizer.

A further aspect of the adaptation process is that if it is applied to a transmission system where the cable characteristics are initially unknown, then the error rate may at first be too high for the adaptation process to converge on the correct equalizer solution. A period of training is therefore required where a data stream that is known by the receiver is initially transmitted to allow the LEQ to reach a converged state where it is providing accurate equalization before it is used to transmit and receive user information.

LEQs are capable of cancelling both pre- and post-cursor interference but suffer from the fact that they will also attempt to equalize dips in the frequency response of the cable, as for example caused by bridge-taps, and therefore increase their gain in this region which increases the noise introduced into the receiver. For the U transmission system, a receiver is required to detect a heavily attenuated signal in the presence of noise, and it is therefore essential to minimize additional noise introduced by the receiver itself. For this reason, an alternative form of equalizer known as a **decision feedback equalizer** (DFE) is typically employed.

4.2.3.2 Decision feedback equalization

The structure of a DFE is shown in Fig. 4.6b. As can be seen, it is essentially an additional stage following an LEQ that attempts to cancel the noise at the output of the LEQ. Its operation works on the fact that the error signal at the output of the LEQ is correlated, which means that the error signal at a point in time is dependent on its previous values. This allows the application of an error feedback filter that uses the past values of the error signal to predict the current error present in the signal at the output of the forward equalizer. The predicted error is then subtracted from the equalizer output prior to the threshold level detector, and the characteristics of the feedback filter are chosen so as to minimise the noise variance at the input to it.

A DFE is only capable of compensating for noise in post-cursor interference because its adaptation to noise is based on past information. This makes it suitable for systems which use line codes that exhibit predominantly post-cursor ISI as opposed to pre-cursor ISI, and is effective in compensating for the post-cursor ISI introduced by bridge-taps in a subscriber loop.

One drawback, however, is that should a decision error occur at the level detector, this error will propagate back through the error feedback filter and affect future decisions, although this effect usually has a short duration due to the finite response of the filter. To offset the disadvantage of error propagation, the DFE has the advantage that as part of the feedback process, it employs the output of the decision level detector which is free from error or noise, and thus does not give rise to the noise-enhancement properties that are present in LEQs. Furthermore, because information that is fed back consists of the data '1's and '0's, the feedback filter in the DFE requires only a simple multiplier compared to the full multiplier required in the digital implementation of a LEQ using a FIR filter.

Typically, both a DFE and LEQ will be used to compensate for ISI, although due to the comparative hardware complexities of the LE, specifically in the provision of hardware multipliers, the design of the LEQ is usually kept as simple as possible with only a few filter taps.

4.2.4 Timing configuration

Like the S transmission system in the previous chapter, the timing requirements in the U transmission system are different at the NT and LT ends. The ISDN subscriber loop is synchronous to the network clock which drives the transmitter in the LT transceiver. As this signal passes around the U transmission system it experiences a phase shift dependent of the length of the U subscriber loop cable.

At the NT, the receiver recovers timing information from the received signal and generates a clock which it uses to sample the received signal prior to the A/D conversion, drive the echo canceller and DFE in the receive path, and also provide timing for the transmitter which sends information back to the LT. At the LT, the receiver also derives its clock from the received data stream from the NT, but although this receive clock is synchronous with the network clock, it will be phase-shifted from the LT transmit clock according to the round-trip LT–NT–LT transmission delay of the loop. The LT transmit and receive paths therefore run from clocks with different phases, which adds complication to the design of the LT echo canceller compared to that in the NT1 where both transmit and receive paths may be driven by the same clock.

4.2.5 Timing recovery

A receive baud clock is typically generated from the received signal by a digital phase-locked loop (DPLL) and is used to sample the received signal prior to A/D conversion and its subsequent compensation by the echo canceller and equalizer. The DPLL is controlled by a phase error signal that represents the difference between the current sampling phase and the desired sampling phase. The equalizers described above are sensitive to sampling phase because an incorrectly sampled signal will represent an error at the threshold-level detector for which the equalizer will attempt to correct. In addition, the DPLL will have the characteristic of a filter which smoothes any high-frequency random fluctuations in the regenerated clock, known as jitter, thus producing a stable clock which is essential to the accuracy of the echo canceller.

Two timing recovery methods, or derivatives of them, are usually found on U interface transceivers. The first is a method called **channel estimation** which recovers timing information from the samples of the impulse response of the transmission system. This information is partially available from the DFE which attempts to compensate for the channel response, and is used to estimate the sampling phase error that controls the DPLL.

The second method relies on synchronization information present in the received information frame to control the DPLL. Synchronization can be performed on the uncompensated receive signal through the use of a known synchronization pattern that is correlated with the received data stream. The recurrent correlation peaks indicate frame synchronization timing which can be used by the DPLL to generate both frame and symbol timing. The benefits of this approach are that the known occurrence of the synchronization pattern in the received signal can be used to train the equalizer instead of a dedicated training pattern, and that the timing recovery is independent of the DFE and is not affected when the DFE suffers a disturbance.

4.2.6 Line codes

The choice of line code for the U transmission system effectively determines its range for a defined BER performance. A low bandwidth code has the advantage that the majority of the signal energy is subjected to lower attenuation than the high-frequency components, thus giving it good range capability, while the NEXT noise is also kept to acceptable levels. A number of line codes have been defined for use in digital subscriber loop transmission systems, some of which are described below. Each of the codes is illustrated in Fig. 4.7.

4.2.6.1 2B1Q

With this code every two bits of data are encoded as one of four equally spaced line signal levels known as **quats**, and represented as +3, +1, −1, and −3 as shown in Fig. 4.7a. (These values represent relative amplitude levels and not the absolute voltage levels.) In a transmitter, the raw data stream is encoded according to the line code rules to generate the line signal, while a reverse process in the receiver processes the line signal to re-generate the original data. Using the 2B1Q line code, the 160 kbps ISDN BRA data rate (2B+D plus 16 kbps overhead and maintenance bits) is therefore halved to a line baud rate of 80 kbaud. As the number of possible combinations of the two data bits is equal to the number of line levels, the 2B1Q code contains no redundancy. The disadvantage of this code is that for the same maximum pulse amplitude, the reduced spacing between each level results directly in a reduced signal-to-noise ratio (SNR) at the receiver when compared with a two level code. However, for long cables the attenuation at higher frequencies increases more dramatically than any degradation in transmission performance caused by a reduced SNR. This indicates that for the case of the 2B1Q code the benefit of a reduced bandwidth code is of more importance in extending the range of the transmission system than is the contribution from having an inherently high SNR.

In general, a multi-level line code increases the complexity and specification of the DFE and echo canceller as they require multi-level arithmetic to be performed. Also, the 2B1Q code does not in itself guarantee a zero average line level (zero DC component), or a suitable number of line signal level transitions for timing recovery in the receiver. Instead it must rely on scrambling the data to achieve these signal characteristics before it is transmitted, although some DC component will remain. Despite this, the baud rate compression of the 2B1Q code results in sufficiently favourable NEXT and ISI performance to make it the line code of choice for BRA ISDN subscriber loop transmission systems in North America and in many European countries.

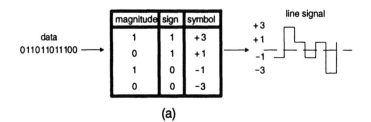

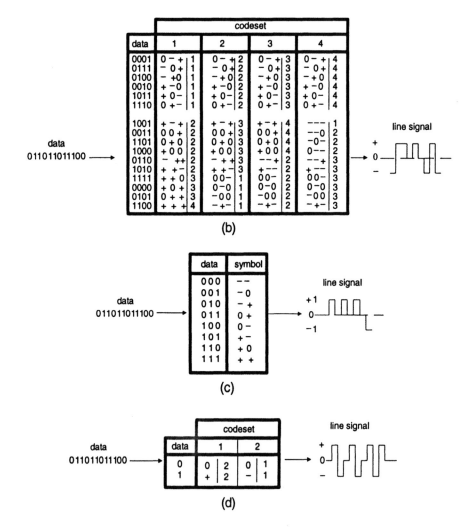

Fig. 4.7 Line codes used for the U transmission system. (a) 2B1Q; (b) 4B3T; (c) 3B2T; (d) AMI.

4.2.6.2 4B3T (MMS43)

This block code converts blocks of four consecutive data bits into one of 27 possible groupings of three baud symbols, where each symbol can have one of three levels, represented as +, 0 and −, as shown in Fig. 4.7b. As there are only 16 possible combinations of input data, the 4B3T code contains redundancy. The 4B3T line baud rate is three-quarters that of the unencoded data rate, or 120 kbaud. Although this is higher than that of 2B1Q, and hence theoretically has a lower range, the three-level 4B3T line code results in a less complex and lower specification DFE, making it simpler to implement than a 2B1Q transceiver. A version of 4B3T code known as MMS43 (modified monitoring state code mapping four bits into three ternary symbols) contains four different sets of codes, as shown in Fig. 4.7b. Switching between code sets is determined by the code set currently in use and the sequence of four bits being coded, the aim being to produce a zero average line level. In Fig. 4.7b, the number of the next code set used is indicated in the table after the current code. MMS43 is used for BRA ISDN subscriber loops in some European countries, primarily in Germany, and in early installations in France which now also use the 2B1Q code.

4.2.6.3 3B2T (SU32)

This code is shown in Fig. 4.7c, and represents a variation of the 4B3T code. As its name suggests, the 3B2T code encodes groups of three data bits as two ternary baud symbols. The line baud rate is therefore two-thirds of the input data rate, or 106 kbaud. As with 2B1Q, the 3B2T code does not have a zero DC component or guarantee the inclusion of timing information. The 3B2T line code was used in early implementations of BRA ISDN subscriber loops in the UK, but has subsequently been replaced with the 2B1Q code.

4.2.6.4 Alternate mark inversion (AMI)

The AMI code represents successive input data '1's as alternate equal amplitude positive and negative pulses, with data '0's being represented as a zero level, as shown in Fig. 4.7d. In this way, the code assures a zero DC component, but requires scrambling of the data input to guarantee sufficient transitions for timing recovery in the coded signal. However, the code provides no baud compression limiting its application to short distance subscriber loops. The AMI code for U transmission systems is used primarily in Japan, but in transceivers that employ a form of duplex transmission called time compression multiplexing (TCM) that eliminates echo and crosstalk. TCM is discussed in Chapter 9 as an alternative to the S transmission system in digital PBX applications. The AMI code is also used in the BRA ISDN S/T transmission system at the user–network interface (Chapter 3).

In comparison, the 2B1Q line code offers the greatest baud reduction, and therefore intuitively concentrates most of its energy at lower frequencies. This assists in minimizing the level of crosstalk which increases with higher frequencies, through the application of low-pass filtering in the receiver. A comparison with the 4B3T code (MMS43) shows that 2B1Q has a combined NEXT and ISI performance which is better by 2 to 3 dB [3], although a 2B1Q transceiver can be expected to be more complex because it has to process a four-level line code instead of a three-level code.

4.3 U TRANSMISSION SYSTEMS

U interface transceiver designs are based primarily around the line code they use. The choice of a particular line code for use in an ISDN subscriber loop depends on several factors. These are:

- the different subscriber loop cable types that exist and their configurations;
- the noise conditions under which they must operate;
- the availability of cost-effective transceivers that conform to recognized standards.

The two dominant line codes in use today are the 2B1Q and 4B3T codes.

4.3.1 The 2B1Q U interface transmission system

The 2B1Q system was originally standardized by ANSI in North America and has subsequently also been adopted with some modifications by ETSI as one of two systems for use in Europe, the other being based on 4B3T. The 2B1Q line coding in Fig. 4.7a is used with line transmit signal levels of +2.5 V, +5/6 V, −5/6V, and −2.5 V which correspond respectively to the 3, 1, −1 and −3 symbol levels.

The 2B1Q system uses a multiframe for transmission and reception of information between NT and LT. A basic frame consists of 120 2B1Q symbols, referred to as **quats** (an abbreviated form for quaternary symbol), which is the equivalent of 240 bits of 2B1Q encoded data. The first 18 bits (nine quats) of this frame contain a synchronization word, while the following 216 bits (108 quats) are allocated to 12 successive frames of the same 2B+D ISDN channels. The remaining six bits (three quats) are used to carry operations and maintenance information. The 240-bit frame is transmitted in 1.5 ms at a bit rate of 160 kbps, which corresponds to a line symbol rate of 80 kbaud.

Eight basic frames are combined into a single multiframe, or **superframe**, and each superframe has a duration of 12 ms. The first basic frame of each superframe has its synchronization word inverted to indicate the beginning of the superframe. The creation of the superframe allows the last six bits in each basic frame to be combined with those from other frames within the superframe, and allocated to a different function according to the bit position and location of the basic frame within the superframe. This permits the available fixed bandwidth of the operations and maintenance channel, referred to as the M-channel, to be more flexibly multiplexed into sub-channels for each function. The framing is illustrated in Fig. 4.8.

Bits M5 and M6 in frames 3 to 8 of the superframe contain a 12-bit cyclic redundancy check (CRC) code for the previous superframe which is derived from all B- and D-channel data and the M4-bits in the operation and maintenance channel. The CRC is calculated when the last M4-bit of the frame in the superframe has been transmitted and inserted into the CRC bits of the next superframe. Should a CRC error be detected in a received frame, the **FEBE** (far-end block error) M-channel bit in the following superframe transmitted is set to 0. Typically, counters within the U transceivers accumulate these bits together with local **NEBE** (near-end block error) bits to allow the network to monitor transmission performance of the U transmission system.

The meaning and use of the allocated M4 and M6 single-bit positions are described in the following sections.

4.3.1.1 The embedded operations channel

The first three bits of the M-channel in each frame, M1 to M3, are used for an embedded operations channel (EOC) which is used by the network to activate loopback modes in the NT1 and request the transmission and reception of corrupted check (CRC) codes between the network and NT1. Each EOC message is 12 bits in length and contains a three-bit address, a one-bit data or message indicator, and an eight-bit information field. The address identifies the destination of the message which presently may either be the NT1 (000), or a broadcast address (111). Other address values will be allocated to repeaters and other equipment used in the subscriber loop transmission system to extend its range. With 24-EOC bits available in eight frames, two complete EOC messages can be contained in each superframe. The EOC channel has a data rate of 2 kbps.

EOC messages are transferred between the network and NT1 using a repetitive command and response sequence initiated from the network. The network continues to send the same command until its receipt is in some way acknowledged by the NT1. Three identical messages, each having the correct address, must be received consecutively at the NT1 before any action defined by the message is taken by it. Under normal circumstances, the NT1 will echo back to the network in the next available frame each occurrence of the received EOC message, where the network treats the reception of three consecutive echoes as the acknowledgment of receipt of the message by the NT1. However, should the NT1

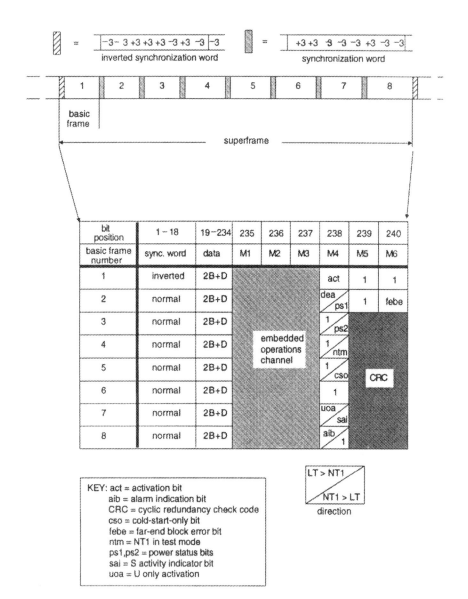

Fig. 4.8 Framing in the 2B1Q transmission system.

not be capable of implementing the action defined by the message, then on the third echo back to the network, it will start to repeatedly send an Unable to Comply message. Also, should the NT1 receive a command with an incorrect address, it will return a Hold State message to the network.

The EOC command messages originating from the LT are as follows.

Operate 2B+D Loopback: The 2B+D channels are looped back in the NT1 toward the network. Any activated TE on the user–network interface side of the NT1 must still be able to maintain synchronization to the NT1 during loopback operation.

Operate B1 or B2 Loopback: Transparent loopback of either B-channel to the network.

Return to Normal: The actions defined by successive EOC commands are maintained until a Return to Normal message is sent to reset them to their initial state.

Request Corrupt CRC: The network requests that the NT1 sends it frames with corrupt CRCs.

Notify of Corrupted CRC: The network informs the NT1 that it is sending frames with intentionally corrupted CRCs.

Hold State: The network sends this message to maintain any EOC controlled operation in its present state.

The messages described above make up only a small number of the 256 possible eight-bit values of the information field in the EOC message. Sixty-four of these values are reserved for non-standard applications, while a further 64 are reserved for internal network use. The remaining 120 values are reserved for future standardization.

4.3.1.2 Activation

The activation of the 2B1Q transmission system involves a handshake sequence of tones and synchronization signals between transceivers at the NT1 and network, and is designed to allow the equalizers and echo cancellers in the transceivers to train themselves to the cable characteristics, and allow the NT1 to become synchronized with the transmissions from the network. Most 2B1Q U transceivers use channel estimation techniques for timing recovery. During the activation sequence, various status bits in the EOC allow the activation sequence at the U interface to be coordinated with the activation at the S/T interface, as both interfaces must be brought up into an activated state such that the user terminal at the far end of the ISDN subscriber loop is also synchronized with the transmissions from the network. Activation at the U interface can occur from either end depending on where the call originates.

Two types of activation can occur at the U interface. A **cold start** of a U transmission system is applied when the transceivers have had their power removed or the cable characteristics may have changed significantly between successive calls. Cold-start activation involves a training period lasting a number of seconds during which the equalizers and echo cancellers in the transceivers fully adapt to the cable characteristics in order to achieve an adequate transmission performance. The resulting activation time is limited by the convergence time of adaptation algorithms, and is defined in standards to be no more than 15 seconds. However, such a long activation time in response to a call establishment request from a local terminal gives rise to an uncomfortable delay before the terminal starts to process the call. As the cable characteristics do not change significantly with time (from one call to another), a shorter activation time can be achieved by using the same filter coefficients in the equalizer and echo canceller from a successful cold-start activation. This quick activation, referred to as a **warm start**, can typically be achieved in less than 300 ms and is used on subscriber loops which are deactivated by the network during periods of no call activity. In a deactivated state, the subscriber loop equipment may enter a sleep state where they consume minimum power.

Figure 4.9a shows the sequence of tones, signals and status bits used across the U interface for the total activation of the subscriber loop from the terminal end. The sequence is initiated with a user request for call establishment at the terminal (TE) which causes it to start activation of the S/T interface with transmission of INFO1 (Chapter 3). Before the NT1 can acknowledge with INFO2, it must first activate the U interface and achieve synchronization with the network. The NT1 U transceiver alerts the LT transceiver to the fact that it is starting U interface activation procedures with a 9 ms burst of a 10 kHz wake-up tone. If the LT transceiver had previously been deactivated and was in a low-power sleep state, then this signal serves to bring it into a fully powered state in preparation for being activated. The wake-up tone is followed by transmission of an SN1 signal which consists of repeated basic frames with no multi-framing, with the contents of both B-, D- and M-channels set to 1. As is the case in the fully activated state, these channels, excluding the synchronization word, are scrambled prior to transmission to maintain sufficient timing information for timing recovery at the far-end receiver.

When the NT1 ceases transmission of SN1, indicated by SN0, it is ready to receive a signal and the LT U transceiver commences with transmission of a signal identical to SN1 called SL1. During these phases, in which either SN1 or SL1 are transmitted, the training of the echo cancellers and equalizers in the transceivers is executed, the duration of which depends on type of activation,

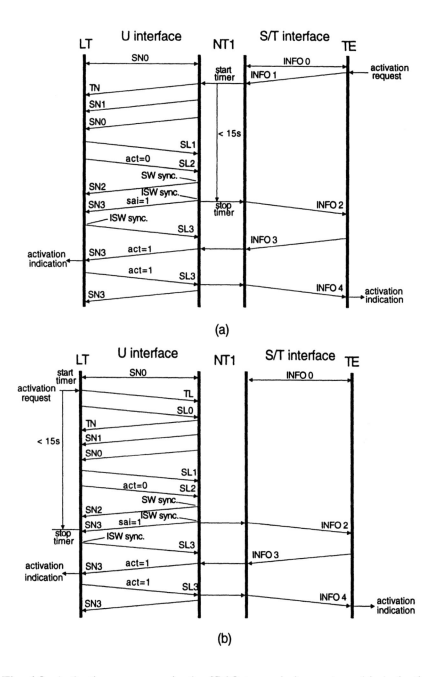

Fig. 4.9 Activation sequences in the 2B1Q transmission system. (a) Activation from the TE side; (b) activation from the network side.

warm start or cold start, is being performed. As an option, some ISDN networks may decide to keep the U transmission system permanently activated, in which case the U interface may not need to support warm-start activations. The NT1 has the ability to declare this to the network by setting the **cso** (cold-start-only) bit in the M-channel of SN3 frames.

At some point in time during the transmission of SL1, the LT transceiver will add the superframe synchronization word to the signal and include valid M-channel information, while the B- and D-channel data are set to 0. This is the SL2 signal and allows the NT1 transceiver to gain basic frame synchronization with the LT transmissions. Having done so, it returns an SN2 signal to the LT which is identical to SN1 and tells the LT that the it has gained frame synchronization, but not yet gained synchronization to the superframe. When the NT1 achieves superframe synchronization, it adds its superframe synchronization word and valid M-channel bits to SN2 to create the SN3 signal. However, as the S/T interface is not yet active, no valid data can be conveyed in the 2B+D channels of SN3 which are instead set to 1. The 2B+D channels only become transparent and contain user or signalling information when the M-channel **act** bit is set to 1 and occurs once the S/T interface is fully activated.

The reception of SN3 by the LT transceiver allows its receiver to gain superframe synchronization at which point it transmits an SL3 signal toward the NT1. SL3 is initially identical to SL2 until its act bit is set to 1 by the LT to indicate transparent 2B+D channels.

At the time the NT1 transceiver achieves full synchronization with the LT, indicated by its transmission of SN3, it can proceed further in the activation of the S/T interface with the transmission of a synchronized INFO2 signal toward the TE, and allows the TE transceiver to become synchronized to the network. At this point, activity at the S/T interface is indicated to the network through the **sai** M-channel bit which is set to 1 by the NT1 transceiver. The TE responds to the INFO2 with an INFO3 signal which then causes the NT1 to set the **act** bit to 1 in the SN3 M-channel toward the network. Finally, the LT transceiver returns SL3 with **act** set to 1 which instructs the NT1 to complete activation of the S/T interface by transmission of INFO4. The SN3 and SL3 signals corresponding to the frame in Fig. 4.8 then continue until the subscriber loop is deactivated.

A similar activation sequence initiated from the network side of the subscriber loop is shown in Fig. 4.9b. The main activation sequence is identical except that it is preceded by a 3 ms 10 kHz wake-up tone, TL, which originates from the network, and **sai** is not set to 1 until INFO3 is received from the TE as this is the first response which indicates any activity at the S/T interface.

Under conditions of testing and maintenance of the U transmission system by the network operator, it is desirable for the U interface to be activated independently of the S/T user–network interface, and thus not be reliant on the presence of a user terminal connected to the S/T interface to permit total activation of the ISDN subscriber loop. This is referred to as **restricted**

activation. The **uoa** (U only activation) M-channel bit is used by the network to indicate to the NT1 if it requires the S/T interface to be activated, and is set to 1 for normal S/T activation and 0 for restricted activation. Similarly, if the U interface cannot be activated for some reason, the S/T interface alone may be activated when initiated from a terminal. If the NT1 does not acquire superframe synchronization after the 15 second activation limit, then it may proceed to send INFO2 and the S/T activation sequence continues as normal. However, as there is no signal from the network to provide timing synchronization, the S/T interface timing is driven from a free-running clock within the NT1.

4.3.1.3 Deactivation

Controlled deactivation of the U transmission system when the subscriber loop is not involved in any call activity can only be initiated from the network side, and only if the U transceivers support warm-start activations. The LT transceiver will change the **dea** (deactivation) bit in the M-channel from 1 to 0 in at least the last three frames it sends before ceasing transmission, which it must do so before commencing transmission of the following superframe. When the NT1 transceiver detects the loss of signal, it too ceases transmission and waits a further 40 ms before it is free to initiate a subsequent activation with transmission of TN. During this time, an activation initiated from the network side is possible. An uncontrolled deactivation will occur at either side of the connection on loss of its received signal or its synchronization for more than 480 ms.

The above descriptions of activation and deactivation sequences are examples which illustrate what happens under normal conditions. Comprehensive activation and deactivation sequences which take into account other circumstances and conditions, typically resulting from errors or failures, are described in terms of state transition tables which form part of the relevant standards discussed in section 4.5.

4.3.1.4 Test and maintenance functions

In addition to the loopback and U only activation features described above, a number of other features have been built into the 2B1Q system to allow the network operator and user to perform tests on the U and S transmission systems for the purpose of fault diagnosis and routine maintenance.

In American ISDNs, tests on the U transmission system from the network can be achieved after the NT1 is placed into a **quiet mode** during which it will not attempt to activate or transmit user information. The trigger signal which initiates quiet mode in the NT1, and other test states, is either a series of DC current pulses or series of half-cycles of a 2 to 3 Hz sine wave. Two methods are proposed because although the DC current method is more simple to implement, it may be incompatible with the power feeding requirements across the U interface. A series of six pulses or half-cycles sent by the network will instruct the NT1 to enter a quiet mode state in which a tester connected to the subscriber loop can perform return loss measurements or impedance tests. Once in the quiet mode state, it remains for a period of 75 seconds unless it:

- receives the same signal, in which case the 75 seconds begins again;
- or it receives a signal of ten pulses or half-cycles, in which case the NT1 exits quiet mode;
- or it receives a signal of eight pulses or half-cycles instructing the NT1 to enter the **insertion loss measurement test** (ILMT) state.

In the ILMT state, a scrambled and framed 2B1Q signal, such as SN1 or SN2, is transmitted by the NT1 for a period of 75 seconds to allow equipment within the network to measure the transmission performance of the subscriber loop cable. At the end of the 75 second test period the NT1 will return to its normal state. However, it can be extended by another 75 seconds on transmission by the network of another eight pulse signal, or can be returned to quiet mode on transmission of six pulses, or returned to normal mode during ILMT on transmission of ten pulses. Quiet mode and ILMT are not defined in the ETSI specification for use on European subscriber loops.

From the user side of the subscriber loop, a user-initiated test mode involving B- and D-channel loopbacks at the NT1 toward the user's terminal will isolate these channels from the network and make them unavailable for call purposes during the test period, but prevent the NT1 from being able to respond to eoc messages sent from the network. Indication to the network that the NT1 is in such a test mode is provided by the **ntm** (NT1 test mode indicator) M-channel bit which is set to 1 under normal conditions and to 0 when in test mode. This test feature is specified for both American and European ISDNs.

4.3.2 The 4B3T U interface transmission system

The 4B3T system uses the multi-codeset line code shown in Fig. 4.7b, and the + ,0 and − symbols are transmitted with corresponding amplitudes of +2 V, 0 and −2 V. Definition of the operations and maintenance channel in the 4B3T system is not as extensive as the 2B1Q transmission system, and no multiframe is defined for the 4B3T system.

The 4B3T frame is shown in Fig. 4.10 and consists of 120 ternary symbols and has a duration of 1 ms at a transmission rate of 120 kbaud. The frame synchronization word is considered to occur at the end of each frame rather than at the beginning, and occupies 11 symbols whose order are reversed in the NT1 to LT direction when compared to the LT to NT1 direction. The 2B+D information in the frame is scrambled using different polynomials in the LT-to-NT and NT-to-

LT directions to ensure that there is little correlation between the transmit and receive signals. Eight blocks of 2B+D information and a single operations and maintenance channel symbol make up the remaining 109 symbols of the frame.

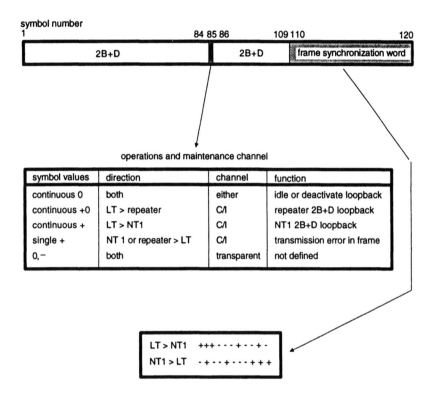

Fig. 4.10 Framing in the 4B3T transmission system.

4.3.2.1 Operations and maintenance channel

The operations and maintenance channel is a 1 kbaud channel that may contain either transparent messages which are not processed by the transceivers but passed directly to their system interface for processing by an attached controller, or continuous symbols whose value represent a loopback command or error information which are processed by the transceivers. Distinction between the transparent messages and command/information (C/I) messages is by means of symbol polarity. The transparent channel uses the − and 0 (or +) symbols to respectively represent 0 and 1 data bits, while the C/I messages use 0 (or −) and + symbols as shown in Fig. 4.10. Priority is given to transparent messages.

The 4B3T frame contains no CRC for error checking. However, an NT1 which detects one or more line code violations in a frame may indicate this back to the LT by setting the C/I symbol in the following frame toward the LT to +.

4.3.2.2 Activation and deactivation

The 4B3T transmission system is activated in a similar way to the 2B1Q system, with a sequence of tones and signals (SIGs) that are related to the frame shown in Fig. 4.10, and which permit the NT1, and subsequently the terminal connected to the S/T interface, to become synchronized to the network and provide the 2B+D physical layer interface between the terminal and the network. Typical activation sequences under normal conditions are illustrated in Fig. 4.11.

SIG 0 corresponds to the idle state of the transmission system where no signal is present on the line. Activation from the terminal side of the subscriber loop is similar to that from the network side, and differs only in the wake-up sequence. SIG 1W and SIG 2W are bursts of 7.5 kHz tones which are used either as wake-up or wake-up acknowledgment signals depending on which side initiates activation, following which SIG 2 is sent from the LT. The SIG 2 signal contains framing information and allows the receiver in the NT1 to gain synchronization to the network. During this period, the NT1 sends SIG 1A to the LT, where SIG 1A is a frame that contains no synchronization word. At such time that the NT1 receiver gains synchronization, it adds the synchronization word to SIG 1A to generate SIG 1, and detection of this change at the LT informs it of the successful NT1 synchronization. The NT1 now proceeds to activate the S/T interface with INFO 2, and receives in return INFO 3 which causes it to send SIG 3 to the LT. SIG 3 informs the LT that the S/T interface is synchronized in both directions. Finally, the LT sends SIG 4H to the NT1 requesting it to establish bi-directional 2B+D channel operation. The NT1 acknowledges by transmitting INFO 4 at the S/T interface and returns SIG 5 to the LT which contains operational 2B+D information from the terminal to the network. SIG 4H then becomes SIG 4 which carries 2B+D information from the network to the terminal. SIG 4 and SIG 5 then continue until deactivation. The maximum warm-start activation time allowed is 210 ms, while the maximum cold-start activation time is 1.5 s. A maximum of twice these times is allowed if a repeater is used in the transmission system. These activation times are significantly quicker than the 2B1Q system and result from faster convergence times at the echo canceller and equalizer due to use of a simplified line code and use of the synchronization word timing recovery.

A deactivation request at the LT that results in it sending SIG 0 (no signal), or a failure condition that causes the NT1 to receive SIG 0 for more than a maximum of 200 ms, results in a loss of synchronization at the NT1 that in turn causes it to deactivate and send SIG 0 toward the LT and INFO 0 toward the terminal. Although the NT1 may not purposefully initiate deactivation, should a failure condition occur whereby the LT receiver loses synchronization, then this too

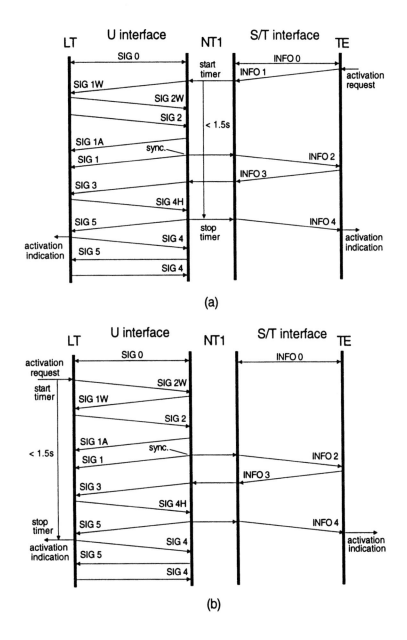

Fig. 4.11 Activation sequences in the 4B3T transmission system. (a) Activation from the TE side; (b) activation from the network side.

results in deactivation. Once deactivated, both NT and LT may enter a low-power sleep state.

4.4 REPEATERS

A number of subscriber loops, particularly those in more rural and remote areas, will exceed the maximum range accommodated by existing standards. In such cases, repeaters may be used to extend the range of the subscriber loop. For example, in North America, a subscriber loop with an attenuation greater than 42 dB must be split into two segments connected with a repeater, where each segment has an attenuation less that 42 dB. Alternatively, a digital loop carrier system (DLC) can be employed to place the LT closer to the subscriber such that only a single segment is required.

When a repeater is used, the activation signals and sequences for each section of the U transmission system in general remains the same with the exception that longer total activation times will be experienced as the DSP training and synchronization with the network must be achieved for each section in a sequential order. However, the overall cold-start activation time of the subscriber loop should not exceed 30 seconds.

4.5 U INTERFACE STANDARDS

The ITU-T has defined two Recommendations which relate to the ISDN subscriber loop connection between the user–network interface and the local exchange. G.960 [4] describes the characteristics of the BRA ISDN subscriber loop digital section as viewed from its interfaces at the T reference point at the user–network interface and the line termination at the LE referred to as the V reference point, while G.961 [5] deals more specifically with the operation of the digital transmission system at the U interface reference point. The distinction between the digital section and its transmission system was shown in Fig. 4.1 at the beginning of the chapter.

The ITU-T Recommendations have a world-wide scope, and G.961 therefore covers all variants of line code that could be considered for use in the U transmission system. These include MMS43 (4B3T), 2B1Q, AMI, AMI using TCM (time compression multiplexing), bi-phase and SU32 (3B2T). However, the G.961 Recommendation is not as complete in its depth and level of detail as the equivalent ANSI and ETSI specifications which are more usually referred to in terms of the operation and performance of the ISDN U transmission systems used in America and Europe.

In America, the NT1 is defined as customer premises equipment which is purchased and maintained by the user. The U interface may therefore be a user–network interface, and as such required standardization at an early stage in the deployment of ISDN to ensure equipment compatibility. The result is that ANSI has led the way in the standardization of the U interface through the ANSI T1.601 standard [6] which defines the use of a 2B1Q transmission system.

In Europe, the NT1 belongs to the network operator, and is installed at the time of subscription and subsequently maintained by the network operator. The user therefore never has direct access to the U interface, and instead connects to the ISDN through the S/T user–network interface.

European ISDNs use both the 2B1Q or 4B3T line code. Through ETSI, a technical recommendation ETR 080 [7] has been defined which covers the licensed use of both of these. The ETSI document exists only as a recommendation which may be adopted by European network operators, and is not an enforced standard in order to respect specific requirements that may exist in the European national networks which fall outside this recommendation. For example, the test loops and conditions which are defined for conformance testing of the U transceiver in different countries may be different in order to use reference loops which more closely represent the population of subscriber loops that exist within a network than do the test loops and conditions defined in the ETSI recommendation. (A further ETSI standard, prETS 300 297 [8], has also been generated for the digital section which is equivalent to G.960.)

Three main differences between the ETSI and ANSI U transmission systems are their performance testing specifications, power supply configurations, and test and maintenance functions.

4.5.1 Performance testing

The U transmission standards and recommendations include test arrangements and procedures which are intended to be used to qualify the performance of LT and NT1 equipment for use in ISDN BRA subscriber loops. According to the requirements of the networks represented by the relevant standard or recommendation, a number of test subscriber loop cables are defined together with simulated noise sources whose signals are injected onto the cable to which the equipment under test is connected. After activation, a **pseudo-random binary sequence** (PRBS) test signal is used to simulate user information in the 2B+D channels and the **bit error rate** (BER) is measured by comparing the transmitted and received data streams.

The ANSI T1.601 standard specifies 12 mandatory test loops which are representative of the subscriber loops found in North America. A further three optional test loops cover more demanding cases which are not so frequently found. The test loops are made up of different lengths of different types of cable with a total length of up to 18 000 feet (5.5 km). The test loops may also include up to three bridge taps. As an example, loop 14 of the ANSI loops is shown in Fig. 4.12a.

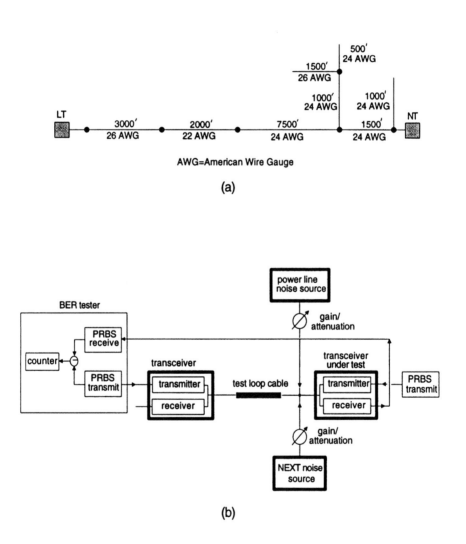

Fig. 4.12 The U transmission performance test system. (a) An example test loop (ANSI loop 14); (b) a typical test system.

The ANSI test configuration is shown in Fig. 4.12b. Noise sources which simulate the presence of NEXT, and noise induced from mains power lines, are injected onto the test loop close to the transceiver under test. These noise sources are filtered such that they have a specific power with respect to frequency which is defined as a 0 dB reference. The gain of the noise signal is then increased or decreased until the BER is equal to the ANSI defined limit of 10^{-7} (one bit in 10^7 is received corrupted), at which point the value of the gain represents the noise margin the transmission system has above or below the 0 dB reference. The ANSI requirement is that the a margin of +6 dB is required on the mandatory loops as well as a zero length loop, while a 0 dB margin is required on the optional test loops. These tests are also performed with added timing jitter to reflect that introduced into the transmission system from the network.

ETSI test procedures make use of eight different test loops, including a zero attenuation loop. Instead of defining a length for each test loop, ETSI specifies a maximum attenuation of 36 dB at a frequency of 40 kHz. The ETSI test configuration is similar to that of ANSI, but has a NEXT noise source with a different-shaped power and frequency characteristic with increasing levels of power toward the lower frequencies. This reflects a significant component of impulsive noise due to the presence of old mechanical switching exchanges and pulse dial telephones still in service in some European networks.

The 0 dB reference noise level of the ETSI noise source is also higher than ANSI in order to reduce the amount of time required to measure the BER performance. To detect one errored bit in 10^7 requires that at least 10^7 bits be transmitted, which at an aggregate data rate of 144 kbps takes just under 70 seconds. To provide an average result, a measurement time much greater than this must be used, and the ANSI standard specifies at least 10 minutes where the PRBS is transmitted in the 2B+D channels a minimum of 13 minutes for a 2B test (128 kbps), and at least 25 minutes for a single B-channel test (64 kbps). The increased ETSI reference noise levels result in a series of fast tests of only 30 seconds on each selected loop where a BER better than 10^{-4} is required with a noise margin of +2.5 dB. The intention with these fast tests is to identify the loop on which the transmission system performs worst, and then perform more rigorous tests on this loop.

4.5.2 Power supply

Figure 4.13 illustrates the method used to provide a power feed across the U interface from the local exchange to the NT1. Unlike the S interface where a four-wire transmission system is available to create a phantom power feed, only two wires are used at the U interface and these are connected to each other through the transformer, which would result in a short-circuit if a DC voltage was applied across the transformer. The two conductors are separated to allow a DC voltage to be applied by splitting each line-side transformer winding into two equal halves and connecting them with a capacitor. The capacitor enables AC signals to pass without attenuation but blocks DC voltages. Components are added to this configuration to protect the circuitry from excessively high voltages caused by lightning strikes.

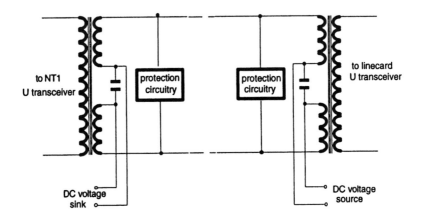

Fig. 4.13 Power feed configuration at the U interface.

In European ISDNs, the U transceiver side of an NT1 for both 2B1Q and 4B3T systems will typically be powered remotely from the network across the transmission cable, thus allowing the network operator to maintain full control over the U transmission system at all times. The S/T user–network interface may under normal conditions be powered locally from the NT1 using a local power source such as mains or batteries, and is backed-up with remote power from the network under emergency power conditions where the local power source fails. When active, the NT1 must consume no more than 500 mW of power from the network, and in a deactivated state must consume no more than 120 mW. Under emergency power conditions when the NT1 is expected to also power the user's designated terminal across the user–network interface, then the power consumption of an active NT1 is allowed to rise to a maximum of 1.1 W[8]. This power is delivered as a DC voltage and current that varies between different ISDNs due to the different safety requirements and subscriber loop configurations. The minimum voltage at the NT1 required for correct operation is 28 V, while the feed voltage at the exchange may vary from network to network from 51 V to 115 V.

---

[8]Slightly higher values of less than 600 mW and 1.3 W respectively were allowed during an interim period until the end of 1994 provided the power is available from the network.

In America, it is expected that remote powering of the NT1 is not provided and both primary power and emergency, or secondary, power may have to be sourced locally at the customer premises. Primary power will usually be derived from a local mains source while emergency power will come from local batteries. The status of power at the NT1 is indicated in both ETSI and ANSI 2B1Q systems using the ps1 and ps2 bits of the M-channel, and Table 4.1 shows the four messages defined by the value of these bits. In particular, the value 00 indicates to the network that both primary and emergency power sources have failed and that the NT1 will shortly cease normal operations. This value is referred to as the **dying gasp** message, for which the American NT1 must have sufficient energy storage to ensure that it continues operation long enough after failure of the power sources to send this status to the network.

Apart from the delivery of remote power, the continuous supply at the exchange of a small amount of current through the subscriber loop helps to prevent the build-up of oxidation at cable joints that could otherwise lead to bad connections. This is known as **sealing current** as it helps to seal the connection.

Table 4.1 Power supply bits of the 2B1Q system

NT1 status	ps1	ps2	Definition
All power normal	1	1	Primary and emergency power supplies are both normal
Secondary power out	1	0	Primary power normal, but emergency power is marginal, unavailable or not provided
Primary power out	0	1	Primary power is marginal or unavailable, emergency power is normal
Dying gasp	0	0	Both primary and emergency power are marginal or unavailable. The NT may shortly cease normal operation

4.5.3 Test and maintenance functions

As a result of the NT1 being considered customer premises equipment in American ISDNs, American operating companies can place the NT1 in quiet mode as described in section 4.3.1.4 in order to perform tests on the subscriber loop using test equipment tapped onto the line either at the local exchange or at some point along the length of the subscriber loop to the NT1. Return loss, insertion loss and impedance measurements can be performed while the NT1 is in quiet mode. In quiet mode, the NT1 is effectively placed in an idle state, ignoring requests for activation from the user or terminal side of the interface.

4.6 U INTERFACE TRANSCEIVERS

A number of BRA ISDN U interface transceivers are available today as VLSI integrated circuit solutions, a selection of which are listed in Table 4.2. As with the S interface transceivers in Chapter 3, the devices are described in terms of their control and data ports, a number of which use the general circuit interface (GCI) standard which is described in more detail in Chapter 6.

In the following section we take a closer look at the 2B1Q U transceiver from National Semiconductor. An in depth discussion of the design and implementation of a 4B3T device can be found in [9].

4.6.1 The National Semiconductor TP3410

The design objective of this 2B1Q U transceiver was to achieve a low-cost, low-power, and low-pin-count single-chip device that can be used at either end of the U transmission system, and has a minimum requirement for external interfacing components. The TP3410 is a mixed analogue and digital design which includes an adaptive analogue hybrid balance filter. This contributes approximately 20 dB of echo pre-cancellation, including the non linear echo components, thereby allowing considerable simplification of the main digital echo canceller which is implemented with a single linear FIR filter [10]. This mixed analogue and digital approach can be contrasted to the predominantly all-digital designs which require significantly more complex DSP, as illustrated by the Motorola MC145472 2B1Q transceiver [11] which implements FIR and IIR linear echo cancellation as well as a memory-based echo canceller for the non linear echo components.

Each design approach has its benefits and drawbacks. The main advantage of the mixed analogue and digital approach is that it can lead to a small, and hence more cost-effective device, but is more susceptible to noise in its analogue circuitry that requires strict design rules to control and limit the affect of manufacturing process variations on the analogue circuit behaviour. The main benefit of the all-digital approach is the ease with which it scales to a smaller geometry manufacturing process, resulting in lower chip size and potential for lower cost, although for this application, which relies heavily on digital filter techniques, it leads to a larger chip size for a circuit with an equivalent function and same device geometries.

A block diagram of the TP3410 is shown in Fig. 4.14, in which the analogue functions are shown in the shaded area. In the transmit direction, 2B+D information is taken from the digital system bus which may be operating in one of a number of time-division multiplex formats that meet the interface requirements of a number of different systems manufacturers. The CRC and M-bits are added and the resulting data stream is scrambled, framed and then encoded into the 2B1Q line code.

Table 4.2 A selection of Basic Rate ISDN 2B1Q and 4B3T U interface chips

manufacturer	device	system	control port	data port
AMD	Am2092	2B1Q	integrated with data port - IOM 2	IOM 2
AT&T	T7264	2B1Q	integrated with data port - K2 bus	k bus
Mietec	MTC 2071	4B3T	integrated with data port - V*	V*
Motorola	MC145472	2B1Q	SCP, or integrated with data port - GCI	IDL or GCI
Motorola	MC145572	2B1Q	As above, with additional parallel uP compatible port	IDL or GCI
National Semiconductor	TP3410	2B1Q	MICROWIRE, or integrated with data port - GCI	GCI, IDL, TDM
Phillips Semiconductor	PCB2390	4B3T	integrated with data port - IOM 2	IOM 2
SGS Thomson Microelectronics	ST5410/11	2B1Q	MICROWIRE, or integrated with data port - GCI	GCI
Siemens Semiconductor	PEB2091	2B1Q	integrated with data port - IOM 2	IOM 2
Siemens Semiconductor	PEB20901+ PEB20902	4B3T	integrated with data port - IOM 2	IOM 2

Note: GCI and IOM are compatible with one another.

Key: IOM - ISDN oriented modular interface (Siemens)
GCI - general circuit interface
IDL - inter-chip data link (Northern Telecom)
SCP - serial control port (Motorola)
V* - (ITT/Alcatel)
TDM - time division multiplex (generic)

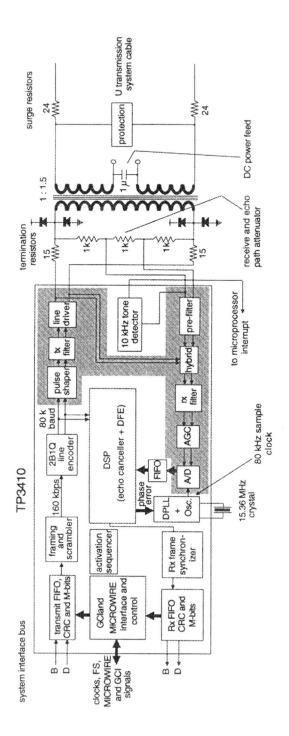

Before the 2B1Q code can be transmitted onto the line, the pulses must be shaped to achieve the transmission performance requirements and meet the transmit pulse mask defined in the standards. A classical solution to this problem is to pass the rectangular 2B1Q pulses through an analogue filter which provides the necessary time-domain shaping. However, as mentioned above, an IC manufacturing process can introduce an unacceptable degree of variability into the characteristics of an analogue filter, and hence the pulse shape, in addition to which the filter introduces a significant amount of ringing which is difficult to control. An alternative solution which is implemented on the TP3410 and overcomes these problems, is a pulse density modulated (PDM) pulse shaping circuit which contains an optimum pulse shape encoded as 96 samples of a three-bit PDM code. Due to the four possible levels of the 2B1Q symbol, the edges of the current symbol pulse are chosen according to the present two-bit data code for that symbol as well as the immediate past two-bit code. The pulse shape is then created by reading out the PDM code through a PDM digital-to-analogue converter. Although each symbol pulse consists of 96 samples, due to pulse symmetry only half of it need be stored as the other half may be created by reading out the PDM code in reverse order. Unwanted high-frequency components of the transmit signal are removed before it leaves the chip by the transmit filter and line driver circuits.

The component values and configuration of the line interface circuit are critical to the transmission performance of the device as they form part of the echo path. In particular, the transformer specification is typically defined by the manufacturer of the U interface device to ensure that it does not introduce non-linearities which cannot be accommodated by the device.

In the receive path of the TP3410, a replica of the transmitted signal is subtracted from the pre-filtered receive signal in the hybrid balance circuit which provides a degree of linear and non linear echo pre-cancellation. The following receive filter band limits the signal and the 13-bit A/D converter makes one sample per baud symbol, which thus simplifies the design of the echo canceller and DFE. Prior to the A/D, an automatic gain control (AGC) circuit adjusts the level of the signal to make full use of the range of the A/D. Both the hybrid balance and the receive filter are implemented as switched capacitor filters.

The DSP block performs the main echo cancellation and DFE functions through the implementation of two transversal FIR filter structures. The filter coefficients are stored in an 80 word RAM (random access memory) while the calculations are performed in two 24-bit ALUs (arithmetic and logic units), one of which performs the filter calculations for both the echo cancellation and DFE, while the other is dedicated to updating the coefficients according to the adaptation algorithm. The ALUs are driven by microcoded instructions stored in two ROMs (read only memories), and each ALU is capable of performing 96 computation cycles per baud.

Although the general architecture of the TP3410 is consistent with that discussed in the early sections of this chapter, three implementation issues are worthy of note. First, a reduction in the size of the FIR filters is contributed by a simple high-pass filter used at the input to the DSP block which limits the long tails of the echo response and received signals. Each FIR filter, one used in echo cancellation and the other in the DFE, therefore requires fewer filter taps. The high-pass filter also reduces the effects of impulse noise whose spectral content is greatest toward the lower frequencies, and also has a beneficial affect on the precursor pulse shape. Second, a simplification of the filter calculations, and thus the ALUs themselves, has been achieved through a shift of the 2B1Q symbol alphabet from (+3, +1, -1, −3) to (+4, +2, 0, −2), and results in all filter coefficients being a power of two which require only binary shift and addition operations on the input signal as opposed to full binary multiplication. The shift of the symbol alphabet can be considered as a DC shift in the output signal and is compensated for by a correction coefficient in the FIR filters. Lastly, a serial test bus is included which can be accessed outside the device and allows the operation of the DSP to be observed in real-time.

The data stream at the output of the DSP is passed through logic circuitry to detect the frame and superframe synchronization words, while the data in between is de-scrambled. The CRC and M-bits are processed while the 2B+D information is buffered and subsequently placed onto the digital system bus.

The TP3410 is designed to be used at both the NT and LT ends of the U transmission system. In NT mode, the DPLL (digital phase-locked loop) generates a 15.36 MHz clock which must become locked to the received line signal during the beginning of the activation sequence, and remain locked while there is a valid signal present. From this 15.36 MHz signal, the 80 kHz sample clock for the A/D is generated, as well as the transmit path timing. The sampling phase error in the received signal, which indicates the difference between the current phase and the optimum phase required to sample in the middle of the pulse, is generated by the DSP block and integrated digitally by the DPLL. The oscillator of the DPLL is a 19-stage ring oscillator whose frequency is controlled by varying the propagation delay equally through each stage of the ring. Depending on the result of the integrated phase error, the DPLL selects its output from the next or previous stage in the ring which causes the phase of the 15.36 MHz output clock to advance or retard by 3.4 ns jumps in order to reduce the magnitude of the integrated phase error. In LT mode, the same DPLL is used to generate a 15.36 MHz clock signal which is phase-locked to the 8 kHz frame synchronization signal, FS, which is generated by the network and is present at the digital system interface. A secondary PLL is implemented on the TP3410 to generate the 80 kHz sample clock for the A/D converter.

Control of the TP3410 and access to its internal registers is either through a dedicated serial MICROWIRE port which can be connected to a microprocessor or microcontroller with a similar port, or through the GCI command/indicate and monitor channels which form part of the GCI digital system bus format. The TP3410 has a power-down mode in which its internal operation is ceased allowing it to consume minimum power. The presence of TN or TL on the subscriber cable is detected with the 10 kHz tone detector circuit that causes an interrupt to be sent to the control microprocessor which commands the TP3410 into a power-up state. Subsequent commands to the TP3410 then cause the activation sequence to be executed.

4.7 THE NT1

An NT1 provides a translation of the physical layer protocols between the S interface and U interface transmission systems of the ISDN subscriber loop. Architecturally, the NT1 is very simple, and consists of an S transceiver and U transceiver back-to-back and interconnected by their serial system interface buses.

Figure 4.15 shows the architecture of two NT1s, one designed with a microcontroller and the other without. The need for a microcontroller depends on the extent to which operation and test and maintenance functions are required in the NT1. Many of the basic functions, such as activation and deactivation, and response to certain eoc messages such as loopbacks, are typically implemented in the U transceiver itself, and may be coordinated by control at the system interface bus between the U and S transceivers. For example, in the NT1 shown in Fig. 4.15a, the GCI system interface bus conveys not only the 2B+D channel information of the subscriber loop, but also an in-band control channel for the purpose of controlling functions between the two devices. Control logic built into each transceiver then handles activation and deactivation sequences and execution of mandatory maintenance functions. Such a configuration is suitable to meet the needs of an NT1 in most European countries based on those mandatory functions defined in the relevant ETSI specifications.

Figure 4.15b shows the architecture of an NT1 with a microcontroller to control additional test and maintenance functions. For example, Fig. 4.15b illustrates the use of opto-isolators to detect the pulses sent to command the U interface into quiet mode when tests are to be performed on the subscriber cable. A microcontroller-based NT1 is typically used in America, where many of these additional maintenance features are required.

The most demanding part of the design of an NT1 is its power supply which must provide high levels of efficiency under emergency power conditions while remaining inside permitted limits for radiated noise. These requirements are particularly stringent in Europe, in which the power from the U interface is required to supply both the NT1 and power to the S interface under emergency conditions. For ISDN in America, it cannot be guaranteed that power is supplied from the local exchange. Instead, the power supply will power the NT1 and

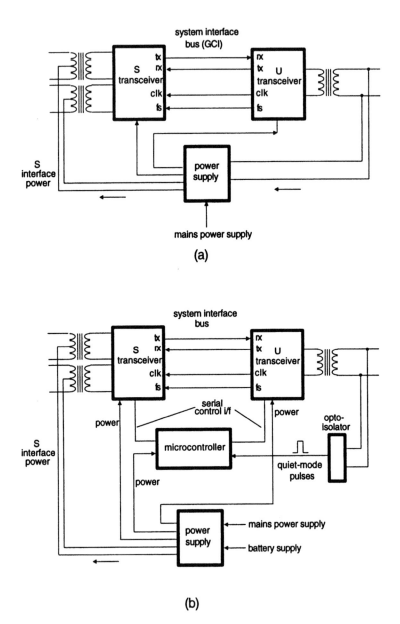

Fig. 4.15 Different NT1 designs: (a) without a microcontroller; (b) with a microcontroller.

deliver power to the S interface either from a local mains source, or a battery when operating under emergency conditions.

The presence of both S and U transceivers in an NT1 make a strong case for their integration into a single device in order to reduce their size and cost. This trend is just starting with devices such as the T7256 from AT&T Microelectronics (now known as Lucent Technologies).

4.8 PAIRGAIN SYSTEMS

Although not part of ISDN, pairgain systems have become an important application area for ISDN U transceivers. Its ability to provide 2B-channels on a normal PSTN copper twisted-pair subscriber loop means that, with the appropriate interfaces, the U transceiver can double the capacity of the subscriber loop from being able to carry a single telephony channel to carrying two telephony channels. In situations where the cost of laying additional subscriber loop cables is prohibitive, this technology can be applied to effectively gain additional pairs, hence the name **pairgain**, by doubling the capacity of existing pairs. Using a pairgain system, a residential user could be provided with a second telephone connection without the need for a installing new cable.

Figure 4.16 illustrates how the U transceiver is applied in a pairgain system. At the local exchange, a box called the **local exchange terminal** (LET) is added and connected to two of the standard PSTN subscriber connections from the exchange. To the exchange, the two interfaces on the local exchange terminal must appear as though the exchange subscriber connections are connected to normal analogue telephones. Each interface is therefore terminated with a **telephony line interface circuit** (TLIC) which mimics this function. Inside the local exchange terminal, the analogue signals are converted to PCM using standard codec/filter devices and the resulting 64 kbps PCM streams interfaced to the U transceiver through the usual system interconnect bus. Each PCM stream occupies one of the two B-channels on the U interface transmission system.

At the user's end of the connection, a **remote terminal** (RT) is provided that converts each of the B-channels back to analogue telephony signals and interfaces them to the two subscriber connections to the user. In this case, each interface must look to the user's terminal equipment as a local exchange, and is therefore provided with the normal subscriber line interface controller (SLIC) functions[9].

As pairgain systems are entirely contained within the subscriber network, their transmission capabilities and the signalling protocols used between the local

---

[9]In most cases the transmission specifications for the SLIC can be reduced as the connection between the remote terminal and the customer premises is likely to be only a short distance.

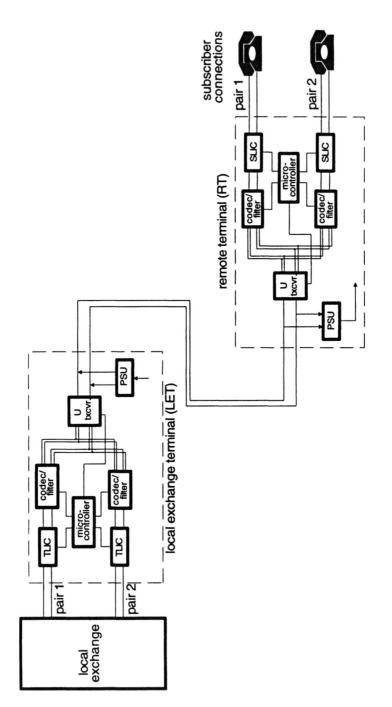

Fig. 4.16 Pairgain system

exchange terminal and the remote terminal need not comply with ISDN standards.

The standard ISDN signalling protocols are too complex and cumbersome for pairgain applications which require only a relatively simple system that in most cases can be handled by a simple single-chip microcontroller device. The DC line signalling states, such as on/off hook, pulse dialling and ringing, must be communicated between the remote and local exchange terminals, as well as operation and maintenance information used exclusively by the pairgain system. This signalling information may be encoded as digital messages and communicated either through the 16 kbps D-channel or through the 4 kbps EOC. In most cases the EOC is used as it is terminated within the U transceiver and does not require additional hardware, for instance a D-channel framing device such as an HDLC controller. DTMF signalling between a telephone and exchange is transported transparently through the pairgain system in the relevant B-channel. In operation, the U transmission system will usually remain activated and allows non-call DTMF signalling, as for example used in the activation and control of network services.

Deployment of pairgain systems in the subscriber loop network requires the RT to be powered from the LET whenever a reliable source of local power is not available. The LET itself is likely to be located somewhere within the local exchange building and so has access to the line feed power supply which is added to the U transmission system in the normal way. At the RT, this power must be used to generate the necessary DC signals and line feed power for the user subscriber connections.

In conjunction with codecs that provide compression as well as conversion, pairgain systems can be used to increase the subscriber loop capability from two speech channels to four or higher. For example, by using ADPCM (adaptive differential pulse code modulation, Chapter 7) codecs, a standard 3 kHz speech channel is converted into a 32 kbps ADPCM stream instead of the 64 kbps output by a normal PCM codec, thereby increasing the capacity of the pairgain system to four voice channels.

REFERENCES

1. McDonald, R.A. (1987) Report of Bellcore Impulse Noise Study. ANSI T1D1.3/87-256.
2. Qureshi, S. (1987) Adaptive equalization (Chapter 12), in *Advanced Digital Communications–Systems and Signal Processing Techniques* (ed. K. Feher), Prentice-Hall, pp. 640–707.
3 Lechleider, J.W. (1989) Line codes for digital subscriber lines. *IEEE Communications Magazine*, **27**(9), 25–32.
4 ITU-T (1993) Digital Section for ISDN Basic Rate Access. Recommendation G.960.

5 ITU-T (1993) Digital Transmission System on Metallic Local Lines for ISDN Basic Rate Access. Recommendation G.961.

6 ANSI (1991) Integrated Services Digital Network (ISDN)–Basic Access Interface for Use on Metallic Loops for Application on the Network Side of the NT (Layer 1 Specification). ANSI T1.601.

7 ETSI (1993) Transmission and Multiplexing (TM): Integrated Services Digital Network (ISDN) basic rate access digital transmission system on metallic local lines. Technical Report ETR 080.

8 ETSI (1993) Integrated Services Digital Network (ISDN); Access Digital Section for ISDN basic rate. prETS 300 297.

9 Szechenyi, K., Zapf, F. and Sallaerts, D. (1986) Integrated full-digital U-interface circuit for ISDN subscriber loops. *IEEE Journal on Selected Areas in Communications*, **8**, 1337–49.

10 Batruni, R., Lemaitre, P. and Fensch, T. (1990) Mixed digital/analog signal processing for a single-chip 2B1Q U-interface transceiver. *IEEE Journal of Solid-State Circuits*, **25**(6), 1414–25.

11 Girardeau, J. et al. (1989) ISDN U transceiver algorithm, development system and performance, in *Proceedings IEEE Globecom'89*, Vol.3, pp. 1957–65.

5

Protocols for ISDN call control signalling

In this chapter we examine the basic control plane (C-plane) protocols for call control signalling which are used both within the ISDN terminal and the network exchange at the ends of the ISDN subscriber loop. These protocols use the D-channel of the physical layer established by the S and U transmission systems to convey call control signalling messages between the terminal and the network. In addition to the basic call control signalling, the last part of this chapter deals with the signalling required to invoke the supplementary services implemented by the network and which enhance the basic call process.

The two layers we are concerned with here are the layer 2 data link and the layer 3 network layer. These layers, along with the physical layer of the subscriber loop, provide a terminal with basic call signalling capability. The data-link layer protocol is the **link access protocol** for the **D-channel**, or LAPD, and is defined in ITU-T Recommendations Q.920 [1] and Q.921 [2], while the network layer protocol, generally known as **call control**, is defined in Q.930 [3] and Q.931 [4]. Extensions to the call control protocol for handling supplementary services are described in Q.932 [5]. The relationship between these layer protocols and the OSI reference model is shown in Fig. 5.1 which illustrates the end-to-end protocols residing in the terminal and the network exchange.

As well as the two layers identified above, operation of the call control functions relies on a management entity which provides services to all layers in the protocol stack. The management entity is required to handle the problems associated with initiating, monitoring and terminating the activities performed by layers and assists the different layers to work together in a coordinated fashion.

5.1 THE LAPD DATA-LINK LAYER

The purpose of the data-link layer is to provide error-free transmission and reception of messages between the network layers on either side of the subscriber loop connection. In the case of call control in the D-channel, packet communication techniques are employed to encapsulate messages from the call

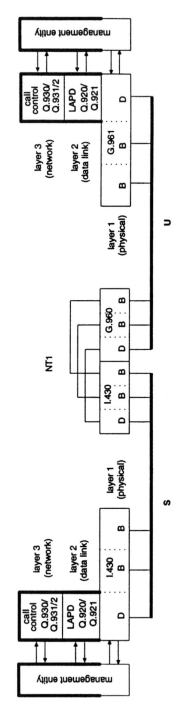

Fig. 5.1 Protocols associated with call control over the D-channel.

control layer in discrete frames which are transmitted and received in the synchronous D-channel. The protocol used for the data-link layer is known as the link access protocol for the D-channel (LAPD).

As up to eight terminals can be attached at the S interface to a single subscriber loop connection, LAPD must distinguish communications between each of the terminals and the network. Each data-link connection between a terminal and the network exchange is a point-to-point connection, although all terminals share the same physical bus and the same 16 kbps D-channel. LAPD distinguishes between connections by allowing the network exchange to uniquely address applications in each terminal through establishment of a **logical connection** between each terminal and the network. These logical connections are differentiated by assignment of a unique address to each terminal which is then included in each D-channel message for that terminal. The creation of multiple logical channels in a single physical channel is a form of multiplexing. But unlike physical multiplexing techniques such as TDM, there is no assignment of a fixed amount of bandwidth to any logical channel. Instead, the entire bandwidth of the D-channel is used by all terminals on a first-come-first-served basis, with the contention resolution protocol described in section 3.2 serving to resolve conflicts caused by multiple simultaneous accesses.

The LAPD data-link layer can be thought of as two distinct sub-layers. The lower of these sub-layers handles the bit-level framing, while the layer above deals with the procedures used to control the establishment of logical connections and the flow of frames across these connections. These procedures also involve interaction with the management entity whose role is mainly in assisting in the management of the logical connections.

5.2 LAPD FRAME STRUCTURE

The structure of the LAPD frame is based on the popular HDLC (high-level data-link control) format used in numerous other derivative protocols, and is shown in Fig. 5.2. The start and end of the frame is indicated by identical **flags** with a bit pattern of one 0 followed by six consecutive 1s and another 0. This pattern is guaranteed to be unique, and not to occur in the remaining contents of the frame between the two flags, through a process called **bit-stuffing**. Bit-stuffing operates on the bit stream at the physical layer between the opening and closing flags, and automatically inserts a 0 immediately after the occurrence of five consecutive 1s. At the receiving end, when five consecutive 1s are detected between opening and closing flags, the 0 immediately following is deleted, thereby providing transparency of the bit-stuffing process to the LAPD layers at either end of the connection. Bit-stuffing is used at the expense of a small increase in the amount of

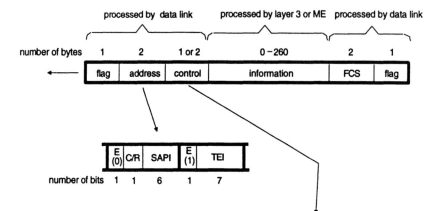

TYPE	NAME	COMMAND/ RESPONSE	ENCODING 1 2 3 4 5 6 7 8							
information transfer	I	C	0			N(S)				
			P			N(R)				
supervisory	RR	C/R	1	0	0	0	0	0	0	0
			P/F			N(R)				
	RNR	C/R	1	0	1	0	0	0	0	0
			P/F			N(R)				
	REJ	C/R	1	0	0	1	0	0	0	0
			P/F			N(R)				
unnumbered	SABME	C	1	1	1	1	P	1	1	0
	DM	R	1	1	1	1	F	0	0	0
	UI	C	1	1	0	0	P	0	0	0
	DISC	C	1	1	0	0	P	0	1	0
	UA	R	1	1	0	0	F	1	1	0
	FRMR	R	1	1	1	0	F	0	0	1
	XID	C/R	1	1	1	0	P/F	0	0	1

Fig. 5.2 LAPD framing.

information that must be transmitted and received, depending on the frequency of occurrence of five contiguous 1s in the transmitted bit stream of the frames.

Error detection and recovery is an important issue in all digital data communication networks, particularly at the physical and data-link layers where errors are more likely to be introduced due to the effects of electrical noise. The last field in the frame before the closing flag is a frame check sequence (FCS) whose value is calculated using a cyclic redundancy check (CRC) technique from the data in the frame between the opening and closing flags without bit-stuffing and excluding the FCS itself. If an error is detected by a receiver, then the frame is simply discarded and the LAPD procedures are left to invoke a request for the frame to be retransmitted if necessary.

In order to multiplex logical connections between end-points, two address bytes are defined for the LAPD frame which contain the service access point identifier (SAPI) and the terminal endpoint identifier (TEI). In addition, these fields contain a command/response bit (C/R) to identify a command and response frame, and an extension bit (E) which is used to indicate the end of the address field. Extension bits are used in general to indicate that a particular field is extended by another byte, but in this case its use is somewhat redundant as the address field is always two bytes in length.

5.2.1 The terminal end-point identifier (TEI)

The TEI of the LAPD frame is an address which is used to identify a point-to-point data-link connection that may be established between a specific terminal and the network. The LAPD protocol and the TEI addressing caters for both the simple point-to-point S bus configuration and point-to-multipoint passive bus configurations in which several terminals can be connected to the user–network interface. Each terminal on a passive bus has the ability to establish a point-to-point data-link connection with the network which is identified with an individual TEI, as shown in Fig. 5.3. With the exception of the broadcast TEI which is used to broadcast a message to all terminals on an S bus, a TEI is typically associated with a single terminal equipment such that a data-link frame that is transmitted from the network over the D-channel can be uniquely addressed to a specific terminal, and a frame transmitted to the network can be identified by it as originating from a specific terminal. A terminal may also have more than one non-broadcast TEI associated with it, where each represents a different point-to-point data-link connection.

The TEI is a seven-bit value split into two ranges and a single broadcast address, as shown in Table 5.1. The two ranges correspond to the two methods used to assign the TEIs to the terminal equipment.

Automatic TEI values are requested by a terminal and assigned by the network through an exchange of messages between the management entities in the

Table 5.1   TEI value assignment

TEI value	Assignment
0–63	Non-automatic TEI assignment
64–126	Automatic TEI assignment
127	Broadcast TEI

terminal and the network. This occurs whenever the terminal is in a **TEI unassigned** state and requires the establishment of a logical data-link connection to transfer information to the network. This is typically the case when a terminal is first attached to an S bus. Once assigned, the terminal is then expected to retain its designated TEI value unless it is disconnected from the S bus or certain error conditions occur causing loss of communications with the network. If the only source of power for the terminal is derived from power source 1 at the S interface, then when the terminal is disconnected from the S bus it is assumed that it loses any TEI values. If the terminal derives its power from another local source, such as batteries, then it must also have the capability of detecting the presence or absence of power source 1 as indication of its state of connection to the S bus, and hence its **TEI unassigned** state. The network is responsible for ensuring that no two terminals connected to the same S bus are assigned the same automatic TEI value, although a terminal may request the assignment of more than one TEI. The management entity procedures used to establish and maintain automatic TEI values are discussed later after we have introduced the different types of messages supported by LAPD for such a purpose.

Non-automatic TEI values are permanently assigned through some hardware feature in the terminal such as a switch. The user is responsible for setting this switch and must ensure that it is uniquely assigned amongst terminals on a passive bus. The network is made aware of the existence on the S bus of a terminal with a non-automatic TEI either from the TEI in the messages sent from the terminal to the network, or if the network makes an audit of existing TEIs using messages which request the terminals to declare their TEIs.

In accordance with the OSI layering rules, the TEI is an address which is known only by the data-link layers on either side of the link, and has no relationship to the telephone number which the network associates with the subscriber loop. Consequently, the presence of an incoming call is indicated by a call control setup message from the network which is broadcast to all terminals connected to the subscriber loop, only one of which may eventually accept the call. The broadcast message contains a broadcast TEI value which all terminals connected to the subscriber loop will receive.

5.2.2 The service access point identifier (SAPI)

In addition to multiplexing connections between different terminals using the TEI, an LAPD frame has a second address called the service access point identifier (SAPI) that is used to identify which network layer entity or management entity is associated with a particular data-link message. For example, as well as the call control entity, a packet layer entity could also be present at the network layer for data communications in the D-channel, and the SAPI is used to identify which entity is to process the message contained in the information field of the LAPD frame. Conceptually, the data-link layer SAP (service access point) lies at the boundary between the data-link layer and the data-link management entity or network layer entity, and provides the point at which the services of the data-link layer are made available to the network layer entities. There are currently four entities identified by the Q.921 Recommendation, each of which is assigned a unique six-bit SAPI value, as shown in Table 5.2.

Table 5.2 SAPI value assignments

SAPI Value	Entity
0	Call control procedures
1	Reserved for packet mode communications using Q.931 procedures
16	Packet communications conforming to X.25 level 3 procedures
63	Management procedures associated with the data-link layer

Together, the SAPI and TEI are known as the data-link connection identifier (DLCI) which uniquely identifies the logical data-link connection between two layer 3 or management entities on the S bus. The use of TEI and SAPI to relate logical data-link connections to the call control and data-link layer and management entities is illustrated in Fig. 5.3.

5.2.3 Addressing layer 3 services

As already mentioned, the DLCI information is known only within the layer 2 data link and remains private to it. However, the services in the layer 3 above the data-link layer on each side of the link must have some means of associating themselves with a particular data-link connection. (The management entity functions associated with the data link are considered to be part of the layer 2). This is achieved through the connection end-point identifier (CEI) which is defined as a combination of the SAPI and a value called the connection end-point

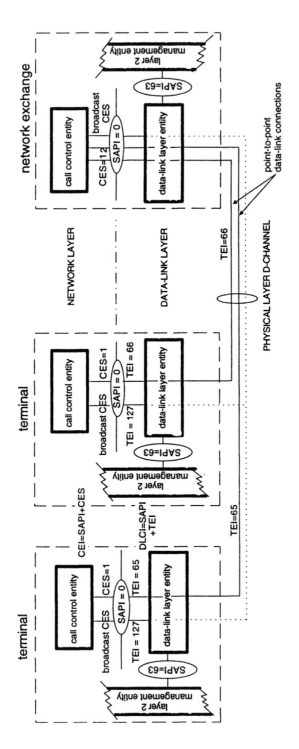

Fig. 5.3 Address parameters on the subscriber loop.

suffix (CES). The relationship between the DLCI and the CEI is therefore through the association of a TEI with a particular CES. The TEI and CES have a one-to-one correspondence, but whereas the TEI is known by both the network and terminal sides of the data link, the CES is allocated separately by the layer 3 in the terminal and the network which have no knowledge of each other's CES value. Each time a new TEI value is assigned to a terminal, a CES value is a allocated to it. The relationship between these variables is also illustrated in Fig. 5.3.

5.2.4 The command/response (C/R) bit

Frames in both directions on a data-link connection are identified as commands or responses by the C/R bit in the first address byte of the frame that also contains the SAPI. The terminal will send commands to the network with the C/R bit set to 0 and responses with the C/R bit set to 1. This procedure is reversed in the network-to-terminal direction.

5.2.5 The poll/final bit

All LAPD frames contain a poll/final (P/F) bit whose function depends on the frame being either a command or a response. In a command frame the bit is referred to as the poll (P) bit which is set to 1 by the data link in order to solicit a response frame from the peer data link. The peer then responds with a response frame in which the bit, now referred to as a final (F) bit, is also set to 1.

5.2.6 Control field

Every LAPD frame contains a one- or two-byte control field that defines the type of command or response frame being communicated. Depending on the type, the LAPD procedures will process the frame accordingly. The commands and responses of LAPD fall into three categories: information (I), supervisory (S), and unnumbered (U). Definition of these frames is by means of the first byte in the control field of the frame, and are listed in Fig. 5.2.

I frames are those used to transfer information between peer layer 3 entities across a point-to-point data-link connection. The procedures used to communicate I frames allow for the full-duplex acknowledged transfer of sequential frames. The U frame format is used in an unnumbered and unacknowledged frame transfer service for the communication of infrequent commands and responses that do not require the same level of reliable data-link transfer as I frames. The S format is used for data-link commands and responses that perform supervisory functions such as acknowledgment of I frames, request re-transmission of I frames, and temporarily suspend the transfer of I frames.

5.3 LAPD PROCEDURES

The LAPD procedures define the way in which frames are processed, based on the data-link layer information contained within the frame.

5.3.1 Numbered I frame operation

LAPD can operate in a number of modes for the communication of frames across the data link, the most usual of which is known as **multiple frame** mode. In this mode, each end of the point-to-point data link has an equal possibility to send multiple frames (a **balanced** frame sending capability), and for each end to send frames as required (an **asynchronous** capability). In this mode, messages that are communicated between peer layer 3 entities are embedded in the information field of I frames. The information field in each I frame can be a maximum of 260 bytes in length.

A sequential frame numbering system is used by LAPD to communicate I frames. The frame numbering mechanism is based on two numbers included in every I frame that reflect the number of I frames transmitted, N(S), and received, N(R), by the end of the data link that sends the frame. The N(S) and N(R) values are incremented for each sequential frame sent and received, and allow I frames received without error to be automatically acknowledged, and frames received in error to be resent. As shown in Fig. 5.2, N(S) and N(R) are represented by seven-bit numbers (0–127) that when incremented from 127 are simply reset to 0 and then further incremented as normal.

Consider the data-link communications between a terminal, designated with suffix t, and the network exchange, designated by the suffix n. When the network side of the data link sends an I frame it will include the numbers $N_n(S)$ and $N_n(R)$, while a frame transmitted by the terminal will contain the numbers $N_t(S)$ and $N_t(R)$. The acknowledgment process works on the basis that when a frame is received, say at the terminal, the data link can compare the received $N_n(S)$ value with that of the last frame it received. Under normal operating conditions it should be one more than the previous value. Receipt of the frame is then acknowledged by the terminal by setting the $N_t(R)$ value used in the next outgoing frame to equal the received $N_n(S)$ value. When received at the other network exchange, it tells the data link there that the terminal has received the same number of frames as it has transmitted. Note that because the transmission of frames is asynchronous, the terminal may not have a frame to send back to the exchange before several frames are received. However, by setting $N_t(R)$ to the last received value of $N_n(S)$, it acknowledges all previously received frames. In order to limit the number of outstanding acknowledged frames, a transmitter will stop when the difference between its own N(S) value and received N(R) value exceeds a parameter known as k. The value of k is fixed according to use and speed of the data link as shown below in Table 5.3.

Should a frame be received at the terminal with a framing error and discarded, then the network exchange will receive an $N_t(R)$ value less than its current $N_n(S)$ value. This then prompts the network side data link to resend the frames with numbers between these values. The reject frame (REJ), which contains N(R), is used to request re-transmission of I frames starting with the frame number contained in N(R), and thus acknowledges transmitted frames prior to this number. This numbered acknowledgment process applies to both network and terminal sides of the data link.

Table 5.3 Data-link k values for different applications

k value	Application
1	16 kbps call control signalling (BRA)
7	64 kbps call control signalling (PRA)
3	16 kbps packet transfer
7	64 kbps packet transfer

5.3.2 Use of RR and RNR frames

A situation can exist during multiple frame mode where one side of a data link sends significantly more I frames than the other. Let us assume that the terminal is sending I frames to the network at a much higher rate than the network is toward the terminal. If the number of outstanding I frames in the data link reaches the limit k before the network has an I frame to send, then the transmitter at the terminal would remain blocked until such time as a frame became available at the network. In this situation, the network can acknowledge the I frames from the terminal using a receiver ready (RR) frame that contains the appropriate N(R).

In the particular case of call control messages in a BRA D-channel, it can be seen from Table 5.3 that the number of outstanding I frames is only 1. This means that unless a receiving data-link entity happens to have an I frame ready to send at the time it receives an I frame, it will be acknowledged by an RR frame.

A data-link entity may become congested at the SAP to the network layer, in which case the receiver not ready (RNR) frame is used to deliberately hold off transmissions from the peer data-link entity until such time as the congestion is cleared. RNR can therefore be used as a means of flow control, and is also used to acknowledge receipt of frames up to the value of N(R) in the RNR frame.

5.3.3 Unnumbered operation

As well as the numbered I frame service, the data-link entity provides a less secure unnumbered information frame service in which no sequence numbering is used and no acknowledgment procedures are provided. Instead, messages for transmission are placed in the information field of a UI frame and sent. Such a service is intended for simple command and response type communications where the responsibility for recovering from lost or corrupted frames is achieved in higher protocol layers. The unnumbered service is typically used for peer-to-peer management or supervision transactions, or for layer 3 communications across the broadcast data link.

5.3.4 LAPD state machine

LAPD is state machine driven by events that occur at its interfaces. As can be seen from Fig. 5.3, these interfaces are the service access points to the layer 3 entity (for example call control), the management entity, and the physical layer. Information is communicated across these interfaces in the form of **primitives**. Primitives were already introduced in Chapter 3 in conjunction with the operation of the S interface transceiver.

The operation of the LAPD data-link entity is described as a state machine which has three basic states as follows:

TEI-unassigned state. In this state, the only TEI assigned is the broadcast TEI which establishes a broadcast data link between the network exchange and the terminals. The broadcast TEI is permanently available in all data-link states and is shared by all terminals connected to the S interface. However, no point-to-point TEI exists in this state. This state is typically entered when a terminal is disconnected from the S bus or has had its power removed. Transition from the TEI-unassigned to the TEI-assigned state involves the assignment of a TEI to a terminal side data link by its layer 2 management entity. This then establishes a dedicated point-to-point logical data link between the terminal and the network.

TEI-assigned state. In this state, a TEI has been assigned and a logical data-link connection is now established between the terminal and network. However, only unacknowledged data transfer is possible using data-link UI frames. The establishment of the I frame service may be requested by a layer 3 entity with a DL-ESTABLISH-REQUEST primitive. If the peer data-link entities are in a TEI-assigned state, a **set asynchronous balanced mode extended (SABME)** command is issued from one side of the data-link connection to the other, which has the effect of zeroing the sequence numbers N(S) and N(R) and state variables in both data-link entities. The data-link entity which receives the SABME frame acknowledges this action by returning an unnumbered acknowledge (UA) frame. Establishment of the service is then acknowledged to the layer 3 which issued the DL-ESTABLISH-REQUEST with a DL-ESTABLISH-CONFIRM primitive, while its peer is informed with a DL-ESTABLISH-INDICATION primitive. Both sides of the data link are then ready to exchange I frames and are said to be in the **multiple frame established** state. If for some reason a data-link entity cannot enter the multiple frame established state, then this condition is indicated to its peer with a disconnect mode (DM) message whenever it is requested to enter the multiple frame established state with a SABME message.

Multiple frame established state. In this state, both sides of the data-link connection can exchange both UI and I frames. Peer-to-peer layer 3 messages are passed to the data-link layer across the SAP using the DL-DATA-REQUEST (layer 3 to layer 2) and DL-DATA-INDICATION (layer 2 to layer 3) primitives. These primitives invoke the I frame service. Alternatively, DL-UNIT-DATA-REQUEST and -INDICATION primitives invoke the UI frame service.

Termination of the multiple frame established state may be caused in several ways. One way is through deactivation of the physical layer when there are no calls in progress on the subscriber loop. The PH-DEACTIVATE-INDICATION/CONFIRM primitives will be received by the data link from the physical layer, and cause it to cease multiple frame operation. Alternatively, it may be achieved by the data link receiving a DL-RELEASE-REQUEST primitive from the layer 3, causing it to issue a DISC frame to its peer. As a result, the relevant DL-RELEASE-CONFIRM/INDICATION primitives will be sent to the layer 3 entities. Termination may also occur as a result of a time-out error condition or persistent deactivation of the physical layer during a call, after which communication with the peer data-link entity is no longer possible. In all cases, the data link returns to the TEI-assigned state.

An additional five states are defined which are used as transitionary states between these three basic states.

5.3.5 Layer management

The layer management functions provides a number of services to the layers which assist in their coordination and management of resources. The services provided to the data-link layer are split into two groups and are implemented by a layer management entity and a connection management entity.

As its name suggests, the connection management entity manages those resources to do with a specific data-link connection such as initialization, negotiation of operational parameters, error processing and the invocation of flow control. Primitives addressed to the connection management entity identify the relevant data-link connection by including its SAPI and CES values. The layer management entity provides the management of resources which have a more global impact to the whole layer, such as TEI management. This involves procedures for the assignment, checking and removal of TEIs. The layer management entity is addressed through the SAPI=63 as illustrated in Fig. 5.3.

5.3.6 TEI administration procedures

Establishment of a data-link connection between peer data-link entities can only take place if a TEI has been assigned to the SAPs within the terminal. For terminals designed to support automatic TEIs, this takes place at some time prior to the establishment of the data-link connection through an exchange of messages between the layer 2 management entity, referred to from now on simply as ME, and its network peer. The terminal side ME communicates with its peer in the network for the purpose of administering TEIs using the unnumbered UI command frames with SAPI=63 and TEI=127. The broadcast TEI is used as this is the only means for the peer data-link layers to communicate prior to the assignment of a point-to-point TEI.

TEI administration consists of three main procedures; TEI assignment, TEI checking and TEI removal.

5.3.7 TEI assignment

A data-link entity which has no assigned TEI, but is requested by the network layer to deliver a message to its peer, will request assignment of a TEI from the ME using a MDL-ASSIGN-INDICATE primitive (notice here that the INDICATE primitive is used as the ME is considered to be a higher level function than the data-link entity, although effectively both reside in layer 2). The primitive contains the CES of the connection for which a TEI is to be assigned. If the terminal is of the non-automatic TEI type, then the ME will simply return to the data-link entity the pre-defined TEI in a MDL-ASSIGN-REQUEST primitive which it uses to associate with the CES of the link to be established. However, if the terminal is of the automatic TEI type, the ME must first communicate with its peer in the network to obtain its TEI.

The message sequence between peer MEs for automatic TEI assignment is shown in Fig. 5.4a. The terminal ME uses the LAPD UI frame service to communicate with its peer by sending and receiving MDL-UNIT DATA-REQUEST and MDL-UNIT DATA-INDICATION frames across the SAP with SAPI=63. These messages contain an ME identifier, a reference number (Ri), a message type and an action indicator (Ai), as shown in Fig. 5.4b.

In response to the MDL-ASSIGN-INDICATE primitive, the ME will send a MDL-UNIT DATA-REQUEST primitive to the data-link entity. The primitive will contain a message with the usual ME identifier, a message type indicating a request for a TEI, and an Ai value indicating that any non-automatic TEI value is acceptable.

In a passive bus configuration with several connected terminals, a situation may arise where more than one terminal sends a TEI request message at the same time. As these messages will all use the broadcast TEI value, the TEI cannot be used by the network to address a message to a particular terminal. This is where the reference number, Ri, is used. Ri is a 16-bit number which is randomly generated by the terminal such that it is unlikely that the messages from different terminals contain the same Ri. The network ME, when responding to a specific terminal with a broadcast UI frame that contains its assigned TEI, will use the corresponding Ri value in the frame in order to specifically address it to that terminal. Other terminals receiving the frame will simply discard it. Although such addressing is not required in a point-to-point S interface configuration, the same procedures are adopted so that a terminal need not have prior knowledge of its connection to a particular type of S interface configuration.

The assigned TEI value is thus returned to the ME of the requesting terminal in a MDL-UNIT DATA-INDICATION primitive. The ME then requests that the data-link entity associate the assigned TEI with the corresponding CES with the MDL-ASSIGN-REQUEST primitive, and the data link enters the TEI-assigned state.

After TEI assignment, the peer MEs and data-link entities may optionally exchange parameter information to ensure that both sides of the data link have compatible capabilities. Parameter negotiation uses the MDL-XID primitives which employ the LAPD XID frame service of the data-link entity.

5.3.8 TEI check procedures

At any point in time, the network may wish to perform an audit of the TEIs it has assigned to terminals connected to the S bus. The network side ME issues an **identity check request** message in which the Ai field contains the TEI value to

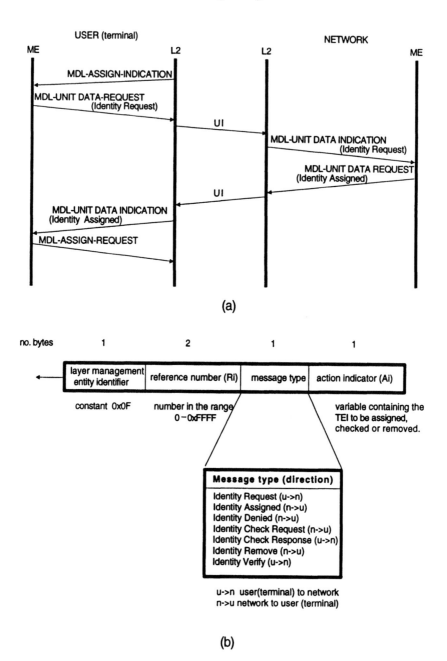

Fig. 5.4 Automatic TEI assignment. (a) The exchange of messages between terminal and network during TEI assignment; (b) the management entity (ME) message format.

be checked. This message is broadcast to all terminals. Under normal circumstances each terminal will contain a different TEI, so either a single **identity check response** from the terminal whose TEI corresponds to that being verified, or no response at all, will be received by the network ME. A one-second timer is set running in the network ME each time a check request is made so as to limit the amount of time a terminal has to respond. Should no response be received by the time the timer has timed-out, the network assumes that the TEI value is free to be assigned to a terminal.

If for some reason more than one terminal has an identical TEI, then the network ME will receive more than one response in reply to its check request. A potential problem here is that the message contained in each response would be identical, and if transmitted at the same time, would appear to the network as only a single response, fooling it into thinking that there was only a single TEI assigned. For this reason, the response messages from the terminals contain the randomly generated Ri value so that if the responses are transmitted simultaneously, a D-channel collision will occur between the responses at the time the Ri values are transmitted. The D-channel resolution mechanism discussed in Chapter 3 ensures that separate responses are received by the network. If the network detects, through use of the TEI check procedures, that a TEI value has been assigned more than once, then it will remove the TEI value in question following which the relevant terminal data links will request assignment of a new TEI value.

The network side ME can also request verification of all TEIs currently assigned to terminals by using the Ai value of 127. When this happens, a terminal will respond with an **identity check response** message in which the Ai field is extended (using the extension bit technique described in section 5.2) to contain all TEI values currently assigned to it.

An option also exists within these procedures to allow a terminal that suspects that multiple TEIs have been assigned, to send a **TEI identity verify** message to the network to request that the network invoke the check procedure for multiple TEI assignment.

5.3.9 TEI removal

Once a terminal is assigned its TEI value it will retain it unless its ME detects one of the following conditions:

- the terminal is disconnected from the S bus (assuming that power is not lost from the terminal);
- the terminal data-link entity has assumed a possible multiple assignment of a TEI value and either (a) removes the TEI value itself, or (b) sends an identity

verify check command to the network following which, if a multiple assigned TEI is found, the network will remove the TEI;

- the network attempts to assign a TEI value which is already in use by the terminal;
- the network specifically requests removal of a TEI.

The terminal ME may, under the above conditions, instruct a data-link entity to enter the TEI-unassigned state by sending it a MDL-REMOVE-REQUEST primitive. This causes its normal data-link procedures to cease and to discard any information it may have in internal queues ready to transmit. Removal of a TEI by the network ME involves it broadcasting an **identity remove** message twice to all terminals connected to the S bus, with the TEI value to be removed in the Ai field of the message. The message is sent twice to reduce the possibility of message loss as UI frame transmissions are not acknowledged. The ME in the terminal with the corresponding TEI will then issue a MDL-REMOVE-REQUEST primitive to its data-link layer.

5.3.10 Data-link parameters

Each data-link connection has a number of parameters associated with it which define the duration of timers, the number of re-transmissions permitted, the maximum number of outstanding I frames allowed, and the maximum number of bytes allowed in the information field of an I frame. These parameters are typically defined as default values by the relevant layer 2 standard, but may be altered according to the requirements of the network through the parameter negotiation procedures implemented by the management entity discussed earlier. The purpose of the timers and their duration, as well as the definition of other data-link parameters, are described below.

5.3.10.1 Timers

Timers are used to impose time limits to the response time allowed for certain procedures. The default values are those defined by the ITU-T Q.921 Recommendation, but may be assigned different values in national ISDNs that reflect the individual requirements of those networks.

T200: This timer sets the limit allowed for the interval allowed between transmission of an I- or S-command frame from the data link and a response from its peer. The default value is set at one second.

T201: This timer is used to set the minimum time between the re-transmission of the TEI identity check message. Its value is also set at one second.

T202: This timer is used to set the minimum time between the re-transmission of the TEI identity request message and has a default value of two seconds.

T203: The value of this timer defines the maximum time allowed without frames of any type being exchanged by the data-link entities. Consequently, a data-link connection between a terminal and the network exchange which is in the multiple frame established state but currently inactive, as for example during the conversation phase of a call, will periodically exchange RR (receiver ready) frames, and provides a confirmation to the data-link layer that the underlying physical layer and cable connecting them is intact. This function is referred to as the **layer 2 monitor function** and is usually initiated on the network side of the data-link connection. The default value of T203 is 10 seconds.

5.3.10.2 Counters

The other parameters which are associated with the data link are values assigned to counters which determine the maximum number of frame re-transmissions allowed, the maximum number of outstanding I frames allowed, and the maximum number of information bytes allowed in an I frame.

N200: This parameter defines the maximum number of frame re-transmissions allowed and has a default value of 3. For example, if one side of a data-link connection attempts to transmit an I frame but receives no response because the S bus has been disconnected, then the data link will, as a result of a time-out of T200, attempt to re-transmit the same frame. It does this three times in succession. If a response is still not received, the data link will inform the ME that an error has occurred using the MDL-ERROR-INDICATION primitive and attempt to re-establish the data-link connection by transmitting a SABME command. However, because the physical connection has been broken, the SABME is not received at its destination and T200 again times-out, causing a re-transmission of the SABME. The re-transmission occurs three times before the data link gives up trying and signals to the layer 3 with a DL-RELEASE-INDICATION that a data-link connection no longer exists. Similarly, an MDL-ERROR-INDICATION is sent to the layer connection management entity.

N201: This parameter defines the maximum number of bytes allowed in an I frame and has a default value of 260 for SAPs that support signalling or packet data communications.

N202: This parameter defines the maximum number of transmissions by a terminal of the TEI identity request message, and has a default value of 3.

5.4 THE NETWORK LAYER FOR BASIC CALL CONTROL

The control of a call in the ISDN across the subscriber loop is performed by a layer 3 protocol known generically as **call control** and defined by the ITU-T Q.931 Recommendation [4]. This network layer protocol generates the necessary messages to set up, disconnect and manage a call with the desired attributes either to or from the network. Together with additional functions built on top of the call control, it is also used to control the invocation of supplementary services.

From an architectural viewpoint, the network layer functionality may be divided into two sub-layers, one which we shall refer to as **protocol control** and the other as **call control**, as shown in Fig. 5.5. The protocol control part interfaces directly to the data-link layer, and is the state machine which takes each call through the logical sequence of states defined for its setup, connection and disconnection phases. As such, the protocol control is application independent and has no knowledge of what the call is being used for. The call control function exists above the protocol control and acts as a pseudo-application layer to provide call processing functions which are specific to a particular application. It may also include the implementation of supplementary service functions.

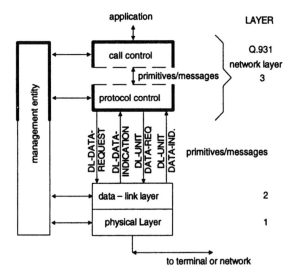

Fig. 5.5 The Q.931 network layer and its relation to other layers in the call control protocol stack.

The network layer makes use of the underlying physical layer and data-link connection to reliably transfer messages between itself and its peer. Communications between the data-link layer and the protocol control are by means of the DL-DATA-REQ and DL-DATA-IND primitives, and these messages are transferred between the terminal and network using the data-link numbered I frame service. An exception to this is the establishment of a call on a passive S bus where the call is offered to all the terminals connected to the S bus through the use of a broadcast message using the unnumbered data-link frame service. These messages are communicated between the data-link layer and protocol control through DL-UNIT-DATA primitives.

The Q.931 Recommendation does not detail any specific functions of a layer ME for layer 3, but in practice it may be used to provide supervisory functions such as error logging and reset for the network layer calls. In the future it can be expected that the ME function will be expanded to support automatic configuration and other network management functions that interact directly with the ISDN.

5.5 Q.931 MESSAGE STRUCTURE

The layer 3 message contains a series of information elements, the first three of which are contained in all messages. These are the **protocol discriminator** (PD), the **call reference** (CR) together with its **call reference length** (CRL), and the **message type** (MT). Depending on the message type, there then follows a number of other information elements (IEs) associated with it. The general structure and different message types of a layer 3 call control message are shown in Fig. 5.6.

5.5.1 Protocol discriminator

The protocol discriminator (PD) is the first byte of the message and uniquely identifies the layer 3 protocol that is to process the message. In the case of Q.931 call control, the PD has a byte value of 0x08 (hexadecimal). Messages that are to be processed according to another protocol, for example layer 3 X.25 procedures, would contain a different PD.

5.5.2 Call reference and call reference length

The layer 3 call control has the ability to establish a number of simultaneous calls, and the call reference (CR) is a value which uniquely identifies the call with which the message is associated.

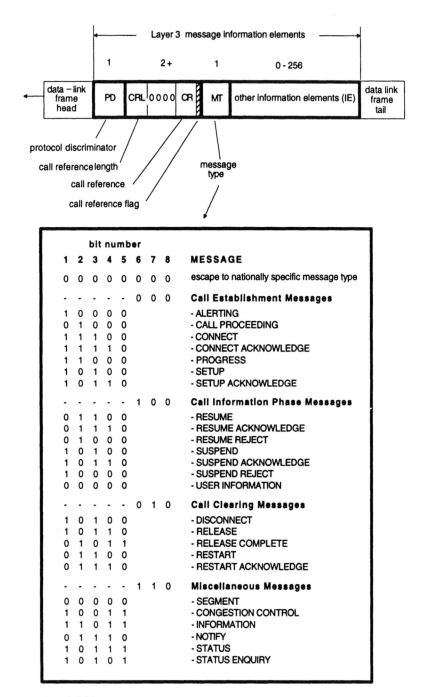

Fig. 5.6 The Q.931 message structure.

A particular CR has only a local significance between the terminal and the network across the subscriber loop, and other mechanisms are used inside the network to associate signalling messages between exchanges with the call. This means that CR value for the call at the calling subscriber loop is allocated independently of the CR value for the same call at the called subscriber loop, and is most likely to be different. A CR is assigned to a call when it is originated and remains until the call is either disconnected or suspended. If a user originates a call, then the terminal call control will assign a CR value, while for an incoming call the network layer assigns the CR. In the event that both the terminal and network exchange simultaneously attempt to set up a call on the same subscriber loop using the same CR, then a bit in the CR called the call reference flag uniquely identifies which side of the subscriber loop originated the CR value, and thus removes any ambiguity or conflict between the CRs for each call. The call reference flag is positioned in the most significant bit of the first CR byte and has a value of 0 in messages from the originating side of the subscriber loop, while messages from the opposite side will have a call reference flag set to 1.

The length of the CR is contained in the first four bits of the preceding byte in the message. All BRA equipment must support a minimum one-byte CR value while the default maximum length of the CR is three bytes. A two-byte length is typically used for PRA interfaces. Additionally, a global CR value of zero is defined which identifies the message as being relevant to all existing CRs associated with a particular data-link connection identifier (SAPI and TEI)[10].

5.5.3 Message type

The message type (MT) identifies the function of the message, and it is this information which is generated by, and controls the operation of, the call state machine in the protocol control for each call. Figure 5.6 lists the main message types defined in Q.931 and their MT values. Bit 8 may be used in the future as an extension bit when more than seven bits are needed to define new message types. More message types are defined for the control of supplementary services and these are dealt with later in this chapter.

5.6 OTHER INFORMATION ELEMENTS

Figure 5.7 shows in tabular form some of the layer 3 messages used to establish a circuit-switched call and their information elements (IEs). The IEs contain information which is communicated between the terminal and the network to

---

[10]Note that the call control messages from multiple calls will use the same data-link connection.

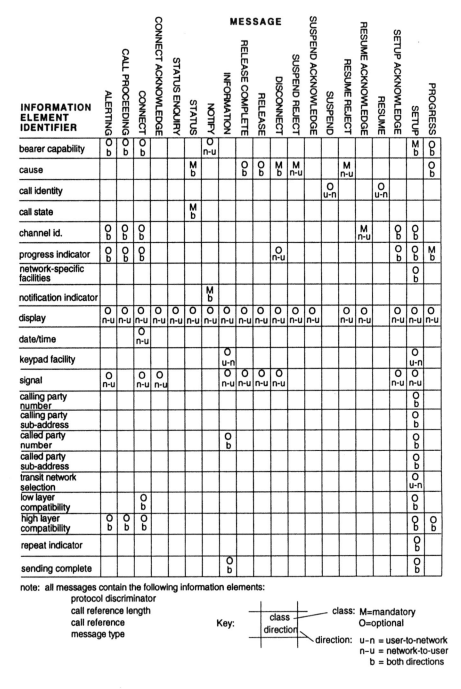

MESSAGE

INFORMATION ELEMENT IDENTIFIER	ALERTING	CALL PROCEEDING	CONNECT	CONNECT ACKNOWLEDGE	STATUS ENQUIRY	STATUS	NOTIFY	INFORMATION	RELEASE COMPLETE	RELEASE	DISCONNECT	SUSPEND REJECT	SUSPEND ACKNOWLEDGE	SUSPEND	RESUME REJECT	RESUME ACKNOWLEDGE	RESUME	SETUP ACKNOWLEDGE	SETUP	PROGRESS
bearer capability	O/b	O/b	O/b					O/n-u											M/b	O/b
cause						M/b			O/b	O/b	M/b	M/n-u			M/n-u					O/b
call identity														O/u-n			O/u-n			
call state						M/b														
channel id.	O/b	O/b	O/b													M/n-u		O/b	O/b	
progress indicator	O/b	O/b	O/b								O/n-u							O/b	O/b	M/b
network-specific facilities																			O/b	
notification indicator							M/b													
display	O/n-u	O/n-u	O/n-u	O/n-u	O/n-u	O/n-u	O/n-u	O/n-u	O/n-u	O/n-u	O/n-u	O/n-u	O/n-u	O/n-u		O/n-u	O/n-u	O/n-u	O/n-u	O/n-u
date/time			O/n-u																	
keypad facility								O/u-n											O/u-n	
signal	O/n-u		O/n-u	O/n-u				O/n-u	O/n-u	O/n-u	O/n-u							O/n-u	O/n-u	
calling party number																			O/b	
calling party sub-address																			O/b	
called party number								O/b											O/b	
called party sub-address																			O/b	
transit network selection																			O/u-n	
low layer compatibility				O/b															O/b	
high layer compatibility	O/b	O/b	O/b																O/b	O/b
repeat indicator																			O/b	
sending complete								O/b											O/b	

note: all messages contain the following information elements:
protocol discriminator
call reference length
call reference
message type

Key:

class: M=mandatory
O=optional

direction: u-n = user-to-network
n-u = network-to-user
b = both directions

Fig. 5.7 Q.931 circuit mode connection messages and their information elements.

define specific actions or attributes associated with the call. Some IEs must be included with certain types of message, and these are identified in Fig. 5.7 with the letter M, indicating that they a mandatory for a particular message. Other IEs may be optionally included, indicated by the letter O, depending on the conditions under which the message is sent. Also, IEs may only be relevant in a particular direction between the terminal and the network exchange as indicated. For other types of call, for example for packet mode connection or non-call associated signalling to control supplementary services, different sets of IEs may be used for the same message, or new messages used to invoke protocol functions outside the basic Q.931 call control.

An IE comprises an **IE identifier** followed by its contents. Both of these may either fit into a single byte called a **single byte IE** or may occupy several bytes in a **variable length IE**, one byte of which will contain the IE identifier while the remaining bytes will be arranged in groups according to their meaning. For example, the bearer capability IE which defines the capability of physical layer channel requested or allocated to a call, is a variable length IE and is identified by the hexadecimal byte 0x04, following which there are a number of fields that contain information such as the transfer capability required and its rate, the transfer mode and its structure, and the protocols used at the data-link and network layers. However, not all of the possible groups, and not all of the bytes within a group, will typically be present in an IE, thus giving the IE a variable length.

The values assigned to IE identifiers and their format are designed to allow the call control protocol to quickly locate and identify those IEs which are relevant to it. The IE identifiers present in a message appear in ascending numerical order, which allows the call control to verify the presence of a particular IE without scanning the entire message. The byte following the IE identifier of a variable length IE always contains the length of the IE, so that from the first IE identifier following the MT in a message, the position of following IE identifiers in the message can easily be located.

By way of an example, the call control SETUP message shown in Fig. 5.8 which is used to request establishment of a call, illustrates the IEs that it would contain in order to request a voice call. The first four bytes contain the PD, CR and its length, and the MT. The CR is a single byte in length and has a value of 1. The CR flag is set to indicate that the SETUP message originated from a terminal rather than the network exchange. The MT value of 0x05 identifies the message as a SETUP.

The first byte to follow the MT will be an IE identifier, which in this case is that of the bearer capability IE. As this is a variable length IE, the following byte indicates how many bytes there are in the remainder of the IE that make up its contents. In this case there are three bytes, where each byte is split up into a number of defined fields containing the necessary information. For example, the coding standard value 00 indicates ITU-T standardized coding. If the value had

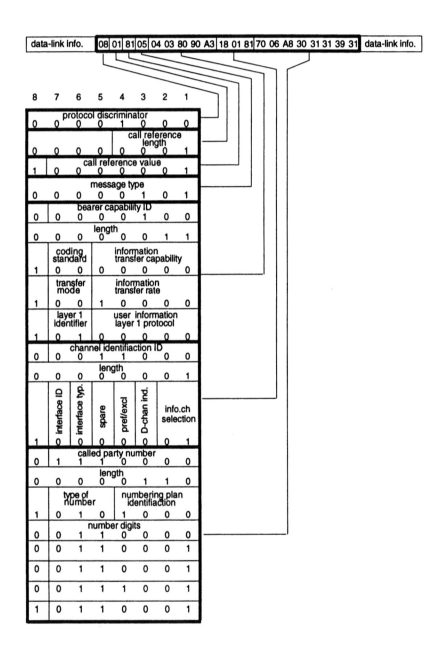

Fig. 5.8 An example of information element detail found in a typical SETUP message.

been 10 instead, then this would have meant that the following elements relating to the bearer capability were coded according to some national standard. As with the other IEs, further options are defined in the Q.931 to represent the possible options available from the ISDN, although not all of the options may be implemented by it. The information transfer capability value of 00000 defines speech capability.

The bearer capability IEs for the SETUP message shown in Fig. 5.8 define the following call parameters:

Transfer mode –	circuit mode
Information transfer rate –	64 kbps
Layer 1 identifier –	Layer 1 user information
Layer 1 protocol –	G.711 A-law

The channel identification IE in the context of this call indicates the preferred use of a particular B-channel for the call. Its single byte content, the only one used from a possible group of seven bytes, is sufficient to define the desired channel as follows:

Interface identification –	implicitly defined
Interface type –	BRA
Preferred/exclusive –	indicated channel is preferred
D-channel indicator –	channel indicated is not the D-channel
Info. channel selection –	B1 channel

The final IE in the SETUP message defines the called party number, and would be included if the terminal making the call had prior knowledge of the complete number before the SETUP message was sent. The following parameters are defined:

Type of number –	national number
Numbering plan identification –	national standard numbering plan
Number digits –	01191 (for example)

The number digits are encoded using the International Alphabet number 5 (IA5) as defined in the ITU-T T.50 Recommendation [6] which is a subset of the commonly used ASCII character set. The called number is a variable length dependent on the numbering plan used. This particular set of IEs are an example of a variable length IE group whose length is determined by use of an extension bit contained in bit 8 of each byte in the group. A 0 extension bit indicates that the byte following is also part of the same group, while a 1 extension bit indicates that the byte is the last in the group and that the following byte is either the first in a new group or a new IE identifier.

5.6.1 Codesets

The coding of IE identifiers for both variable length and single-byte IEs allows a maximum of 133 identifier values to be assigned. Provision is made in Q.931 to extend this to eight different codesets, each with a minimum of 133 IE identifiers. The default codeset which we have assumed so far is codeset zero, while codeset five is reserved for national use, codeset six for local network specific IEs, and codeset seven for user specific IEs. Use of the remaining codesets is as yet undefined.

Changing from one codeset to another involves one of two procedures which are similar to the shift procedures used on a typewriter or keyboard to change between upper-case and lower-case letters. The first procedure is referred to as a locking shift procedure whereby the inclusion of the single-byte locking shift IE in the IEs of a message indicates that all the IEs following it have been encoded according to the new codeset defined in the locking shift IE. The call control protocol that processes the message therefore interprets the IEs according to the new codeset until another locking shift IE is detected, indicating a change to another codeset. The action of the locking shift procedure is therefore to make the change to the new codeset permanent until another codeset is indicated with another locking shift IE.

The alternative is the non-locking shift procedure, which also uses a single-byte IE (but with a different IE identifier) to indicate the new codeset to be used, but is used to interpret only the IE immediately following it, after which the IEs are interpreted according to the original codeset until another codeset shift IE is detected. The action of the non-locking shift procedure is therefore not permanent and changes the active codeset only for the single IE that follows.

5.7 PROTOCOL CONTROL

The protocol control state machine uses the MT for the control of a call through the logical states required to set up, tear down and manage a call.

5.7.1 Point-to-point call setup and tear-down

Figure 5.9 shows the sequence of Q.931 messages exchanged across the subscriber loop of both the calling party and called party in order to set up and tear down a call between two ISDN telephones, each on a point-to-point BRA S bus. The sequence of messages represents a typical successful call setup and tear-down procedure and is used here primarily to introduce the basic Q.931 messages and illustrate their use. Other sequences are possible depending on the type of call, the way it is established, and to account for the possible occurrence of errors. For simplicity, Fig. 5.9 shows only the peer-to-peer layer 3 interactions and assumes that the underlying data-link connection and physical layer have already been established. As already mentioned, the Q.931 messages are communicated between the layer 3 call control and the data-link layer in the form of DL-DATA-REQUEST and DL-DATA-INDICATION primitives which instructs the data link to transfer the messages using the numbered I frame service in the physical layer D-channel.

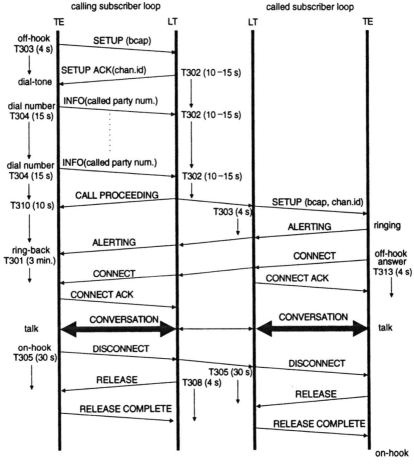

note: TXXX (Y) - layer 3 timer number (XXX) with Y timeout value

Fig. 5.9 Layer 3 peer-to-peer messages for call set up and tear-down.

The user initiating the call goes off-hook which causes the telephone to send a SETUP message containing a new call reference to the network exchange. The SETUP message will also contain information elements which will inform the network of the bearer capability requested. Having established that the network can support the call, the exchange returns a SETUP ACKNOWLEDGE message to the telephone containing identification of the B-channel to be used for the call in the channel identification IE. In some cases, the telephone may indicate a preference of the B-channel(s) it wishes to use in its outgoing SETUP message to the network. Receipt of the SETUP ACKNOWLEDGE message by the telephone is an indication that the network requires more information in order to establish the call, in particular the number of the called telephone. Receipt of SETUP ACKNOWLEDGE will prompt the telephone to issue a dial-tone which may either be generated locally by the terminal if it has the capability, or alternatively may be generated in the exchange and sent in the selected B-channel to the terminal.

The series of INFORMATION messages which follow each correspond to a single dialled digit which together make up the called party telephone number. When the last digit has been sent, the exchange starts to establish a circuit through the network to the destination subscriber loop, and indicates to the telephone that the call is being processed with the CALL PROCEEDING message. The transmission of the telephone number a digit at a time toward the network is referred to as **overlap sending**. Should the telephone contain a pre-programmed telephone number, then it may be sent **en-bloc** in the called party IE in the initial SETUP message. All the information needed by the local exchange to start to establish the network circuit to the chosen destination is therefore contained within the SETUP message, and so the local exchange acknowledges the SETUP message with CALL PROCEEDING instead of SETUP ACKNOWLEDGE.

At the called subscriber loop, the local network exchange informs the called telephone that an incoming call is present with a SETUP message containing its own call reference, the bearer capability and the channel identification. If the latter two parameters are compatible with the capability of the terminal, the called telephone activates its ringer to alert the user of an incoming call and indicates to the network that it has done so with an ALERTING message. The alerting condition propagates back through the network to the calling subscriber loop whose local exchange sends a corresponding ALERTING message to the calling telephone. This prompts it to activate a ring-back tone to the user as an indication that the called telephone is ringing. When the called user answers the call, the telephone generates a CONNECT message which is sent to the network. The CONNECT message instructs the exchange to connect the assigned network circuit to the designated B-channel of the subscriber loop, which it acknowledges with the CONNECT ACKNOWLEDGE message to the terminal. As a result, the telephone connects the B-channel to its voice circuits and the subscriber loop call enters the active state. If the terminal equipment can answer a call automatically without user intervention, as for example a fax machine can, then the CONNECT message is sent out immediately by the called terminal without a preceding ALERTING message.

To complete the connection at the calling subscriber loop, the CONNECT message sent from the network arrives at the calling telephone, indicating that the called user has accepted the call. Receipt of the CONNECT message causes the telephone to connect its voice circuits to the designated B-channel, if it has not already done so, to enable tone indications from the network exchange, and enter the active call state. A conversation may then take place between the two telephones. On receiving a CONNECT message, the telephone may optionally send a CONNECT ACKNOWLEDGE message back to the network. This message has no effect on the network, but is included to maintain a degree of symmetry between the network and terminal call control 3 protocols.

When the conversation is finished, either user may go on-hook first. In this example, we assume that the calling user goes on-hook as shown in Fig. 5.9. Call tear-down is initiated with a DISCONNECT message which, when received at the network exchange, prompts it to disconnect the B-channel from the network circuit, and start to clear the network circuit to the called subscriber loop. Once the network has disconnected the B-channel, it sends a RELEASE message to the terminal, requesting that the B-channel and call reference be released and thus made available for a future call. Completion of this is acknowledged to the network with a RELEASE COMPLETE message.

The disconnect state of the calling subscriber loop is communicated across the network to the local exchange of the called subscriber loop, which sends a corresponding DISCONNECT message to the called telephone to indicate that the end-to-end connection is being cleared. The terminal responds with a RELEASE message whose action is confirmed with a RELEASE COMPLETE message from the exchange.

5.7.2 Broadcast call setup

The call procedures must be capable of establishing a call to a terminal which is connected to a point-to-multipoint S bus (passive bus) configuration as well as the point-to-point configuration. The passive S bus configuration poses the problem of which terminal should receive the incoming call. If we assume that the called subscriber loop in Fig. 5.9 has a point-to-multipoint S bus, then the SETUP message from the network is broadcast to all terminals using the broadcast data-link connection. First, each terminal checks the compatibility of the call attributes defined in the SETUP message with its own capability, and those terminals unsuitable to receive the call take no further action. Second, where each terminal on the S bus can be identified by a different telephone number[11], and which was included by the calling user when placing the call, then the broadcast SETUP message will contain the number which allows only that terminal to respond with ALERTING to the network and then CONNECT the call. Lastly, where several connected terminals exist which cannot be distinguished using the previous means, then each may send an ALERTING message to the network. The first terminal to then send a CONNECT message is awarded the call with a CONNECT ACKNOWLEDGE message from the network, while the remaining terminals are returned to their idle states with RELEASE messages from the network.

5.7.3 Layer Initialization

The setup and tear-down procedures described above assume that an underlying physical layer and data-link connection in multiple frame mode have already been established and are available for use by the layer 3. For a BRA ISDN subscriber loop, the choice of leaving these layers established or not during periods of no call activity is dependent on the network operator or administration. For example, in America the physical and data-link layers may remain active, while in Europe the physical layer may be deactivated and the data-link layer returned to the TEI-assigned state. These layers on both sides of the connection must be initialized into the correct state before the peer layer 3 entities can communicate. The initialization procedures are not described in the Q.931 Recommendation and may vary according to the application.

To illustrate the problems posed by layer initialization, we examine how the layers in the C-plane of a terminal may be initialized for an outgoing call toward the network exchange. The layer initialization sequence is shown in Fig. 5.10a. It is assumed that the calling terminal prior to the off-hook condition has a deactivated physical layer and its data-link layer is in the TEI-assigned state. When the protocol control generates a SETUP message, the physical and data-link layers will not be in a position to support the transmission of this message to the network. The SETUP message must therefore be stored at the layer 3 while it requests the establishment of a multiple frame mode data-link connection with the DL-ESTABLISH-REQUEST primitive. However, the data-link layer is also unable to send a SABME frame to the network in order to establish multiple frame operation until the physical layer is activated. This can be achieved either from the management entity[12] or the data-link layer. In the case of the data-link layer, a physical layer activation is requested with a PH-ACTIVATION-

---

[11]Assignment of different numbers to terminals on a passive S bus is an ISDN supplementary service called multiple subscriber number (MSN).

[12]The management entity will be made aware that the terminal has gone off-hook, and this can be used to start an activation of the physical layer from the management entity.

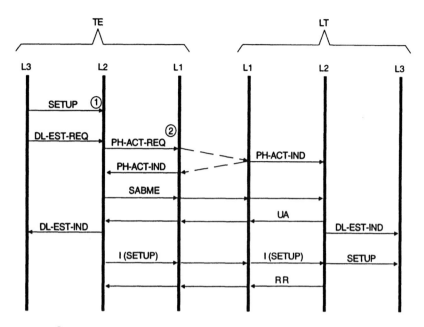

① The SETUP message is stored here for later transmission

② Activation of the physical layer may also be done through the management entity

(a)

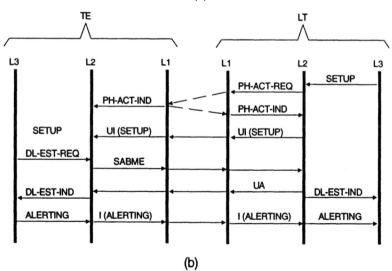

(b)

Fig. 5.10 Layer initiazation sequences for: (a) a user originated call; (b) a network broadcast (far-end) call setup.

REQUEST primitive, and is shown in Fig. 5.10a. Successful activation of the physical layer is then indicated to the data-link layer with a PH-ACTIVATION-INDICATION primitive, and the data-link layer proceeds to send a SABME frame. If the data-link was initially in a TEI-unassigned state, then the TEI assignment procedure must take place prior to sending the SABME frame. The network acknowledges the SABME with a UA frame and the data-link enters multiple frame mode which it indicates to the layer 3 Protocol Control with a DL-ESTABLISH-INDICATION primitive. Finally, the SETUP message can progress down through the data-link and physical layers and out to the network to request establishment of the call. All the layer 3 messages destined for the network which are generated while the multiple frame data-link connection is not available must be stored at the layer 3 and then sent once it has been established. At the network side of the connection, the local exchange experiences a more orderly in-sequence initialization of its layers starting from its physical layer followed by the data-link layer and finally the layer 3 call control.

The layer initialization for a terminal on a point-to-multipoint S bus with an incoming call from the network will be different, as the initial SETUP message delivered by the network will use the layer 2 broadcast data link to all terminals. The multiple frame mode of the data-link layers is therefore not established until the terminals need to respond with either an ALERTING or RELEASE COMPLETE messages, as shown in Fig. 5.10b.

5.7.4 Establishing a packet mode call

The Q.931 procedures are also used to establish packet mode calls in the ISDN. Depending on the implementation of ISDN packet services they are considered to be either inside or outside the ISDN. Q.931 can be used to establish either an ISDN circuit-switched connection to an access port of the packet network outside the ISDN, or an ISDN packet-switched connection to packet handling facilities that reside within the ISDN.

In the case of a circuit-switched access connection, the ISDN is simply used to provide a physical circuit connection between the packet terminal and an access port to the packet network. The B-channel therefore transports X.25 packets from the terminal to the access port transparently, and the Q.931 call procedures used to establish the connection between the terminal and the access port are the same as those used for a voice call described above. From a signalling and service viewpoint, the ISDN has no knowledge that a packet network is being accessed.

However, for connection to packet network services that are considered to lie inside the ISDN, call signalling identifies the use of packet transfer mode instead of circuit mode. In addition, connection to a packet handler facility can be made either using a B-channel or the D-channel. In the case of the D-channel, X.25 packets are multiplexed along with other Q.931 messages in the same 16 kbps channel. A B-channel connection is established using a sub-set of the Q.931 voice call procedures. In particular, procedures for call proceeding and overlap sending are not used, and ALERTING is no longer used as calls are answered automatically.

Once a logical connection to a packet handler is established across the D-channel, data packets are exchanged and processed by a layer 3 X.25 packet layer protocol. Although these packets share the D-channel with call control messages, they are distinguished at the data-link layer by a separate data-link connection having the same TEI but a SAPI of 16. The procedures for accessing packet services using the ISDN are detailed in Recommendation X.31 and are discussed further in Chapter 8.

5.7.5 Other Q.931 messages

A number of other Q.931 message types are defined whose purpose are to provide supporting functions to the layer 3 call control process. A brief description of these messages is provided here for completeness.

PROGRESS: The PROGRESS message is used to indicate the progress of a call connection towards its destination when the connection involves inter-working with another network, for example an analogue PSTN. The message contains a progress indicator IE which indicates the type of connection and the availability of progress tones in the B-channel chosen for the call connection.

STATUS: The purpose of the STATUS message is to communicate the current state of the protocol control to its peer. A STATUS message is typically returned to its peer when the protocol control of one side of the connection receives an unexpected message for its current state. The returned STATUS message will contain its current state and a cause IE indicating that an incompatible message has been received. As a consequence, both sides of the connection will attempt to re-align their states, which typically results in the call being cleared and the peer protocol control entities returned to their null state.

STATUS ENQUIRY: On the occurrence of certain procedural errors, the protocol control may wish to be informed of the current state of its peer, in which case it sends a STATUS ENQUIRY message. The peer then responds with a STATUS message.

RESTART and RESTART ACKNOWLEDGE: The RESTART message is used by Q.931 to request that all current calls that have been established through a particular channel be reset to their null state. For calls which have been established over channels other than the D-channel, for example data calls to a packet data network, the channel identification IE is also included. Confirmation that the peer has taken this action is received with the RESTART ACKNOWLEDGE message. The RESTART message contains the global call reference as its action affects all calls.

USER INFORMATION: During signalling to establish a packet mode call, a terminal may send USER INFORMATION toward the network which it then conveys to the far-end terminal. In this way, terminals can exchange information directly with one another during the call signalling phase, and may be used for example as a means of exchanging terminal capabilities or proprietary signalling.

CONGESTION CONTROL: This message is sent by either the terminal or network as a means of local flow control of USER INFORMATION messages. The CONGESTION CONTROL message indicates either a receiver ready or receiver not ready status.

5.7.6 Q.931 timers

The protocol control procedures of Q.931 contain numerous timers which are used to limit the response time taken by the peer Q.931 to messages sent to it across the user–network interface, and also the response of the underlying data-link layer to the primitives sent by the layer 3. The timers have a typical duration of between 4 seconds and several minutes depending on the purpose of the message or primitive sent. Should a timer time-out before an expected response is received, then corrective action is taken by Q.931. Persistent time-outs will result in the call being automatically terminated.

Some of the layer 3 timers that are used during a call setup and tear-down procedure are shown together with their Q.931 recommended time-out values in Fig. 5.9.

5.7.7 Error and exception handling

The more complex a system is the higher is the likelihood that something will go wrong! In comparison with the PSTN subscriber loop, ISDN represents a significant jump in complexity, and as a consequence has had to be designed with features and mechanisms to make it robust to possible errors and failures. A number of such features and mechanisms are present within Q.931 as follows:

- response timers that prevent the call control waiting indefinitely for a response from its peer;
- a cause IE used in messages such as RELEASE and DISCONNECT to indicate the reason for a call being terminated;
- the STATUS message containing cause and call state IEs that may be returned to the peer call control when a message sequence error occurs or a message is detected which cannot be understood. Action to be taken on receipt of a STATUS message is dependent on the system implementation and will most likely lead to termination of the call.

As well as exception conditions that occur for peer-to-peer communications at the network layer, a failure may occur at one of the layers beneath that makes such communications either temporarily or permanently unavailable. Should the network layer call control be informed of a failure at the data-link layer by receipt of a DL-RELEASE-INDICATION, then calls that are currently in a stable state are maintained while those in an overlap sending or receiving state are internally cleared. In addition, a timer, T309, having a typical value of 90 seconds is started. If a DL-ESTABLISH-CONFIRM primitive is not received from the data-link layer prior to T309 timing out, then any remaining calls are terminated.

5.8 CALL CONTROL APPLICATION

The top of the ISDN C-plane protocol stack is the call control sub-layer of the layer 3. Its main purpose is to provide an interface between the underlying layer 3 protocol control state machine and the application specific hardware of the terminal or exchange. The call control sub-layer will typically be responsible for processing the call reference and other information elements, and implement further processing required for supplementary services.

Communication between the call control and protocol control of the layer 3 is achieved with primitives that are related to the messages and the states of the protocol control. Whereas the names, coding, content and format of the protocol control messages are fully defined, only the names of the call control primitives appear in the Q.931 in the protocol control SDL diagrams. These primitives are not more explicitly defined so as to maintain flexibility in the way the application interface to the ISDN protocol stack is implemented.

5.9 SUPPLEMENTARY SERVICES

Supplementary services enhance the basic call setup and tear-down service defined in Q.931, and provide an ISDN with a number of additional services that significantly improve the flexibility of the basic call service and give the user access to associated call information from the network. Supplementary services utilize a special set of messages, as well as the IEs in these and the other messages already discussed, to activate and provide information to the procedures which implement the services. Table 5.4 gives a list of common supplementary services together with a brief description of the service. Although these services are in the main applied to voice calls, some of them may also be applied to calls involving other types of information transfer such as data or video, and as time goes on, more services may be defined as necessary to meet the evolving needs of the user. Typically, an ISDN may not offer all of the services described in Table 5.4, but may instead offer a sub-set of services based on the prevailing user requirements, the ability of the network operator to provide the service, and the regulatory conditions that influence the provision of the service. For example, in Europe a minimum sub-set of supplementary services have been defined for provision in the Euro-ISDN, these being multiple subscriber number (MSN), direct dialling in (DDI), calling line identification presentation (CLIP), calling line identification restriction (CLIR), and terminal portability (TP).

There are two main aspects to the subject of supplementary services. The first is the definition of the service itself, which includes a procedural description of how the service operates and its interaction with other existing services, and secondly the general mechanisms by which the services are controlled. The descriptions of supplementary services are currently provided in the ITU-T Recommendations I.250 through to I.257, while the methods used in the control of supplementary services is provided in Recommendation Q.932. Additionally, Q.931 also contains the definition of those procedures which support the implementation of supplementary services and that also have an impact on the operation of the basic call control. The detailed definitions of individual ISDN supplementary services are published as part of the ITU-T Q-series Recommendations.

In the remainder of this section on supplementary services we describe the underlying Q.931 features that support these services, and the different procedures by which the services are controlled. These provide the foundation on which the existing and future supplementary services are built.

5.10 Q.931 SUPPORT FOR SUPPLEMENTARY SERVICES

Support provided by the basic call control in Q.931 for supplementary services are call rearrangement through the use of the SUSPEND and RESUME messages, and user-to-user signalling procedures which allow users to transfer call related information between themselves during the different phases of a call.

Table 5.4 ISDN voice supplementary services

Service	Description
Number identification	
Direct dialling in (DDI)	A user may directly call another user connected to an ISDN PBX
Multiple subscriber number (MSN)	Multiple numbers may be assigned to a single user–network interface
Calling line identification presesntation (CLIP)	The calling user's number is provided to the called user
Calling line identification restriction (CLIR)	Used to restrict the presentation of the calling user's number to the called user
Connected line identification presentation (COLP)	The called user's number is provided to the calling user
Connected line identification restriction (COLR)	Used to restrict the presentation of the called user's number to the calling user
Malicious call identification (MCID)	The source of an incoming call is identified and registered by the network
Sub-addressing	A sub-address may be used to select a particular terminal on a passive S bus using the sub-address IE in a message instead of using part of the telephone number as in DDI and MSN
Call offering	
Call transfer (CT)	Allows an active call to be transferred to a third party
Call forward on busy (CFB)	A call is forwarded to a third party during it's establishment if the called user is busy
Call forward on no reply (CFNR)	A call is forwarded to a third party during it's establishment if there is no reply from the called user after a defined time.
Call forward unconditional (CFU)	A call is unconditionally forwarded to a third party during it's establishment
Line hunting (LH)	An incoming call to a particular number is distributed to a group of interfaces
Call deflection (CD)	The called terminal automatically deflects the incoming call to another terminal
Charging	
Advice of charge (AOC)	The user paying for the call is informed of the charges at the end of the call, during the call, or during call establishment

Table 5.4 *Continued*

Service	Description
Credit card calling (CRED)	Allows a user to pay for a call using a credit card
Reverse charging	Allows a user to reverse charges on a call on request
Freephone	Reverse charging is automatically made based on the number called
Call completion	
Call waiting (CW)	A busy user is notified that an incoming call is present
Call hold (CH)	Allows a user to interrupt an existing call (to answer a waiting call for example) and then subsequently re-establish the original connection
Completion of calls to busy user (CCBS)	A call is automatically requested to a previously busy user once the line is free
Terminal portability (TP)	Allows a user to move a terminal from one socket to another on the same interface during the active state of the call
Multiparty	
3-party service (3PTY)	Allows a user in an active call to establish a second call to a third party and switch as required between the two. Privacy is maintained between the two calls
Conference call – add on (CONF)	Allows a user to simultaneously communicate with other users who may also communicate among themselves
Meet-me conference	Allows a user to arrange for a call between more than two participants by each participant dialling into a pre-arranged number
Community of interest	
Closed user group (CUG)	Allows users to form groups. Access to and from the group may be restricted
Additional information transfer	
User–user signalling	Allows terminals or users to exchange call related information

5.10.1 Call rearrangement

The protocol control procedures of Q.931 allow the physical layer and the data-link layers of a connection to be rearranged during a call that is already in progress. By rearrangement it is meant, for example, that a terminal can be disconnected from the S bus and another terminal connected in its place. Alternatively, a user may suspend a call present on one terminal and resume it on another terminal in a different location on the same passive S bus. Both cases cause a change of state of the call being rearranged which must be implemented by the protocol control in Q.931. Call rearrangement is used in the terminal portability (TP) supplementary service.

Rearrangement of a call is initiated by a user from a terminal which sends a SUSPEND message to the network containing the call reference and optionally a number sequence called the call identity which is entered by the user and used by the network to identify the call when it is resumed. If the network accepts the suspension of the call, it acknowledges it with a SUSPEND ACKNOWLEDGE message. At this point, the call control in both the network and the terminal considers the call (and its corresponding call reference) to be released and thus enters the null state. However, the local exchange maintains the call identity number for future call resumption, and reserves the B-channel used by the suspended call until it is resumed or cleared.

The modifications that take place to the state of the suspended call take place only in the subscriber loop of the user which originated the suspension. No state changes take place in the subscriber loop at the other end of the connection, although its terminal is notified that the call has been suspended by a NOTIFY message with an appropriate indication contained in its information elements.

Once a call has been suspended, the originating terminal may if necessary be disconnected from the S bus thereby disconnecting the data link and deactivating the physical layer.

The suspended call may then be resumed from a newly connected terminal by the user instructing it to send a RESUME message to the network together with a new call reference, and the call identity if it was used during the suspension of the call. If the underlying physical and data-link layers were deactivated during the rearrangement, then they must first be established before the RESUME message can be sent to the network. If the call identity is not recognized by the network, or the suspended call no longer exists, either because it was cleared by the far end or the call was not suspended in the first place, then the network rejects the request with a RESUME REJECT message with an appropriate cause information element. Otherwise, the request is accepted by the network which sends a RESUME ACKNOWLEDGE message to the terminal indicating the B-channel which it had reserved when the call was suspended and to be used again when the call is resumed. The far-end subscriber loop is then notified with a NOTIFY message indicating that the user has resumed the call. The sequence of messages is illustrated in Fig. 5.11.

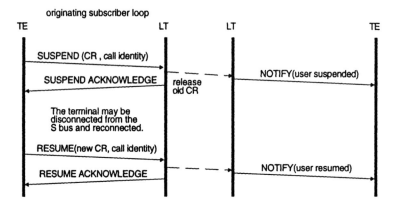

Fig. 5.11 Layer 3 peer messages used for call rearrangement.

5.10.2 User–user signalling

User–user signalling (UUS) is a supplementary service which allows users to exchange call related information during different phases of a call. Once the UUS service has been requested by the calling terminal and accepted by the called terminal (see the following section), the **user–user** IE is then used to transfer user or terminal information either in the call control messages between the user's terminal and the network during call setup and clearing phases, or in the USER INFORMATION message during the call establishment phase (between the ALERTING and CONNECT messages) and the active phase of the call. This information is then passed unaltered through the networks' signalling channels to the destination subscriber loop where corresponding Q.931 messages are used to transfer it to the destination user's terminal. Any end-to-end protocol processing of this information, such as acknowledged information transfer, takes place solely between the communicating terminals. The network acts merely as a transparent connecting medium and has no knowledge of the content or meaning of the information.

Support for UUS by the basic call control is required because the user–user IE may be communicated in basic call control messages according to the state of the call. The user–user IE consists of the user–user IE identifier, length, a protocol discriminator which identifies the way in which the information is to be processed in the receiving terminal, followed by the information itself which may be up to either 35 or 131 bytes in length depending on the network implementation. Information greater in length may be split over several USER INFORMATION messages through the use of the **more data** IE, and the network may implement flow control over these messages by using the CONGESTION CONTROL message.

5.11 CONTROL OF SUPPLEMENTARY SERVICES

A particular supplementary service supported by the ISDN must first be activated by the terminal before it may be used. This can be done on a per-call basis during the different phases of the call, or outside the call process, depending on the type of service being activated.

There are two types of protocol used for the invocation and control of supplementary services. These are **stimulus protocols** and **functional protocols**. Although both of these types of protocol have the same purpose, the functional protocol is perhaps the more important of the two in that it is universally defined in the ITU-T Q-series Recommendations for all current supplementary services, and therefore provides the basis for widespread implementation. The stimulus type of protocol, however, is considerably simpler than the functional type, but does not define a unique coding for the identification of the supplementary service being controlled. Instead, the semantics for the service are chosen by the equipment manufacturer or network operator, which consequently impacts the ease of interworking between the terminal and exchange. However, because standards are still in definition for some supplementary services implemented using the functional protocol, stimulus protocols have provided a means of early implementation of these services. Both functional and stimulus protocols may co-exist in an ISDN.

5.11.1 Stimulus protocols

The stimulus protocols are designed to be simple in their implementation and assume that the intelligence associated with processing the procedures which implement the supplementary service are contained in the network rather than the terminal. The terminal need not have any knowledge of the service being invoked or its state, and simply provides the means through which the user can directly communicate and control the service entity within the network. The stimulus protocols are therefore only suited to services which are invoked in the user-to-network direction. Two stimulus protocols have been defined, one called the keypad protocol and the other the feature key management protocol.

5.11.1.1 Keypad protocol

As its name suggests, the keypad protocol is driven by the user interactions with a keypad that is part of an ISDN terminal, and through which the user selects a particular service to be invoked, and supplies additional information where necessary to complete the invocation. The information from the keypad is transferred to the network in the **keypad facility** IE of either the SETUP or INFORMATION messages, and the network uses the **display** IE in the INFORMATION message to indicate to the local or remote user the service being invoked and the need for the local user to enter more information where necessary. The keypad information is sent to the network as a number of IA5 characters using either overlap sending or en-bloc transmission. As previously mentioned, the coding of this information is network or equipment dependent. The keypad protocol may only be used to invoke supplementary services while a call exists with a valid call reference, that is during the establishment, active and clearing phases of a call.

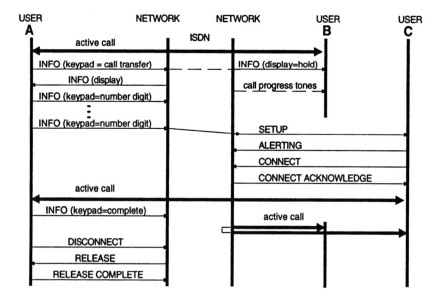

Fig. 5.12 Call transfer example using the keypad protocol.

Figure 5.12 illustrates the use of the keypad protocol to invoke a call transfer service. Initially an active call exists between users A and B, and user A then wishes to transfer the call to user C such that a connection then exists between users B and C. User A makes a request to transfer the call, for example, by entering an access code as a sequence of keys on the keypad of the terminal. The terminal then sends an INFORMATION message to the network with a keypad facility IE that contains the en-bloc access code which identifies the call transfer service. The call to user B is then placed on hold by the network and user A is prompted by an INFORMATION message with the display IE, whose contents are displayed on the terminal, to enter the number of user C to whom the call is to be forwarded. The number of user C is then transferred to the network in the keypad facility IE of further INFORMATION messages. When the dial-plan is complete, the network proceeds to connect the call in the usual way and establishes an active call between users A and C. User A subsequently completes the call transfer by pressing a key on the keypad which causes another INFORMATION message to be sent to the network with a keypad facility IE requesting completion of the forwarding process. The network then associates the original call to user B with the new party C, and tears down the connection to user A.

5.11.1.2 Feature key management protocol

The feature key management protocol is based on the use of the **feature activation** and **feature indication** IEs to invoke a particular supplementary service. In the user-to-network direction, the feature activation IE may be contained in a SETUP or INFORMATION message to activate or deactivate a specific service, while in the opposite direction the feature indication IE may be contained in any of the main Q.931 call control messages shown in Fig. 5.9 to indicate to the user the status of the service. The feature activation IE contains a **feature identifier** number which the network uses to identify the relevant supplementary service in a **service profile** for the user. The service profile is held by the network and typically contains a one-to-one association between the feature identifier numbers and specific supplementary services. For example, the call hold service may correspond to feature identifier number 3, whereas the call transfer service may have a feature identifier number 4. The service profile is set up by the network operator according to the services made available to the user[13]. To overcome the problem of terminals on a passive S bus addressed by a single network telephone number, but requiring different service profiles, a **service profile identification** IE may be included in a message from a terminal to uniquely identify its service profile. The service profile identifier consists of a series of IA5 characters and is assigned by the network operator. Feature identifier numbers are also allocated to the supplementary services by the network operator and may therefore differ between ISDNs.

As well as activating a particular service, the feature activation IE may also be used to deactivate it, either by re-sending the same identifier in a toggle on–off

---

[13]In the USA, a procedure exists to download a user's service profile from the network exchange to the terminal so as to avoid the necessity of 'exchange' specific terminals.

type of action, or by selecting an identifier which has been explicitly defined by the network operator as a deactivation for that service. In response to the user sending a message with a feature activation IE, or at any other time, the network may return a message with a feature indication IE that indicates the status of the service as deactivated, activated, pending, or waiting for more user input from the user.

As several calls may exist at the same time across an ISDN user–network interface, the activation of a service that is to be associated with a particular call is achieved through the use of the same call reference for the INFORMATION messages that contain the feature activation IE. The activation of services which are not associated with a particular call is achieved with messages containing the dummy call reference value of zero.

5.11.2 The functional protocol

The functional protocol differs from the stimulus mode protocols described above in that the user terminal must have the ability to process the user-end procedures associated with the services it is capable of controlling. This means that the user terminal is capable of understanding the semantics of the protocol elements contained within the IEs of messages it receives from the network and can process them without the need for user intervention. It is from this ability that the term **functional** is coined in the name of the protocol.

The control of services by the functional protocol makes use of either separate messages dedicated to the control of a particular service, such as the HOLD and RETRIEVE messages which are defined for use in the call hold service, or the general purpose FACILITY message which contains the **facility** IE. In addition the facility IE may be included in the normal Q.931 call control messages, typically in the call establishment and call clearing phases, although the definition of a separate FACILITY message for the control of supplementary services does allow the activities associated with normal call control to be easily separated from those associated with supplementary services. The benefit of the facility IE approach over the use of separate messages is that it does not lead to the generation of a host of messages which are required to control the services, particularly when more services may be defined in the future that require new messages. Consequently, the use of the facility IE simplifies the implementation of services within the terminal and the network, in particular their software implementation.

From the functional viewpoint, the supplementary services implemented by the functional protocol may be viewed as applications which are distributed through the ISDN in the terminals, PBX equipment, network switching exchanges and so on, and which are required to interact with one another in order to implement the services. Protocols to achieve this in a distributed open systems environment have been defined by the ISO in compliance with their OSI-RM, and subsequently mirrored in the ITU-T X.200 series of Recommendations. In the terminology of these Recommendations, the entities which process the services are known as remote operations service elements (ROSE) which exchange application protocol data units (APDUs) according to a ROSE protocol.

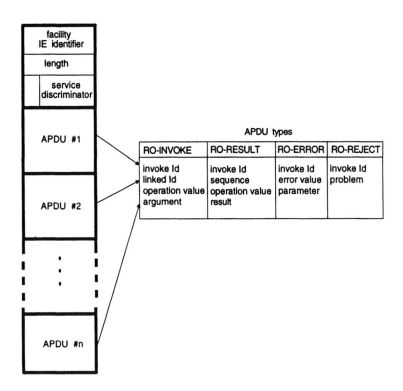

Fig. 5.13 The facility information element.

The facility IE, as shown in Fig. 5.13, contains a number of APDUs which define the actions to be taken by the peer ROSE service elements in the processing of the supplementary services. The APDUs can be one of four types as shown in the diagram and are sufficient to invoke the service, confirm its activation, report errors, and reject its request. The operation value within the RO-INVOKE and RO-RESULT APDU defines the particular supplementary service being activated. Successful activation of a service during the active phase of a call would involve the user terminal sending a FACILITY message to the network with the facility IE that contains a RO-INVOKE APDU. Notice that several services may be activated by the same FACILITY message by simply stringing together RO-

INVOKE APDUs for each service being invoked. If the user is successful in activating the service, the network returns another FACILITY message with the corresponding RO-RESULT APDU in its facility IE. This is illustrated in Fig. 5.14a.

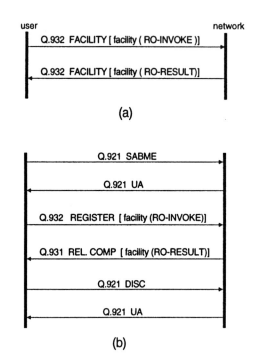

Fig. 5.14 Invocation of supplementary services: (a) during the active phase of a call; (b) outside a call.

The invocation of services which do not require the presence of an active call, for example the call forwarding services, can be performed while no call exists through the use of the REGISTER message. Prior to a user sending a REGISTER message, the underlying peer physical and data-link layers across the subscriber loop to the network exchange must first be established if not already done. The terminal then sends a REGISTER message containing, among other elements, a call reference (CR) and a facility IE with the appropriate RO-INVOKE. Subsequent information, if required, is transferred between the terminal and network in further FACILITY messages which use the same CR value. When the interaction is complete, a RELEASE COMPLETE message, which may also contain a facility IE, is sent by either side, thus releasing the CR. Figure 5.14b shows a service invoked by a single REGISTER message and assumes that the activated service does not require the continued existence of an active data-link connection, which is therefore disconnected.

5.12 SYSTEM MANAGEMENT

It is intended that the ISDN user–network interface support a number of management functions that allow comprehensive configuration control and diagnostic procedures to be carried out by the network in conjunction with the attached terminal across the user-network interface. These functions will extend the basic layer management functions that provide specific support to a particular layer, for example the TEI management functions provided to the data-link layer, into more generic functions, such as:

- fault management;
- configuration management;
- accounting management;
- performance management;
- security management[14].

These management functions across the user–network interface will align with those to be provided within the network itself, and thus provide a powerful management infrastructure for the ISDN as part of the future **intelligent network** (IN). The general aspects of the user–network management protocol, such as its architecture, are defined in ITU-T Recommendation Q.940 [7].

Figure 5.15 shows the generic model of a management entity. It consists of two parts; a communications component and system management component. The communications component is associated with the D-channel through which peer management entities communicate. Layers 1 to 3 of the communications component are those of the call control stack already discussed in this chapter. At the top of the communications component is the application layer in which resides the system management application entity (SMAE) whose function it is to provide access between its local system management application process (SMAP) and other management application processes. Peer management application processes

---

[14]Access to management functions and information requires security measures in order to preserve the integrity of the system. One way in which this may be achieved is through the use of the calling line identity service to provide authentification of the entity attempting to access it.

will communicate across connections established for them by their respective SMAEs.

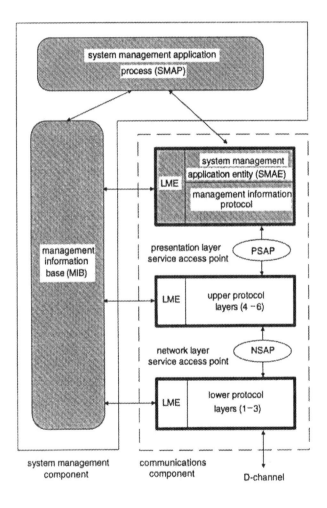

Fig. 5.15 The system management architectural model.

For strict compliance with the OSI-RM, the SMAE should be attached to a presentation layer protocol entity (layer 6) through a presentation layer service access point (PSAP). However, this requirement would create an unnecessary overhead for simple terminals such as telephones, as they would need to implement transport, session and presentation layer protocols (layers 4, 5 and 6) in order to support only a limited set of management functions required by a terminal. In this case, the SMAE can instead be attached to the network layer through a network layer service access point (NSAP) provided that a convergence function replaces the layers 4 to 6 with a minimal set of services required to interface the application layer directly to the network layer.

The system management component consists of the system management application process and the management information base (MIB). The SMAP is the process which performs the management functions. Invariably, this management process will be required to perform a task based on the value of one or a number of parameters or variables that are contained in the MIB. The contents of the MIB may be accessed either by the application process or by the layer management entities (LMEs).

REFERENCES

1 ITU-T (1993) ISDN User–Network Interface Data Link Layer – General Aspects. Recommendation Q.920 (I.440).
2 ITU-T (1993) ISDN User–Network Interface – Data Link Layer Specification. Recommendation Q.921 (I.441).
3 ITU-T (1993) ISDN User–Network Interface Layer 3 – General Aspects. Recommendation Q.930 (I.450).
4 ITU-T (1993) ISDN User–Network Interface Layer 3 Specification for Basic Call Control. Recommendation Q.931 (I.451).
5 ITU-T (1993) Generic Procedures for the Control of ISDN Supplementary Services. Recommendation Q.932.
6 ITU-T (1993) International Alphabet number 5. Recommendation T.50.
7 ITU-T (1988) ISDN User–Network Interface Protocol for Management – General Aspects, **VI**(11), Blue Book. Recommendation Q.940.

6

The ISDN terminal

This chapter provides an introduction to ISDN terminals, and focuses on the implementation of ISDN call control functions and how these translate into hardware and software components, sub-systems and their interconnection. All ISDN terminals that establish a circuit-switched connection across the ISDN must contain at least a set of functions required to process basic ISDN call control. These functions can therefore be thought of as a core set of functions that are generic to ISDN terminals.

The terminal will contain an application function that processes and conditions the user information. It is from this capability that the terminal gains its identity, for example a telephone, videophone, fax machine, and so on. Together, the terminal and its application is the user's focal point of his or her ability to communicate with other people because the terminal determines what information is communicated and how it is communicated.

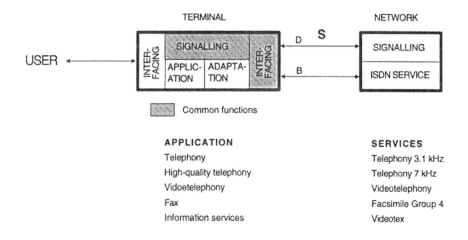

Fig. 6.1 ISDN services and applications.

Figure 6.1 presents a functional partitioning of an ISDN terminal, and shows some of the appropriate applications and the corresponding network services they might use. An ISDN terminal performs the following functions.

- **Signalling.** Signalling is required between the terminal and the network to request and provide a connection to other terminals connected to the network.
- **Application function.** For the terminal to be a useful communications device, it will host an application that will define how the terminal is used and what network services it is required to access.
- **User interface.** A user interface is required both for signalling functions and the application which may be resident within the terminal.
- **Network interface.** The S user–network interface.
- **Data rate and format adaptation.** A conversion process is required between the information generated by the user or application and data that is suitable for transportation across the ISDN.

Of these functions, the signalling and user–network interface are common elements found in all types of ISDN terminal, while application specific function will vary according to the application itself. Much of this stems from the fact that all ISDN terminals, regardless of their application, must have the ability to establish calls to the network through an exchange of signalling messages in the D-channel of the ISDN user–network interface, as described in Chapter 5. This ability requires a minimum set of hardware and software functions which are typically implemented by a set of core terminal hardware and software components, which include the following:

- an S interface transceiver (described in Chapter 3) that provides access to the physical layer D-channel;
- an HDLC controller that provides layer 2 framing for call control messages in the D-channel;
- a microprocessor system that can control the HDLC controller and S interface transceiver, and is capable of running call control software which processes the call control messages exchanged by the terminal and the network;
- call control software that implements the necessary data-link layer (layer 2) and network layer (layer 3) protocols required for call control signalling (described in Chapter 5).

Figure 6.2 illustrates the hardware components of an ISDN telephone and their interconnection. The telephone is chosen here as a specific example as it is the most common type of telecommunications terminal in use today, and perhaps the simplest type of terminal which can be considered for use with the ISDN. As such, much of its complexity is due to the call control processing requirements of ISDN. The hardware components and sub-systems which constitute the core functions defined above are enclosed inside the dashed area of Fig. 6.2, while those outside implement the application (codec/filter for speech communication) and user interfacing (keypad and display) to the telephone.

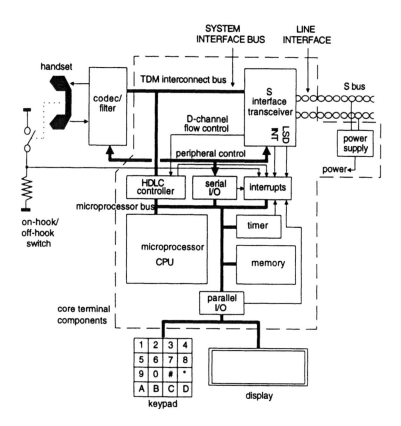

Fig. 6.2 Core components that form part of an ISDN telephone.

6.1 THE S INTERFACE TRANSCEIVER

The S interface transceiver provides the physical layer framing and timing functions for the B- and D-channels across the ISDN user–network interface. These functions are typically implemented as a single IC and include the activation and deactivation sequences required to achieve synchronized transmission and reception with the network. The transceiver will have a control interface through which activation and deactivation, as well as other features and functions, may be controlled by a device within the terminal such as a microprocessor or microcontroller.

In an ISDN terminal, the S transceiver will be responsible for synchronizing its transmissions to those it receives from the network. In this role, it controls the timing and synchronization of the TDM interconnect bus at the system interface, and generates the clock and frame synchronization signals for devices which interface to it, such as the codec/filter and HDLC controller.

A more detailed description of the S interface was provided in Chapter 3, in which an example of an S transceiver was described in section 3.9.

6.2 THE HDLC CONTROLLER

Call control messages carried in the D-channel between the terminal and network are framed according to the HDLC format used by the call control data-link layer called LAPD. An HDLC controller is used to provide hardware support for building an HDLC frame for transmission according to the format shown in Fig. 5.2 in the previous chapter. It will also check the integrity of received frames and extract the relevant information from the frame for further layer 2 and layer 3 processing by the microprocessor system. As a minimum, the HDLC controller will:

- provide a parallel-to-serial and serial-to-parallel conversion of data between its microprocessor bus interface and its HDLC interface, which in this case connects to the D-channel of the serial system interface bus;
- automatically add and remove the opening and closing flags (F) of a frame;
- perform bit-stuffing to preserve the integrity of the opening and closing flags;
- calculate and insert the frame check sequence (FCS) of a transmitted frame and check the FCS of a received frame.

Figure 6.3 illustrates a the major functional blocks to be found within a typical HDLC controller along with its interconnection to the other main components within an ISDN terminal. The following description indicates the way in which the controller would operate when interfaced to the D-channel across the system interface bus.

6.2.1 HDLC operation

Control logic in the HDLC controller has the responsibility of controlling the transmit and receive framing in conjunction with the various flag and CRC registers. For example, in the transmit path, when no frame is being sent, the control block will enable the fixed **idle** pattern (which consists of 11111111) into the TxBuffer such that the TxOut line will remain in the logic 1 state in the D-channel time-slot during idle periods. When the HDLC controller is instructed by the CPU to send a frame by setting an appropriate bit in its control register, the first address byte of the transmit frame will be fetched from memory and stored in the TxRegister while the contents of the flag register (01111110) is loaded into the TxBuffer and sent. When the last bit of the TxBuffer has been transmitted, it is loaded from the TxRegister and the following byte of the frame is fetched from memory and again stored in the TxRegister. As the contents of TxBuffer are transmitted, the CRC value excluding the opening flag is calculated as the serial data is clocked into the D-channel of the TDM interconnect bus through the bit-stuffing logic. (The bit-stuffing logic is disabled during transmission of flags, abort and idle characters in order to maintain their bit patterns.) This process continues until the final byte in the transmit frame from the memory has been transmitted, at which point the two bytes of CRC are transmitted followed by the closing flag of the frame.

In the receive path, the reverse process takes place. As frames occur asynchronously in the D-channel, in order to trigger the reception process the HDLC controller must first detect the presence of a frame by recognition of its opening flag. As successive bytes of the frame are received, they pass through the RxBuffer and RxRegister and to the microprocessor bus interface where they are loaded into memory, either under control of the microprocessor itself or automatically using direct memory access control. These techniques are elaborated on further in section 6.2.3. When the flag detector detects the closing flag, the calculated CRC value is compared with the last two bytes of the frame corresponding to its CRC, and if different, an error bit is set in the status register that can be subsequently read by the microprocessor. This would then initiate any necessary error recovery procedure implemented in software.

An additional feature that is often included in HDLC controllers is address filtering. Here, the receiver contains a number of registers, the contents of which are compared with the address field in the incoming frame. If a match is found then the remainder of the frame is received as normal, while a mismatch would result in the frame being ignored and the HDLC receive state machine being reset to detect the opening flag of the next frame. Address filtering is of little use in a point-to-point S bus configuration where under normal conditions all messages from the network will be addressed to a single terminal. However, it is useful in passive bus configurations with more than one terminal connected to the S bus, where it can be used to filter out those messages with a TEI value that matches that of the terminal, and prevents the microprocessor in the terminal from spending processing time determining the validity of the message, although this overhead is usually very low. As a terminal may have at least two TEI values to which it must respond (its own TEI and the broadcast TEI), an equivalent number of registers should be available in the HDLC controller.

6.2.2 HDLC interfacing to the TDM interconnect bus

The devices which provide serial information into the B- and D-channels of the ISDN BRI are interconnected within the terminal by a synchronous serial time division multiplexed (TDM) bus. Details of several multiplexing schemes are discussed later in section 6.8.3. The outputs of devices connected to the TDM bus are equipped with a tri-state driver that puts the output into a high impedance state unless it is driving the bus during the time-slot allocated to that device. A time-slot controller circuit in each device may be used to indicate the correct time-slot.

The HDLC controller interfaces to a 16 kbps channel within the TDM bus that in turn is interfaced to the D-channel of the S interface by the S transceiver. Access to this D-channel time-slot is different in the transmit and receive directions due to the fact that collisions can occur on the S bus D-channel from the transmissions of multiple terminals connected to it in the passive S bus arrangement see section 3.2). D-channel flow control in the transmit direction is therefore necessary and is performed by the S transceiver with a hardware signal which enables the HDLC controller to transmit into the D-channel with the least likelihood of it causing a collision with that of another terminal. This is the DEN, or D-channel enable, signal shown in Fig. 6.3. In the receive direction, a collision is not possible, and indication of the receive D-channel time-slot on the TDM bus may be provided either by the S transceiver with an additional control line similar to DEN, or by a time-slot circuit within the HDLC controller.

6.2.3 HDLC interfacing to the microprocessor system bus

Data in the address, control, and information fields of an HDLC frame will typically reside in the memory of the microprocessor system, and are transferred between memory and the HDLC controller as required to maintain the transmission or reception of an uninterrupted frame. For example, to maintain a D-channel bit rate of 16 kbps during transmission or reception of a frame, a byte of data must be transferred between memory and the HDLC controller on average every 0.5 milliseconds. This transfer can either take place under control of the microprocessor, or semi-autonomously by means of a direct memory access (DMA) controller.

6.2.3.1 The FIFO buffer bus interface

In reality, the transfer of each byte individually under microprocessor control introduces a significant amount of processing overhead. A preferred arrangement which reduces this overhead is the use of a buffer called a FIFO (first-in-first-out) in which several bytes can be stored in preparation for transmission, or in which several bytes are stored after reception and then transferred to memory as a block. As its name suggests, the FIFO is a sequential buffer which stores information and outputs it in the order in which it is received.

The FIFO size in an HDLC controller is typically between 32 and 64 bytes and is double-buffered. This means that the FIFO is organized in two sections such that, in the case of the transmit FIFO, while one section is transmitting information the other section may be filled by the microprocessor with the next data from memory to be transmitted. Similarly for the receive FIFO, as one section is filled with information received by the HDLC controller, the other section may be emptied by the microprocessor and stored in memory. A full receive FIFO buffer or empty transmit FIFO buffer will result in an interrupt to the microprocessor system instructing it to empty or fill the respective buffer. In this way, frames that are larger than the FIFO size can be transmitted and received by the HDLC controller.

6.2.3.2 The DMA bus interface

Instead of performing data transfer between memory and the HDLC controller under program control, a DMA bus controller, once programmed with the correct source and destination information, can autonomously perform this task. The microprocessor need only be interrupted when a complete frame has been transmitted or received.

DMA relies on the ability of the microprocessor to relinquish control of the microprocessor bus to another bus controller device such as the DMA controller. Control of the microprocessor bus is requested and acknowledged using hardware control lines between the HDLC/DMA controller and the microprocessor. In addition, the DMA controller will contain address registers which can be programmed by the microprocessor that indicate the start and end address of a frame in memory to be transmitted, and the start and maximum permitted end address for a frame to be received. When active, the DMA controller will request control of the microprocessor bus whenever the HDLC controller has need to transfer information in order to maintain the uninterrupted transmission of a frame, or maintain frame reception without overflow and loss of received information. The benefit of DMA transfer is that it is performed automatically by hardware instead of software instructions, and is therefore significantly faster and more efficient on microprocessor processing resources. DMA can transfer a single byte at a time, or a block at a time if used in conjunction with a FIFO buffer.

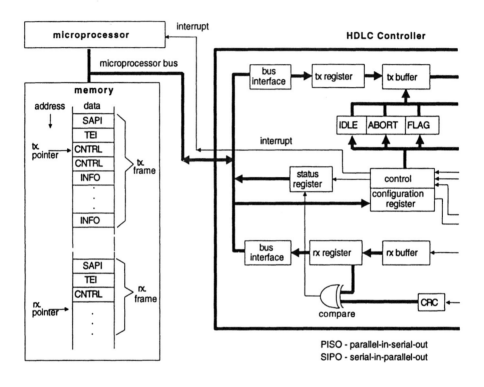

Fig. 6.3 Functional block diagram of the HDLC controller and its interface to memory and the serial TDM interconnection bus.

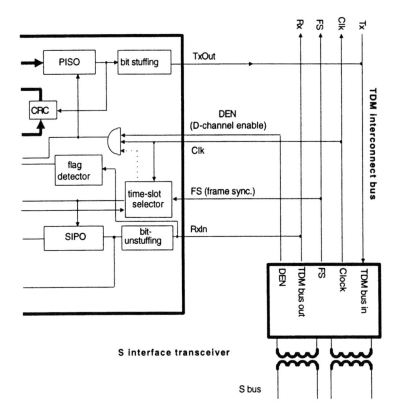

Fig. 6.3 *(cont)*

Table 6.1 A selection of HDLC controllers for ISDN applications

Device	Number of channels (full-duplex)	CPU bus interface	TDM bus interface format	Remarks
Siemens HSCX SAB 82525	2	64 byte FIFOs per channel	Yes	Automatic I and S frame control field processing for window size $k = 1$
Zilog Z16C31	1	32 byte FIFOs + DMA	Yes	Contains 8 general purpose I/O lines
Hitachi HD64530	1	DMA	No	Supports LAPD procedures
STM ST5451	1	64 byte FIFOs	Yes (GCI)	Contains GCI controller
Motorola MC145488	2	DMA	Yes (IDL)	Contains 2 timers
Motorola MC68606	1	DMA	No	Implements LAPD procedure functions for multiple logical links
Plessey MV6001	1	DMA	No	
AMD Am79C401	1	32 byte Rx FIFO 16 byte Tx FIFO	Yes	Integrated with USART and dual-port memory controller
AMD Am79C30	1	8 byte FIFOs	Yes (GCI)	Integrated with S interface and codec/filter
Siemens PEB2085	1	64 byte Rx FIFO 64 byte Tx FIFO	Yes (IOM-2)	Integrated with S transceiver. Automatic I and S frame control field processing for window size $k = 1$
Intel 29C53A	1	32 byte Rx FIFO 32 byte Tx FIFO	Yes (SDL)	Integrated with S transceiver
Hitachi HD81501	2	DMA	Separate B-channel ports	Integrated with S transceiver and an embedded microprocessor system with LAPD firmware
Yamaha YM7303	1	DMA or 4 byte RxFIFO 4 byte Tx FIFO	Yes	Integrated with S transceiver. Supports LAPD procedures

6.2.4 HDLC controllers

Table 6.1 lists some examples of commercially available HDLC controllers that may be used in ISDN terminal applications. The first seven devices in this list contain just the HDLC function, while the last four devices are examples of more highly integrated parts. The Siemens PEB2085 and Intel 29C53A both combine an HDLC controller integrated with an S transceiver, while the AMD Am79C30 integrates three main components of an ISDN telephone, the HDLC, S transceiver and codec/filter into a single device. In the last two devices, the Hitachi HD81501 and the Yamaha YM7303, the HDLC controllers support elements of the LAPD procedures in hardware as well as framing functions, and are integrated together with an S transceiver into a single device. In the case of the Hitachi 81501, the LAPD control procedures are implemented on the device by a microprocessor system with embedded firmware.

HDLC controllers intended for ISDN or other telecommunication applications will typically incorporate a serial data format interface compatible with a TDM bus format, and will permit D-channel flow control and perform time-slot control.

6.3 THE MICROPROCESSOR SYSTEM

A microprocessor system is an essential part of an ISDN terminal due to the amount of call control protocol processing required, the need to process application information, and the need to provide an effective user interface. Today, the implementation of these functions is software intensive. While it is conceivable that in time much of the common protocol functions may become implemented in hardware VLSI circuits, the wide and varying range of programmable features and functions of ISDN terminals will generally require an implementation which is flexible, and one which currently is ideally suited to the programmable nature of a microprocessor.

A microprocessor system consists of three main components: the microprocessor or CPU (central processing unit) itself, memory, and input/output (I/O) which includes peripheral functions such as the hardware timer and interrupt controller shown in Fig. 6.2. Systems built from discrete versions of these components are too large, too costly, and consume too much power to be effectively embedded in equipment such as ISDN terminals. However, the trend towards ever increasing levels of higher integration for VLSI semiconductor circuits has brought many of these functions into a single device of which there are two types which we identify here as a **microcontroller** and a **highly integrated microprocessor**.

The microcontroller typically has a 4-, 8- or 16-bit internal microprocessor architecture, and integrates memory and many essential supporting functions, such as I/O, interrupts, timers and other peripheral functions, into a single-chip device. The constraint usually lies in the amount of on-chip memory which may typically be in the range 0–512 bytes of variable data storage or RAM (random access memory), and 0–32 kbyte of program code or fixed data storage known as ROM (read only memory). More recent devices also include small amounts of EEPROM (electrically erasable read only memory) designed to retain parameters such as configuration information while the device has no power applied to it.

The microcontroller has its origins in embedded control applications that typically involve simple sequence control operations requiring a low computational performance but a high control ability involving bit manipulation and logic computation. From here it has evolved upwards in performance and complexity to meet applications such as communications and motion control, which require a high computational ability.

The highly integrated microprocessor device has evolved from the opposite end of the performance spectrum where a higher computational ability and large amounts of memory are required for embedded applications such as image or graphics processing, communications and robotics.   These devices typically consist of a 16- or 32-bit CPU with on-chip complex high performance peripherals. As the memory requirements of the applications, and hence the devices, are typically much larger than can be practically integrated into a single-chip device, the memory is provided externally to it and accessed through an external microprocessor bus.

The prevailing trend is that the functionality of the embedded microprocessor is determined more and more by the specific application for which it is intended, and as a consequence of this, the boundary between the microcontroller and the integrated microprocesser has become increasingly vague in recent years. Today, these microprocessor systems are tailored with peripherals that target them at specific application areas, such as communications, industrial control, automotive, and so on.   For an ISDN terminal, the typical requirements are described as follows.

Parallel I/O: These ports provide a programmable parallel data interface to functions such as the keypad and display. Individual port lines may usually be programmed as either input, output or bi-directional, and may include special features, such as comparators and interrupt capabilities[15] that simplify interfacing to external devices.

Serial I/O: There are two types of serial I/O. The first is more commonly known as a **UART** (universal asynchronous transmitter and receiver) and is typically used to communicate information to other electronic systems or terminals that

---

[15]For example, a change of state at an input may be indicated to the microprocessor with an interrupt which relieves it from having to periodically check the state of the port in order to detect a change.

contain a similar UART over distances of up to 30 m depending on the data rate. For example, in an ISDN TA, the UART could be used at the R interface to connect a data terminal or computer.

The second type of serial I/O is more simple than the UART and is used as a means of interconnecting peripheral devices to the microprocessor within the same system. Specific examples are described later in section 6.7.

Interrupt control: All peripherals, including the I/O, will in general be associated with an interrupt facility that will alert the microprocessor to the occurrence of an event at the peripheral, such as a change of state of an input port, a timer timing-out, or receipt of a new frame by the HDLC controller. The source of each interrupt is uniquely identified to the microprocessor using a number of techniques, each of which will cause the microprocessor to automatically jump to a pre-defined memory location where a software interrupt service routine is located to provide any processing required as a result of the interrupt.

Timers: A hardware timer relieves the microprocessor from performing timing functions as counted loops in software. Most timers are capable of performing both timing and counting functions, and will interrupt the CPU when a timer has timed-out or a count value has been reached. One use of a hardware timer in an ISDN terminal is as a time base to implement the protocol timers specified in the call control protocol layers.

In addition, a hardware timer may also be used as a system safety feature known as a **watch-dog**, whereby a time-out causes the microprocessor to be reset to a known state if a software routine fails to periodically reset the timer, thereby indicating that normal program execution by the microprocessor has ceased and the system has crashed.

Memory: The memory requirements for basic ISDN call control software will vary according to its implementation. A 'bare-bones' implementation that offers basic call setup and tear-down functions, and written in assembly language, can be less than 64 kbyte, 80 to 90% of which would be program code (stored in ROM) and the remainder data storage (RAM). A more extensive implementation including supplementary services, and written in a high-level language such as C, could be 100 kbyte to 120 kbyte of program code or more in size, and require around 16 kbyte of data storage.

The limiting factor in the integration of memory into single-chip microprocessor systems is the complexity and size of RAM data storage compared with ROM. While devices are beginning to appear with upwards of 100 kbyte of program store, they are more normally equipped with sizes up to 32 or 64 kbyte, and up to 4 kbytes of RAM. Of course, devices with these amounts of integrated memory are significantly more expensive than their discrete counterparts, and most ISDN terminals are designed with microprocessor devices with external memory.

A small amount of non-volatile memory which can be erased on demand, such as an EEPROM (electrically erasable programmable read only memory), may also be found in some systems to provide storage of user programmable information such as pre-programmed telephone numbers, and initialization settings which need to be stored if power is removed from the terminal.

Power management: Most of today's microprocessor devices are implemented using a CMOS (complementary metal-oxide of silicon) semiconductor process. A property of a CMOS transistor is that it only consumes power when it is switched, and for a clocked system such as a microprocessor this means that its power consumption is roughly proportional the frequency at which it is clocked. Many microprocessors have the ability to put themselves into an idle mode where it ceases to fetch and execute instructions, and instead lies dormant until an external event such as an interrupt signal causes it to wake up. In such a dormant state, the clock oscillator of the CPU is typically stopped, or if the microprocessor contains circuitry which needs to be kept ticking-over, is reduced to the lowest frequency sufficient to maintain these functions. Stopping the CPU clock altogether typically reduces the power consumption to that determined by the leakage current of the device, which is typically less than 1% of that in its normal operating mode.

6.3.1 Integrated microprocessor devices

Table 6.2 lists a selection of integrated microprocessor devices designed for telecommunication and ISDN applications, and illustrates the peripheral functions that can typically be found on such devices. All devices contain one or more HDLC controllers.

The TMPZ84C711A is perhaps the most unusual of the four devices in that it is designed to implement as far as possible all functions associated with the physical layer, data-link layer and network protocol control layer of the D-channel signalling protocol. A second microprocessor system may implement higher layer and application functions if necessary, and interface to the TMPZ84C711A through a parallel port. As such, the TMPZ84C711A integrates the S transceiver (apart from the line driver and receiver which must be added to the system as a separate device), a dual channel HDLC controller and a microprocessor system that can support 64 kbyte of external memory.

The basic 64 kbyte logical address space is also present on the HPC46400, but can be extended to address half a megabyte of external physical memory through a technique known as **bank-switching**, thus allowing a system with sufficient memory to accommodate both signalling and application firmware. The bank-switching technique uses port lines of a parallel I/O port to select, under software

Table 6.2 A selection of integrated microprocessor devices with functions for ISDN applications

	Feature	Toshiba TMPZ84C711A (see note 1)	Zilog Z80181	NSC HPC46400	Motorola MC68302
CPU	Architecture	8 bit, Z80 CPU	8 bit, Z180 CPU	16 bit, HPC core	16 bit, MC68000/ 68008 CPU
	External addressing range	64 kbyte	64 kbyte basic, 1 Mbyte with on-chip memory management unit none	64 kbyte basic, expandable to 544 kbyte with bankwitching	16 Mbyte + DRAM refresh controller
	On-chip memory	none		256 bytes RAM	1152 bytes dual-port RAM
	Low power modes	2 idle 1 stop	Halt and low power modes	Halt and Idle modes	Yes
GENERAL PURPOSE PERIPHERALS	Parallel I/O ports	2× 8-bit ports	2× 8-bit ports	1× 16-bit port (see note 1) 1× 8-bit port 1× 8-bit input port	1× 16-bit port 1× 12-bit port
	Timers	6 dedicated 2 general purpose	2 general purpose	3 general purpose	2 general purpose
	Watchdog	Yes	No	Yes	Yes
	UART	1 (see note 3)	2	1	1 (see note 2)
	Serial control port	No	Yes	Yes (MICROWIRE)	Yes (SCP)
TELECOM/ISDN SPECIFIC PERIPHERALS	HDLC channels (full duplex)	2 channels	1 channel	2 channels	3 channels
	DMA	4 channels	2 channels	4 channels	6 channels
	Serial time slot formatter	No (see note 3)	No	Yes SLD, IDL, TDM GCI (see note 4)	Yes GCI, IDL, TDM
	Integrated S transceiver	Yes (not including the line drivers)	No	No	No

Notes: 1. Some general purpose I/O lines are multiplexed with other peripheral functions
2. The UART and HDLC controllers are implemented as general purpose serial controllers which can be programmed for a number of different protocols. If the controller is programmed as a UART, then it is not available as an HDLC controller.
3. Serial data at 64 kbps, from a codec/filter for example, can be transferred to and from a B-channel at the S interface through a general purpose serial port.
4. The serial data format conforms to the GCI specification although no processing is performed on the MON and C/I channels of the GCI.

control combined with external logic, one of a number of banks of memory which share the same logical address space. However, the instructions which perform the bank-switching itself, as well as other hardware dependent data structures such as the system stack and interrupt vectors, must reside in a common area of memory which is the same for all banks. For example, within a 64 kbyte address space, the lower 32 kbyte of address space can be shared by a number of banks, of which only one bank is active at a time, while the upper 32 kbyte remain the same regardless of which bank is selected.

Although the Z80181 is based on the same core microprocessor as the TMPZ84C711A, and therefore also has a 64 kbyte logical address space, it can address a physical memory of 1 Mbyte as it incorporates a memory management unit (MMU) that provides automatic hardware support for the address translation.

In contrast, the MC68302 is based on the MC68000 CPU core which provides 16 Mbyte of linear addressing which obviates the need for any memory extension mechanisms. This device is equipped with three multi-purpose serial controllers, each of which can be configured as an HDLC controller, which make it suitable for high performance terminal adapter applications.[16] Furthermore, the MC68302 also supports cheaper dynamic RAM (DRAM) memory which is beneficial where the application demands large amounts of memory, and are typically found in slot-in PC or workstation terminal adapter designs.

6.4 HARDWARE AND SOFTWARE PARTITIONING

Figure 6.4 illustrates the way that the hardware and software components of an ISDN telephone map into the layers of the OSI reference model. As is typical in many communication systems, the lower layers of the protocol stack are primarily implemented in hardware, while the upper layers are implemented in software that runs on the microprocessor system embedded within the terminal. Those hardware and software functions which are above layer 7 are not considered part of the communications system, but are instead part of the application. For ISDN call control, we are concerned only with the first three layers of the OSI-RM, where the hardware/software boundary typically lies halfway in layer 2 at the interface to the HDLC controller. As most HDLC controllers provide only hardware support for basic framing functions, this leaves the LAPD procedures to be implemented in software. A small amount of **driver** software is required below the boundary to implement the interrupt service routines and control interfaces between the hardware and the call control software.

---

[16]For example, one HDLC channel can be assigned to D-channel signalling, while the two other HDLC controllers can each be assigned to the two individual B-channels on an ISDN BRI interface that would allow the terminal adapter to handle two simultaneous data calls.

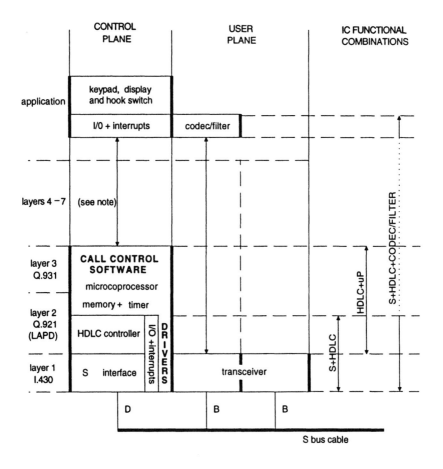

Fig. 6.4 Hardware and software components of the ISDN telephone mapped into the layers of the OSI-RM. Note: for applications other than telephony, any protocol processing required of B-channel data in layers 3–7 may also be implemented as software running on the microprocessor system.

Figure 6.4 also indicates some of the popular hardware function combinations illustrated by the devices listed in Tables 6.1 and 6.2. An integrated S+HDLC device, for example, the Siemens PEB2085 and the Intel 29C53A in Table 6.1, contains the main communication components required to terminate the ISDN D-channel and is a suitable combination for both a telephone TE and PBX (NT2) linecard designs. An even more integrated telephone solution is provided by the S+HDLC+CODEC/FILTER device, as for example in the AMD79C30 also listed in Table 6.1. For data terminal designs that require B-channel HDLC data capability, one or two additional HDLC controllers are needed, and if integrated may be combined with the microcontroller or microprocessor device, as for example in the National Semiconductor HPC46400 and Motorola MC68302 listed in Table 6.2.

6.5 POWER SUPPLY

Power may be delivered to a terminal across the S interface either by a phantom circuit superimposed upon the signals of the transmit and receive lines of the S interface (power source 1), or by separate power lines that are part of the S bus cable (power source 2). The power supply block shown in Fig. 6.2 is responsible for translating the nominal DC voltage (40 V in Europe and 48 V in America) available across the S interface into a regulated power supply for devices in the terminal. The power supply must be capable of achieving high conversion efficiencies when only a low power is available at the S interface. For example, up to 1 W is available at the S interface to power the terminal under normal operating conditions. At this level of input power, conversion efficiencies of around 80% can be achieved to deliver up to 800 mW to the terminal. However, when a designated terminal operates under **restricted** power conditions, it is allowed to consume a maximum of only 380 mW from the S interface, and when idle, only 25 mW with which the terminal is expected to be functional to a level which allows it to identify a call request from the network. The efficiency of converters at such low input powers is typically much less than 80% because the power they consume in relation to that which they convert becomes significant. At 25 mW input power, efficiencies of around 50% are typical, which means the terminal then has a power budget of less than 13 mW with which to maintain an idle mode operation.

The restricted power condition is indicated to the terminal power supply by a polarity reversal of the S interface supply voltage, and under this condition, all terminal functions which are not needed in the provision of the basic telephony service may be disabled. If a restricted power condition occurs during a basic voice call, then operation of the power supply should be so as to allow the call to continue without interruption. A non-designated terminal which normally draws power from the S interface must draw no power when it is in restricted mode. In addition, it must be ensured that a terminal which is powered locally by a mains power supply, but the power is not switched on, does not adversely load the S bus so as to cause its impedance to fall outside that specified by the I.430 specification. This is particularly important for a point-to-multipoint configuration where a terminal that adversely loads the S bus will likely cause other terminals to suffer poor transmission performance.

The design of a terminal for restricted power operation is a demanding task which requires detailed attention to the control of the power consumption of individual devices within the terminal. Most terminal components are fabricated from a low-power CMOS semiconductor process, and possess circuitry that can be powered down when not in use.

6.6 THE TERMINAL CODEC/FILTER

A codec/filter performs conversion of speech signals between analogue and digital domains. Unlike an analogue subscriber loop where this conversion takes place at the local exchange, in an ISDN subscriber loop it must now take place in the telephone in order to interface digital data to a B-channel at the S interface. As telephony over the ISDN must interwork with the PSTN, the conversion process used is the same.

A voice signal for telephony is band limited to the frequency range 300 to 3400 Hz, the highest frequency considered for the purposes of sampling and digitization is 4 kHz. The Nyquist sampling criterion dictates that the sample rate must be twice the highest frequency, or 8 kHz, if no information in the analogue signal is to be lost by the conversion process which is essential for reconverting the digital data back into an analogue signal at the far end of a connection. Each sample of the analogue speech waveform is converted into eight bits of digital data, which when transmitted in a serial binary format produce a data stream having a data rate of 64 kbps (8 bits, 8000 times per second).

The serial data format for voice is known as pulse code modulation (PCM) because the amplitude of the analogue voice signal is effectively modulated by a train of pulses (the sampling process), each of which is converted to serial digital data, or code, that represents the sampled analogue voice signal. The process of digitization causes distortion to the signal known as quantization noise, which is particularly detrimental to low amplitude samples of the speech signal. However, observing the amplitude distribution of a typical speech signal, we find that the majority of the signal spends its time close to the origin where the quality of the signal is degraded the most by the quantization noise. To overcome this problem, the conversion of a voice signal to PCM also involves a stage of nonlinear encoding designed to reduce the impact of quantization noise at lower amplitudes and increase it at higher levels, such that the impact of the quantisation noise is spread evenly over the permitted amplitude range of the signal. Two encoding laws are used, one called **A-law** which is used primarily in Europe, and the other called **µ-law** which is used mainly in North America and Japan, and both are defined in ITU-T Recommendation G.711. The A-law or µ-law encoding and decoding normally takes place in the codec.

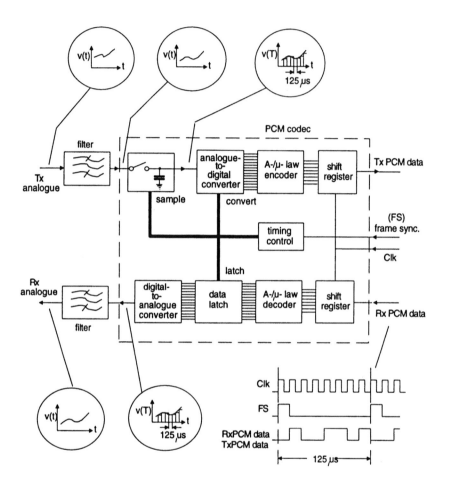

Fig. 6.5 Operation of the PCM codec/filter.

The digitization process of an analogue signal in the codec/filter combination of a line circuit is illustrated in Fig. 6.5. In the transmit direction, a new sample of the filtered speech signal is taken at the rate of 8000 times a second, or once every 125 µs, and converted to a digital value with at least 14-bit linear resolution by the analogue-to-digital (A/D) converter. The A- or µ-law converter then translates this value to an eight-bit PCM code which consists of one polarity bit plus seven bits representing 1 of 128 amplitude levels. The timing control for the codec is derived from a frame synchronization signal, which occurs once every 125 µs, and a clock signal which is usually some multiple of 64 kbps. The PCM data is then passed into a digital shift register and output as serial data. The reverse process takes place in the receive direction.

In a telephone, the features of a codec/filter are typically enhanced with telephony specific functions such as an additional microphone input, a loudspeaker output, acoustic feedback control for hands-free operation (where a loudspeaker and microphone are used instead of the handset), and a tone and ring generator. The typical features of a such a codec/filter are illustrated in the block diagram in Fig. 6.6 and described below. Programmable control of these features is provided by the microprocessor through the codec's control interface.

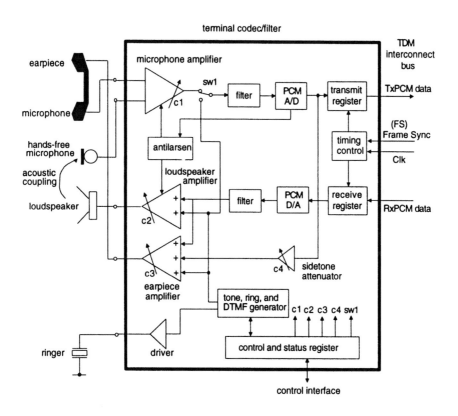

Fig. 6.6 Block diagram of a typical terminal codec/filter.

6.6.1 Sidetone circuit

The normal acoustic path between the mouth and ear of a person using a telephone is significantly attenuated due to the earpiece of the handset being placed over the ear, thus making it difficult for the user to hear what he or she is saying, particularly in the presence of high levels of background noise. The solution to this is to take the signal from the mouthpiece of the handset and feed it back to the earpiece such that the user can hear his or her own voice as well as the far-end talker signal. This near-end feedback signal is referred to as **talker sidetone**. In an analogue telephone, talker sidetone is introduced by deliberately de-tuning the telephone hybrid to allow a portion of the analogue speech transmit-path signal into the receive path. An ISDN telephone, however, will not contain a hybrid with which such an analogue signal feedback can be achieved, so a feedback path, typically with a programmable attenuation, is included to provide the talker sidetone signal to the earpiece amplifier.

6.6.2 Tone generation

A tone generator, under control of the microprocessor, provides audible indications to the user. Typically, this would include generation of a ringing signal which is sent via a low impedance drive circuit to an external piezo-electric ringer on receipt at the telephone of a SETUP message from the network, indicating the presence of an incoming call.

For ISDNs that offer the 3.1kHz audio or speech service, the network will provide in-band tones in the B-channel for indications such as dial-tone and ringback. The availability of these tones is indicated to the telephone by the network through the progress indicator IE included in the SETUP ACKNOWLEDGE message. Should these in-band tones be available, then the terminal must connect the codec/filter to the relevant B-channel at its TDM interconnect bus on receipt of the SETUP ACKNOWLEDGE message, thus allowing the user to hear the tones. Should these indicator tones not be available from the network, then they may be generated locally within the codec on receipt of appropriate call control messages from the network.

The ability to generate DTMF tones is also of benefit when interworking with PSTNs and analogue terminal equipment that are controlled by DTMF signalling, such as voice-messaging systems and automatic call distribution systems.

6.6.3 Hands-free

Use of a telephone during conversation without a handset is referred to as **hands-free**. Instead of the handset, a separate microphone and loudspeaker, which are typically built into the body of the telephone, are used. As well as freeing the user's hand, hands-free operation also allows third parties to participate in the telephone conversation at the same terminal.

Hands-free operation poses a problem in that the attenuation of the acoustic path between the loudspeaker and microphone is reduced compared to a handset placed close to a user's mouth and ear. Consequently the higher sound levels produced by the loudspeaker may cause a significant portion of the received signal to be acoustically coupled into the microphone, thus creating an echo signal. If hands-free operation is also used at the far-end telephone, then there is also a possibility that the gain of the signal loop, created by the received signal at each telephone being coupled back into the transmit path, is greater than unity and therefore causes unstable oscillations to occur which are heard at both telephones as a loud howling sound.

There are two methods used today to implement hands-free operation, and both work to eliminate the presence of an echo signal. The first method is echo suppression. The hands-free block shown in the terminal codec in Fig. 6.6 suppresses the occurrence of an echo signal by attenuating either the transmit or receive path according to whether the conversation participants are speaking or listening. This approach is based on the fact that a telephone conversation is for the most part a simplex communication where one side is talking while the other side is listening. The hands-free circuit therefore controls the gain of the loudspeaker and microphone amplifiers such that when the far end is talking the loudspeaker amplifier has a high gain while the microphone amplifier has a low gain. When the near end is talking, the microphone amplifier is set to a high gain while the loudspeaker gain is set low. When both users are talking or silent, attenuation is added to both transmit and receive paths. Switching between these conditions is dependent on the signal levels of the transmit and receive paths, and is controlled by the hands-free circuit which transitions smoothly between the different conditions so as not to generate audible clicks. The difference between the high and low gain settings is typically around 40 dB.

This technique is the one most commonly found in telephones with hands-free capability as it provides an adequate performance and is relatively cheap to implement. However, the enforced half-duplex operation and its susceptibility to be falsely triggered by background noise, can disrupt the flow of conversation, particularly if more than two people are involved. Hence, for conference facilities, and in particular videoconferencing, the above technique can be enhanced through the use of echo cancellation. The principle is the same as that discussed for the U transceiver in Chapter 4. Here, the codec subtracts from the transmit signal a copy of the received signal which has been modified by a filter whose characteristics closely match the acoustic impulse response of the room, and hence the transmission path between the loudspeaker and microphone. The echo cancellation significantly reduces the gain of the acoustic feedback path, and therefore only reduces the echo signal rather than the talker's signals, thus making the interaction between participants of the conversation much more natural.

6.6.4 Loudspeaker output

The ability of third parties to listen-in to a telephone conversation, often referred to as **loudhearing**, may be provided by a loudspeaker within the telephone. A variable gain amplifier may also be included in the codec to directly drive the loudspeaker. Variable gain amplifiers, which can be controlled by the microcontroller through the control interface, may also be provided in the transmit and receive path of the handset in order to adjust the characteristics of its earpiece and microphone to provide the correct signal levels.

Table 6.3 A selection of terminal codec/filter devices

Feature	Siemens PSB 2160 Audio Ringing Codec Filter (ARCOFI)	Mitel MT8994/5	SGS-Thomson ST5080	AMD79C30A
PCM	Programmable A-law and μ-law or 16-bit linear PCM	μ-law (MT8994) or A-law (MT8995)	Programmable A-law or μ-law	Programmable A-law or μ-law
Variable gain transmit and receive amplification	Yes, programmable	Yes, programmable	Yes, programmable	Yes, programmable
Sidetone control	Yes, programmable	Yes	Yes, programmable	Yes, programmable
Tone generation	Programmable DTMF tone and ringing	DTMF, tone and ringing	Programmable DTMF tone and ringing	Programmable DTMF tone and ringing
Hands-free	No (Included in the PSB2165)	Yes	Partial (AntiLarsen circuit)	No
Loudspeaker output	Yes	Yes	Yes	Yes
TDM interconnect bus	SLD or IOM-2	ST-BUS	PCM or GCI	Proprietary (Has on-chip S/T transceiver and HDLC controller)
Control interface	SLD or IOM-2	Microprocessor port	Serial port (MICROWIRE)	Microprocessor port

Acoustic feedback from the loudspeaker to the microphone of the handset may also cause loop instability, as in the case of hands-free operation, particularly when the handset is in close proximity to the loudspeaker, as for example when a user hangs-up and replaces the handset on the telephone prior to the hook switch being depressed. However, the degree of control and attenuation required is not as demanding as in hands-free operation.

Table 6.3 lists some commercially available terminal Codec/filter devices and their features.

6.7 PERIPHERAL SERIAL CONTROL PORT

Many microcontrollers and microprocessors designed for embedded control applications are equipped with a simple serial control port capable of transferring a byte or word command to a peripheral device equipped with the same port, and reading back status information from it. This type of serial control interface is preferable for the interconnection of a microprocessor to peripherals that require only low-speed control and monitoring, and for which the connection of a full microprocessor bus is an 'overkill' and costly in terms of printed circuit board space and pin count on the peripheral device. The S transceiver and codec/filter in the telephone are ideal candidates for such a control port as they only require mode initialization followed by infrequent commands for activation and deactivation, or to enable and disable various functions.

The principle by which a serial control port operates is based on a transfer of information between two shift registers[17]. The microprocessor and its peripheral are connected such that their two shift registers are configured as a ring. This scheme permits full-duplex communications such that, at the same time that a bit of data is shifted out of one shift register to the input of the other shift register, a corresponding bit of data is transferred in the reverse direction. Therefore, as a byte of data is transmitted in one direction, a byte is received in the opposite direction. To provide an orderly control over the transfer of information, one of the two devices acts as a master, which means that it determines when information transfer is initiated, while the other operates as a slave. However, situations do arise where a slave device must initiate a request to send information to the master, for example the S transceiver whenever it changes its operational state. This is indicated by means of a hardware interrupt to the microprocessor which then initiates a transfer.

The serial control port transfers information synchronously to a clock signal generated by the master and connected to all slave devices. In addition, multiple

---

[17]A shift register is a basic building block function that provides parallel-to-serial and serial-to-parallel data conversion.

slaves can be supported if each is equipped with a chip-select input that can be individually selected by parallel I/O lines of the microprocessor.

The operation of this type of interface is mostly manufacturer specific and relies on a compatible port existing both within the microprocessor system and the interconnected peripheral devices. The following sections describe some typical examples of serial control port.

6.7.1 MICROWIRE (National Semiconductor)

MICROWIRE is National Semiconductor's serial control port that is implemented on the majority of its microcontroller products as well as peripherals such as ISDN transceivers, programmable codec/filters, LCD display drivers and EEPROMs [1]. Both 8-bit and 16-bit data transfer operations are supported at speeds of up to 5 Mbps.

In addition to the mode of operation described above, MICROWIRE can operate with a single serial data line that serves as both input and output, and allows half-duplex communications between the master and slave. As a consequence, the direction of the data transfer (master-to-slave or slave-to-master) is determined by the type of transaction to be performed across the MICROWIRE. Under normal conditions, the data direction is set to allow the master (the microconroller) to transmit to the slave (the peripheral). If for example, the master wishes to read the contents of a register in the slave, it would first send a byte to the slave that indicates its intent to perform a read operation and also identify the register to be read. Both master and slave then reverse the direction of their serial data port line to allow the slave to send the contents of the register back to the master.

6.7.2 SPI/SCP (Motorola)

The SPI (serial peripheral interface) is a general purpose serial control port implemented on many of Motorola's microcontroller products [2]. The SPI supports eight-bit data transfers at speeds of around 1 Mbps.

A subset of the SPI known as SCP (serial control port) is used by Motorola on their communications specific microprocessor (MC68302) and S interface transceiver (MC145474) for the exchange of status and control information [3]. SCP supports full-duplex four-bit and eight-bit read and write operations at speeds of up to 4 Mbps. The eight-bit read and write operations are similar to those described above for MICROWIRE where the first byte identifies the type of operation and the register to be operated on, followed by a second byte that contains the data to be written to, or that is read from, the relevant register. The exception in this case is that, as SCP only supports separate SO and SI lines, the operation can also be made as a single 16-bit transaction. The MC145474 also contains a number of four-bit registers, and the SCP implements the four-bit read and write operations as single eight-bit transactions where the first four bits identify the register and the type of operation, and the last four bits contain the data.

6.8 THE TDM INTERCONNECT BUS

The serial B- and D-channels of the ISDN user–network interface are extended inside the terminal on a serial time-division multiplexed (TDM) bus which interconnects those devices responsible for providing information to and from the B- and D-channels of the S interface. In addition, the TDM bus may also be used as a means of transferring control and status information between connected devices. The channels of the TDM bus are organized into a frame which is repeated once every 125 µs and is synchronized to the 8 kHz network timing clock.

Control of the physical timing of the TDM bus is determined by a bus master which ensures that the TDM bus timing is synchronized to the network clock. In our ISDN telephone example, the S transceiver is the TDM bus master which synchronizes the TDM bus timing with the clock and frame synchronization recovered from the S interface signal. The remaining devices attached to the TDM, such as the codec/filter and HDLC controller, are slaves to the bus timing. Typically, the TDM interconnect bus consists of the following four signals:

Frame Sync (FS). This is the frame synchronization pulse which occurs every 125 µs at the beginning of every frame. Together with the clock signal, a device connected to the TDM bus can identify the presence of each channel according to its specific order within the frame. The Frame Sync signal is generated by the TDM bus master and output to the connected slave devices.

Many of the devices that can be attached to a TDM bus will transmit and receive information into a time-slot or channel that immediately follows the Frame Sync signal. It may therefore be necessary with these devices to delay the Frame Sync signal that it receives to correspond with the time-slot to which it is assigned. This function is provided by a programmable **time-slot assignment** circuit that may either exist separately, or may be integrated into the device itself.

Clk. The clock signal, Clk, is used to transfer the serial information into and out of a device and onto the TDM bus. In conjunction with the Frame Sync signal, it determines the current bit position within the frame. The clock frequency may have a frequency within a typical range of 256 kHz to 4 MHz depending on the number of channels provided. Like the Frame Sync signal, the Clk signal is generated by the TDM bus master and synchronized to the 8 kHz network clock.

Tx and **Rx**. These signals contain the serial data input and output between a device and the TDM bus. Depending on the implementation, the Tx and Rx signals may be separate, or be multiplexed onto a single signal line.

Although many of the features of different implementations of the TDM bus are similar, each is different, having been defined by different IC and equipment manufacturers. Examples of such buses are IDL, SLD and ST-BUS which are described in the following sections. However, equipment manufacturers invariably wish to maintain a degree of flexibility in their designs, and as a consequence may require the ability to use a device without being constrained by the incompatibility of the TDM interconnect bus. A manufacturers' standard bus has therefore been defined called GCI, the general circuit interface, and has been adopted by numerous manufacturers of ISDN IC devices.

6.8.1 The IDL bus

IDL [3] is the interchip digital link used by Northern Telecom, Motorola and other manufacturers. It is a four-wire bus and has separate Tx and Rx signals. The basic frame structure contains 20 bits made up of the 2B+D channels of the BRA interface (18 bits per frame) and two 8 kbps channels known as the A- and M-channels (one bit each per frame) that may be used to transfer control and status information between devices. The frame format is illustrated in Fig. 6.7a. The eight bits of each B-channel are transmitted consecutively within a time-slot, while the two D-channel bits are separated and appear after each B-channel time-slot. The time between the end of the frame and the beginning of the next frame may be used by the output of other devices connected to the IDL bus, or remain unused.

6.8.2 The SLD bus

The subscriber line datalink (SLD) [4] bus was developed by Intel and is used in devices by numerous other manufacturers, including Siemens. It is an example of a three-wire bus where the Tx and Rx are combined on a single connection. The SLD signals and basic frame format are illustrated in Fig. 6.7b. In addition to the Tx/Rx and Clk signals, a Direction signal indicates the direction of the Tx/Rx line. In addition, the rising edge of the Direction signal marks the beginning of each 125 µs period and thus also operates as a Frame Sync signal.

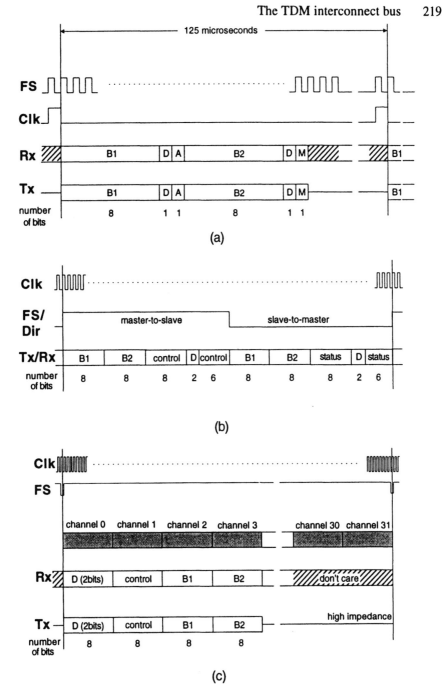

Fig. 6.7 Examples of different TDM interconnect bus formats. (a) IDL; (b) SLD; (c) ST-BUS.

Each SLD frame consists of four channels, each of 64 kbps. The first two are allocated to the two B-channels of the S interface. As the Tx and Rx signals are combined, the transmit and receive SLD frames must be multiplexed to make use of only a single line. For the first half of the 125 μs period, this line has a master-to-slave direction, while during the second half it is reversed such that a frame can be transmitted in the slave-to-master direction. The contents of the third channel in the SLD frame depend on the direction of the Tx/Rx line. In the master-to-slave direction, it may contain control information, while in the slave-to-master direction it may contain status information. The last channel in either direction contains the D-channel and additional control or status information according to the direction.

The SLD Clk signal has a frequency of 512 kbps in order to accommodate the total of eight channels (four in each direction) within the 125 μs time period.

6.8.3 The ST-BUS

The ST-BUS is a four-wire bus developed by Mitel Semiconductor for the system interconnection of their terminal and linecard components [5]. Each frame of the ST-BUS consists of thirty-two 64 kbps channels, as illustrated in Fig. 6.7c. The use of these channels depends on the type of device connected to the bus. For example, the MT8930 BRI S bus transceiver from Mitel makes use of four consecutive channels starting from the time it receives its frame synchronization input signal. As well as the 2B+D channels, a control channel also exists, through which devices may be controlled and their status examined. For a PRI transceiver, an additional ST-BUS is dedicated to control and status information, where each channel is used to provide control and status information for the corresponding B-channel.

The ST-BUS employs a Clk signal of 4.096 Mbps that results in two Clk cycles for each bit position in the ST-BUS frame.

Mitel employs a novel daisy-chain approach to the assignment of time-slots on the ST-BUS, obviating the need for a more complex programmable time-slot assignment circuit. A synchronization signal is relayed from one device to the next according to the order in which the time-slots are assigned by the devices connected to the ST-BUS. Thus, the synchronization signal is used as input to the first device, and when it has finished transferring information to and from the bus, its sends the synchronization signal to the second device, and so on.

6.9 THE GENERAL CIRCUIT INTERFACE (GCI)

The GCI [6] is an industry standardized serial TDM interconnect bus that allows telecommunication specific ICs from different manufacturers to be interconnected

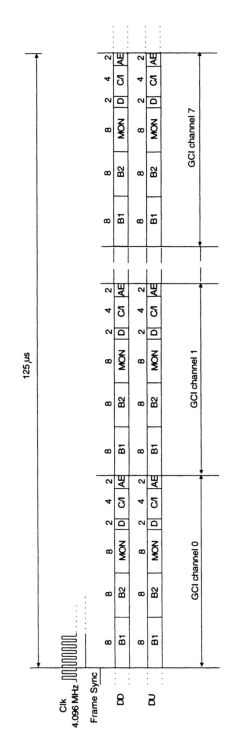

Fig. 6.8 GCI multiplex frame format.

and controlled across a single interface. It was defined by four equipment manufacturers, Siemens, Alcatel, Italtel and GPT, and tends to be used predominantly by telecommunication equipment manufacturers in Europe. Because of Siemens' involvement in the definition of GCI, it is incorporated in many of the telecommunication specific devices made by the semiconductor division of Siemens, although they give it the name **ISDN oriented modular** (IOM) interface [7]. Alcatel also have a bus related to GCI known as V* [8]. However, GCI is an open standard and is used or implemented by numerous other equipment and semiconductor manufacturers.

GCI was primarily defined as a component interconnect bus for PBX and local exchange switch linecard applications. As such, the GCI frame structure consists of a multiplex of eight basic GCI frames, where each basic frame consists of four 64 kbps time-slots, as illustrated in Fig. 6.8. Each basic frame consists of two B-channels, a 16 kbps D-channel, and a monitor channel (MON) and command/indicate channel that are used in the control of devices attached to the GCI bus. The A- and E-bits are used as handshaking bits for the transfer of information in the MON channel. With eight basic GCI frames in a single multiplex, up to eight BRA devices, such as S or U transceivers, can be connected to a single GCI bus.

The GCI bus consists of four signal lines. Two lines, the data upstream (DU) and data downstream (DD), contain the serial data previously referred to as Tx and Rx, while the usual frame synchronization (Frame Sync) and data clock (Clk) are the remaining two lines. The data clock has a frequency of 4.096 MHz, twice that of the bit rate on DU and DD.

6.9.1 The GCI special circuit interface T (SCIT)

In addition to linecard applications, the GCI also defines a version of the interface called the **special circuit interface T** (SCIT) for terminal applications. SCIT uses the same basic set of signal lines in its bus as GCI, but employs a different arrangement of time-slots to better meet the requirements of terminal applications.

To illustrate the use of SCIT, Fig. 6.9 shows a simplified diagram of an ISDN data terminal whose components are interconnected with the SCIT bus. An S interface controller is interconnected to a D-channel HDLC controller, a data interface and a data compressor. The purpose of the compressor is to reduce the bandwidth of the data at the data interface such that it can be conveyed in a single 64 kbps B-channel. Communication between the data interface and the compressor is by means of additional **inter-chip** (IC) channels on the SCIT bus.

Like the GCI multiplex, SCIT contains MON and C/I channels for the control of devices connected to the SCIT bus. The microprocessor is able to send and receive information through these channels by means of a GCI controller, which in this case is integrated into the HDLC controller as it has a direct interface to

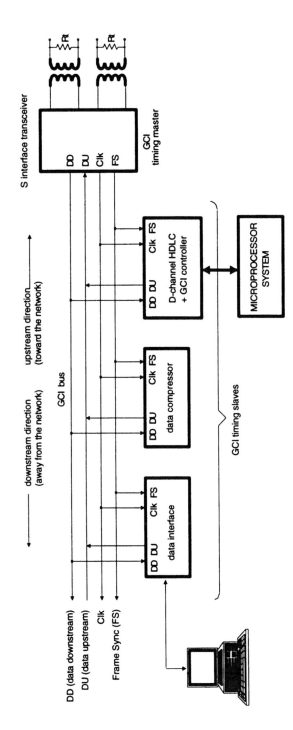

Fig. 6.9 The GCI-SCIT bus used in an ISDN data terminal application.

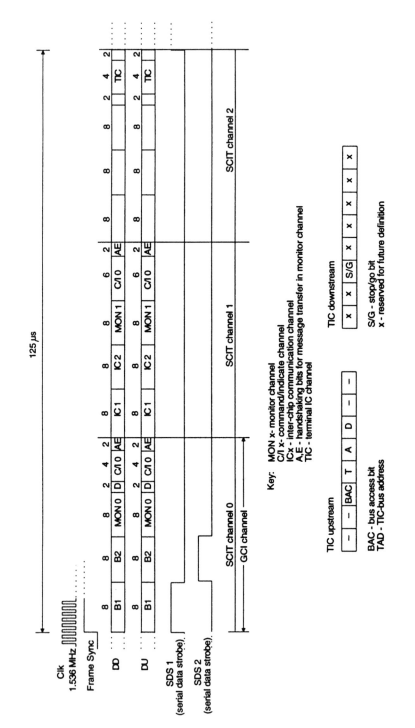

Fig. 6.10 The GCI-SCIT bus format for digital terminals.

the microprocessor system. Note that the GCI controller does not have to be the same device as the GCI bus timing master from which the frame synchronization and clock signals are output.

6.9.2 The SCIT bus format

Figure 6.10 shows the GCI-SCIT frame format. It consists of 12 eight-bit time-slots organized into three groups of four. Each group is referred to as a SCIT channel, the first of which, SCIT channel 0, has a format identical to the basic GCI format. However, the formats of SCIT channels 1 and 2 are unique to SCIT. In addition, the SCIT bus also contains two data strobes, SDS 1 and SDS 2, which indicate the presence of the B1-channel and B2-channel time-slots. These may be used to enable the data from devices such as voice codecs that have only a simple serial PCM interface.

SCIT channel 0 is dedicated to the transfer of the 2B+D ISDN channels between the S transceiver and devices **downstream** of it. The term downstream is used in a relative sense to indicate devices further away from the network in their connection to the subscriber loop than the S transceiver. The monitor channel, MON 0, and command/indicate channel, C/I 0, with its associated handshaking bits A and E, are used to communicate control and status information between the GCI controller and the **upstream** S transceiver. Similarly, the term upstream here refers to the fact that the S transceiver is closer to the network, from a connection point of view, than the HDLC controller.

Channel 1 is used for inter-chip (IC) communication using the IC1 and IC2 channels, and the associated MON 1 and C/I 1 channels provide communications between the GCI controller and the corresponding devices. The data terminal application in Fig. 6.9 is used here to simply illustrate how the inter-chip communication channels are used. The data interface generates data at speeds up to 128 kbps which is transferred from the data interface to the compressor in the two IC channels. The IC channels exist only internally on the SCIT bus within the system and are not visible externally from the S interface. The compressor then reduces the data to a maximum of 64 kbps which may be transported in one of the B-channels of SCIT channel 0 and across the S interface to the network.

Several devices may be connected to the SCIT bus and communicate with the GCI controller using the same MON and C/I channels. Each device therefore has a unique address to which it responds, and the C/I channel in the upstream direction may be used to provide device specific interrupt information to the GCI controller to identify the source of information generated by each device. The GCI controller may therefore exercise control over a number devices on the SCIT bus. In addition, devices may communicate with one another using the IC channels, as illustrated above. To avoid conflict in defining which device has the right to transmit information on the DU and DD lines, the devices make use of these lines according to their own position relative to upstream and downstream directions. For example, in Fig. 6.9 the HDLC and GCI controller are situated downstream of the S transceiver and will read the MON 0 and C/I 0 channels from the DD line. However, the same device is situated upstream of the data interface and data compressor, and so outputs information into the MON 1 and C/I 1 channels in order to control these devices. The S transceiver itself is always in the upstream position. Therefore, the devices connected to the SCIT bus, apart from the S transceiver, must be capable of switching their DU and DD pins between input and output according to the channel present on the SCIT bus.

6.9.3 Command/indicate (C/I) channel operation

The purpose of the C/I channel is to communicate real-time status information and maintenance commands such as activation, deactivation and loopback requests. The commands are defined in the GCI specification as four-bit codes that are transferred in the C/I channel[18]. A simple information transfer method is used, where data in the C/I channel is persistently transmitted in each frame until a device has a new command to send or new status to indicate. No acknowledgment or error checking procedures are provided, so devices receiving the C/I channel will therefore typically ensure that the same data has been received at least twice before taking any action as a simple means of error protection.

6.9.4 Monitor (MON) channel operation

The MON channel is used to convey commands and status information other than those mentioned for the C/I channel. Its purpose is to provide an interface through which the programmable functions of a device may be configured by the GCI controller.

Multi-byte messages may be conveyed in both directions in the MON channel, with each byte being acknowledged through a handshaking protocol that uses the E-bit on the transmit line and the A-bit (acknowledgment bit) in the reverse direction[19]. Figure 6.11 illustrates the use of the E- and A-bits to transfer a three-byte message.

Transmission of the first byte in the message follows a slightly different procedure from the following bytes. When no message is being transferred the E- and A-bits are inactive and can be read as a logic 1 level on the SCIT bus. The

---

[18]Although the coding of the commands is defined in the GCI specification, some manufacturers' implementations deviate from this definition.

[19]On the Alcatel V* interface, the equivalent A- and E-bits are respectively used as a transparent service channel and to enable the monitor channel.

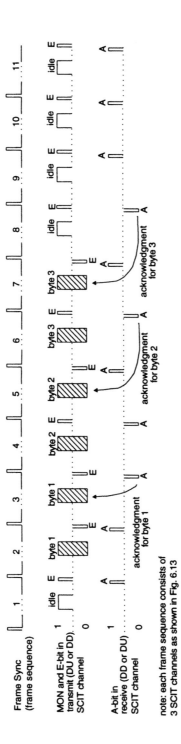

Fig. 6.11 GCI monitor handshake protocol.

device transmitting the message will place the first byte in the MON channel and activate the E-bit by pulling it to a logic 0 level, as shown in frame 2. It continues to transmit the same first byte with the E-bit active for at least another frame, and remains for as long as the A-bit in its receive direction stays inactive. The receiving device is notified of the transmission of the first byte by the active E-bit, and as a means of ensuring correct reception, also reads the following MON channel byte to ensure they are the same. This method of sending and receiving the same byte twice as a simple means of error checking is also employed for subsequent bytes. The receiver acknowledges correct reception by setting the A-bit active (low) as shown in frame 3, and keeps it active for one further frame. The transmitting device, sensing the acknowledgment, then places the second byte into the MON channel in frame 4 and signals its presence by deactivating its E-bit. This in turn causes the receiver to set its A-bit inactive during the following frame.

The transmitter again re-transmits the second byte in frame 5, this time with an active E-bit, and the receiver makes its acknowledgment in frame 6. Instead of the transmitter placing a new byte in the MON channel when it detects an inactive to active transition of the A-bit, as it did after the transmission of the first byte in frames 3 and 4, the transmitter now transmits a new byte whenever it detects an active-to-inactive A-bit transition, as shown in frames 5 and 6. Under normal data transfer conditions, the E- and A-bits therefore alternate in their polarity for each successive frame.

When the message has been completely transferred, the transmitter reverts to transmitting successive frames with inactive E-bits, two or more of which indicate that the MON channel is idle.

Control over the flow of information from the transmitter to the receiver can be achieved by the receiver holding the A-bit active, during which the transmitter will continue to re-transmit the last byte in the MON channel. When the receiver is ready to receive further bytes of the message, it releases the A-bit to its inactive state and the transfer of bytes resumes. The reception of two or more successive inactive A-bits during transmission of a message is an indication from the receiver that it is aborting reception of the message, and causes the transmitter to attempt to re-transmit the message.

The messages themselves contain both address and data information relevant to programming the registers within a device. The purpose of the registers depends on the application of the device and is therefore not defined as part of the standard. However, the first four bits of the first byte define an address which can fall into one of four ranges that are defined in the GCI standard and used to identify the type of device to which and from which messages are transferred. Also, as the GCI bus is an 'open' definition, a GCI-SCIT bus may have devices from different manufacturers connected to it, each of which may require different software drivers. Consequently, the GCI does define one particular message which allows a GCI controller to identify the type of interface device connected to the bus and its manufacturer.

6.9.5 Terminal IC (TIC) channel operation

Most of SCIT channel 2 in Fig. 6.10 is reserved for further definition, apart from the last time-slot, which contains four bits that perform the function of D-channel access control in the downstream direction, and the access control of multiple D-channel controllers in the upstream direction. These four bits are known as the terminal IC (TIC) channel.

Up to eight D-channel devices may be connected to the SCIT bus, and the TIC channel in the upstream direction provides access and contention resolution for these devices on the SCIT bus. The upstream TIC channel contains four bits as shown in Fig. 6.10, three of which represent the address of each D-channel (HDLC) controller on the SCIT bus, while the fourth bit is the bus access control (BAC) bit which indicates if the upstream D-channel is busy or not.

A D-channel controller wishing to transmit information first examines the status of the BAC bit. If the D-channel is already occupied (BAC = 0) the device waits until it is free (BAC = 1). It then starts to transmit its address into the TIC channel and at the same time read the condition of the DU line so that it can compare it with the address bits which it sends. If it is the only device requesting an access to the D-channel, then the information the device reads from the line will match the address bits which it has sent. However, if another D-channel controller is also requesting D-channel access at the same time, it too will be attempting to transmit its address into the TIC channel. As the address for each device is unique, there will be contention of the line where transmission of a logic 1 by one device and a logic 0 by another devices will result in a logic 0 appearing on the line. Therefore, only the device with the lowest logical address will read back its address correctly from the line, giving it priority over devices with a higher address. The device which successfully reads back its address immediately asserts a bus occupied condition by setting the BAC bit to 0, and starts to transmit its information in the D-channel.

In the downstream direction, the TIC channel contains a single stop/go bit (stop = 0, go = 1) which indicates the availability of the D-channel at the S interface. If the D-channel becomes unavailable during transmission due to a contention on the S interface, then the HDLC controller must abort its transmission in the D-channel.

6.10 IMPLEMENTATION OF CALL CONTROL SOFTWARE

A large part of the D-channel call control signalling within a terminal is implemented as software that runs on the terminal's microprocessor system. Today, those functions of the call control which lay above the HDLC controller (the lower part of layer 2) are predominantly implemented in software, while those below are mainly hardware based and interfaced through device specific software drivers into the higher software layers of the system. The large number of functions that exist at the upper layers of the protocol stack, particularly the supplementary services, and the regional variations in their specification, will mean that software will continue to remain the most practical means of implementation.

6.10.1 The multi-tasking executive

A popular approach to the software implementation of a layered communication system is to use a multi-tasking operating system or **executive** to provide the infrastructure and common services that are required by the functions within the system. The name executive is typically given to versions of multi-tasking operating systems that are embedded within equipment and provide coordination and execution control for multiple tasks within a system. In a layered communications system, the protocol entities that exist in each layer can be thought of as separate tasks, where, according to the layering principles discussed in Appendix A, each task is responsible for processing specific parts of protocol messages that are communicated between peer entities on either side of the communications link. Each task is a **state machine** driven by the messages which propagate up and down the layers of the communication stack.

In addition, other, control oriented, messages are used to control and coordinate specific actions within and between the layers. For example, these would include the activation and deactivation primitives found at the physical layer, as well as timer **time-out** indications within a layer.

Figure 6.12 illustrates a layered communication system. Each task, or state machine, is driven by messages that it takes as input from a message queue. The state machine will process the message according to its current state and the type of message received. This may result in other messages being sent to the next layer up or down, and relevant timers being activated or deactivated. Finally, the task may transition to another state before retrieving another message from the queue and repeating the process. In effect, each task is a continual loop which consists of reading a message and processing it.

At the hardware dependent levels, the devices will indicate to the microprocessor system with an interrupt signal when an event has occurred, such as a message being received by the HDLC controller, or a change of state in the S transceiver. This causes the normal program execution to be interrupted and a specific interrupt service routine to execute which will read the message from the device and place it in the relevant queue before returning execution to the program. In the output direction, a message will be delivered to a device driver task whose execution will read a message from the queue, interpret it and output information to the device accordingly.

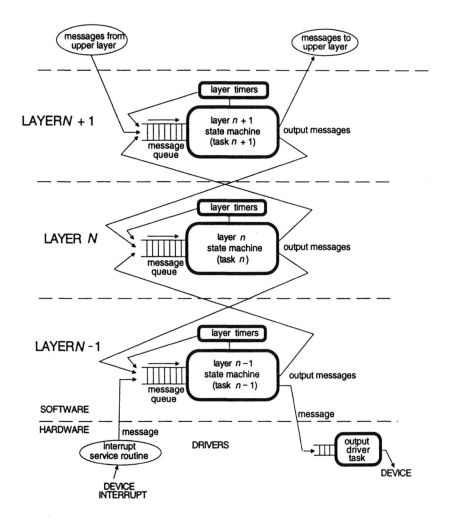

Fig. 6.12 A node of a communications system consisting of layered state machines.

From this basic mode of operation, some general requirements can be defined for the coordinated execution of tasks within the communications system implemented by a software executive running on a microprocessor system. The main attributes of an executive are described below.

6.10.1.1 Multi-task execution

A microprocessor executes instructions in a serial manner, and can therefore process only one task at a time. Consequently, the executive must provide some means of allowing each task to perform some work as required. A task may be defined as ready to do some work if it has a message in its queue. However, there may be more than one task in such a state, although it is possible for only one of them to be executing on the microprocessor at any one time. Therefore, an execution queue is required which is managed by the executive and defines the order in which the ready tasks execute. The process of ordering the ready tasks on the execution queue is known as **scheduling**, and may be done according to some priority rules. For example, if the task that has just become ready has a higher priority than all other tasks currently on the execution queue, then it will get placed at the front of the execution queue and will therefore be the next task to execute. Alternatively, if all tasks have an equal priority, they will execute on a first-ready-first-to-execute basis.

When the currently executing task has processed a message, one of two things may happen. Firstly, if its message queue contains another message, then it ceases to execute, but gets placed by the executive back onto the execution queue, typically behind the other tasks already waiting to execute (assuming they all have the same priority). Secondly, if no more messages are present, the task ceases execution and is placed into a waiting state where it remains until another message arrives in its message queue, and at which point the executive returns it to the execution queue.

As described, a task may therefore be in one of three basic states: **executing**, **ready to execute** and **waiting**. These states and the transitions between them are illustrated in Fig. 6.13.

Many executives provide different types of scheduling schemes. The scheme described briefly above is known as **non-preemptive** scheduling on the grounds that a high priority task that becomes ready to execute during the current execution of a lower priority task does not cause the executive to immediately suspend execution of the lower priority task and to install and execute the higher priority task. Instead, the higher priority task will be placed at the head of the execution queue, but must wait until the lower priority task has completed processing its current message before the task switch can occur. In this way, a message is completely processed by a layer before another message in the same or another layer can be processed. Adherence to this non-preemptive scheduling is important in order to ensure that the consequence of a message be affected before the next message is processed, as the way in which the next message is processed may well depend on these previous actions.

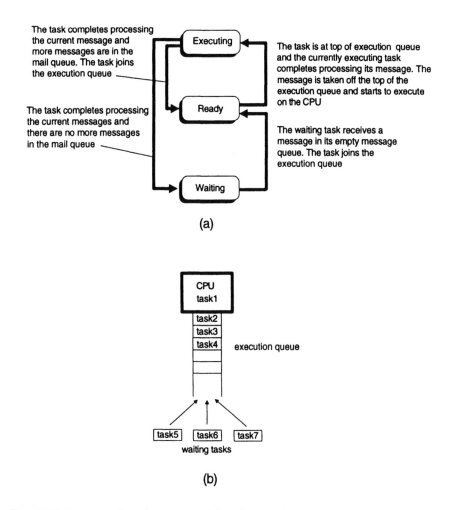

(a)

(b)

Fig. 6.13 An example task state transition diagram for a multi-tasking executive.

Other forms of scheduling are **preemptive**, which does allow the execution of a low priority task to be interrupted, or **preempted**, by a higher priority task, and **time-slicing** which allows tasks of equal priority to share the execution bandwidth of the microprocessor through allocation of a fixed execution time. Preemptive scheduling is typically found in systems that are driven by strictly real-time applications such as robotics and automation, while time-slicing may be used for the execution of background or supervisory tasks while the main tasks are idle.

6.10.1.2 Inter-task communication

Tasks represent layer entities within a communications system, and are implemented as state machines that are driven by receipt of messages from other tasks. Each task is associated with a message queue that stores messages until such time as the task can process them. The message queue, known as a **mailbox**, is shown in Fig. 6.12. A task successively reads messages from the mailbox until an absence of messages forces it into a waiting state. Similarly, while processing a message, a task will have the ability to send messages to the mailboxes associated with other tasks, normally those representing the layers immediately above and below the sending task.

In general, the mailbox concept makes no constraints on the format or content of the messages themselves. Some executives also provide facilities for the communication of an event occurrence, such as an interrupt signal, that has no information associated with it, but will also cause its task to transition into a ready to execute state. Such facilities are useful in the support of interrupt service routines for hardware device drivers.

6.10.1.3 Timers

Each task may have a number of timers associated with it to perform the response timing described in Chapter 5. A timer expiry may result in an appropriate time-out message being sent to the task's mailbox such that when it executes and processes the message, the appropriate action can be taken.

It is unlikely a microprocessor system will provide sufficient hardware timers for the total number of timers required by all the tasks that make up the communications system. Instead, an executive may support the implementation of numerous software-based timers driven from a single hardware timer. The hardware timer is used to generate a clock tick with a period equal to the highest resolution required by all the timers, typically in the range 1 to 10 ms for Q.931- and Q.921-based timers. The active software timers are implemented simply as software counters which are updated on each time-out of the hardware timer. During the updates, each counter is checked to see if it has exceeded its time-out value, and if so a message that identifies the timer and its time-out condition is sent to the task's mailbox. In this way, numerous 'soft' timers can be supported from a single hardware timebase.

6.10.1.4 Memory management

Certain resources used by an executive and its tasks require the use of system read/write memory, or RAM. As not all resources will ever be in use at the same time, a smaller memory requirement can be defined if the available memory is allocated dynamically from a pool rather than being fixed. An executive will therefore provide a memory manager to provide allocation of memory as requested by the executive or its tasks, and to recover the memory once it has been used.

6.10.2 Example – Telenetworks Adaptable ISDN Software

A number of ISDN software packages are commercially available today, either from semiconductor manufacturers who license the software for the design-in support of their ISDN chip-set, or from independent software companies. The following example is used here to illustrate how such software is implemented around a multi-tasking executive, and the typical features and functions it can provide.

The Telenetworks Adaptable ISDN Software package is a software implementation of the standardized data-link procedures, network and call control functions for the BRA and PRA ISDN D-channel call control signalling[20] [9]. The operation of the software is based on a multi-tasking executive called MTEX that was written specifically to support communications-oriented tasks, and contains the four main functional features described in the previous section [10]. Each layer entity within the protocol stack corresponds to a task, and each task comprises a number of modules arranged as shown in Fig. 6.14. Those modules outlined in bold represent the core functions provided by the software.

6.10.2.1 Architecture

The layer 2 (L2) and layer 3 (L3) procedures are designed around the same state machine engine, and each consists of three modules. The main task loop is found in the Lx interface module (Lx means either L2 or L3, as the principle is the same for both layers). Here the task's mailbox is read and the resulting message checked for basic validity. A function call is then made to the state machine which is implemented in the module Lx. Based on the current state of the task and the type of primitive received, the task will call one of the routines contained in the module Lx_SDL which processes the information according to one of the SDL branches defined in the relevant standard[21]. As a result of this action, the task

---

[20]This description is based on revision 4.0 of the Telenetworks software.
[21]The Telenetworks software package can be used in both terminal and network side applications. Although Fig. 6.14 illustrates the terminal call control protocol stack, a network side stack would look very similar. However, as the layer 3 network side SDLs are sufficiently different from the terminal side SDLs, a different L3_SDL module is used.

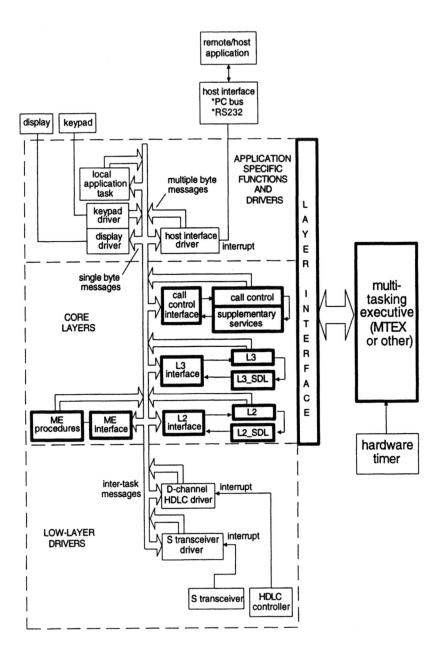

Fig. 6.14 Architecture of the Telenetworks adaptable ISDN software package.

may change its current state, mail messages may be generated and sent to the mailbox of other tasks, and timers may be activated or deactivated. Program execution then returns to the main loop to pick up another message from the mailbox.

Application specific call control functions are implemented in a separate module from the network layer procedures, which is where most of the customization work will be done to interface the protocol stack to a particular application. Much of the work done at this level involves the processing of information elements within the protocol messages, and a library of generic functions are provided to assist in building up the information elements in messages sent to the network protocol layer below, and to help search for and decompose those in messages received from it. At this level there are also significant regional differences to the implementation of supplementary services, which has resulted in a range of application specific and country or region specific modules that supplement a generic call control module and call control interface.

The management entity (ME) provided as part of the software package implements those management functions that are defined to support the data-link and physical layers. As such, it provides TEI assignment and management, parameter negotiation, error handling, and is also involved in physical layer activation and deactivation. It is expected that these functions will be expanded at some later time, for example to allow comprehensive error monitoring by the local exchange to ensure performance to quality of service parameters. As yet these functions are undefined, but the ME task provides a framework on which they may be implemented in the future. The implementation of the existing functions within the ME module is divided into two parts: the layer management and the connection management. Layer management is responsible for TEI management, while the connection management provides parameter negotiation and error handling for each of the data links established by layer 2.

Drivers exist for a number of popular HDLC controllers and S transceiver devices, although writing a driver for a new device is relatively straightforward. In general, a driver will consist of three parts: hardware initialization, a service request (SRQ) task and an interrupt service routine (ISR). An SRQ task is similar to any other task in that it possesses its own mailbox, and performs a continuous loop reading the mailbox and processing the message. In the case of an HDLC driver, the message read from the mailbox will have been sent from the layer 2, and will contain a message for transmission in the D-channel. In the receive direction, a hardware interrupt will be generated when a message from the D-channel has been received, and the ISR associated with it will check the validity of the message, mail it to the layer 2 mailbox, and reinitialize the hardware to receive another frame. It is desirable to keep execution time of the ISR as short as possible in order to minimize interrupt latency and maintain multi-tasking operation. The initialization of device specific registers to establish its correct operational mode may be performed in a variety of places, but most typically as part of the MTEX initialization or as the initial part of the service request task before it enters its main loop that reads messages from its mailbox.

Figure 6.14 shows that all tasks except for the lower layer device drivers interface to the multi-tasking executive, MTEX, through a layer interface, called LIF. Provision of a single interface between tasks and the executive has a benefit in that should MTEX be substituted for another executive, then only a single module, the LIF, need be modified, rather than each task. In this way, the Telenetworks software may also be supported by other executives instead of MTEX. However, a disadvantage of this approach is slower execution speed, as each software function call from a task to the executive must be made through another function call within the LIF module. This conflicts with the driver ISR requirement to keep execution speed as quick as possible. As a result of this, the lower layer drivers are allowed to interface directly to the executive instead of through the LIF module.

6.10.2.2 The layer 2 and layer 2 tasks

The layer 2 and layer 3 tasks are based on state machines and are driven by the messages that arrive at their mailboxes. Because of the similarities between them, the layer 2 task can process messages according to Q.921 LAPD, X.25 LAPB and the rate adaptation protocol V.120 described later in Chapter 8. The layer 3 task processes messages according to the protocol procedures within Q.931.

Before a message can be processed by the layer 2 task, the correct logical data link must be identified, and the current state of that data-link loaded into the layer 2 state machine. The state of each data-link is stored in a data structure known as the data-link control block (DLCB), and in addition to the state of the LAPD data-link connection, contains other information such as the current status of the numbered I frame state variables, sequence numbers $N(S)$ and $N(R)$, and operational parameters such as k and timer time-out values. The layer 2 task maintains a list of active DLCBs, one for each active data-link connection, and each time a message is processed, the list is searched to identify and load the correct DLCB for the corresponding logical data link into the state machine. The correct DLCB is identified by matching the DLCI in the message to be processed with that stored in the DLCB. When a message has been processed, the status is updated in the DLCB before the next message is processed. The DLCB also contains a parameter which identifies the protocol with which the message is to processed, and allows the layer 2 task to support a number of data-link protocols, in this case LAPD, LAPB and V.120.

An identical operation is found in the layer 3 task. As it is concerned with processing only the protocol procedures of Q.931, it uses the call reference (CR) and message type (MT) to drive its state machine. In the same way that the layer 2 task can handle multiple logical links within LAPD, the layer 3 task can control multiple overlapping calls where each call is identified by its call reference value which is used to identify an NLCB (network layer control block) that contains the status of each call.

6.10.2.3 MTEX

The multi-tasking executive, MTEX, provides task scheduling, mail management, memory management and timer management services to the tasks that make up the call control protocol signalling stack, including the drivers and the application that interface to it. MTEX consists of three modules, one of which contains the user definition and initialization of system resources such as task control blocks[22], mailboxes and timers. The remaining two modules contain the core MTEX functions, including the definition and initialization of memory management functions.

When used to support the ISDN D-channel call control signalling, MTEX employs a non-preemptive scheduling scheme such that once a task starts to process a message it continues until it has completed its execution[23]. In addition, the tasks that go to make up the call control stack are given the same priority such that those tasks that are ready to execute get an equal opportunity to process the messages waiting in their mailbox. This has the effect of smoothing the flow of messages up and down the protocol stack, and easing the need for large message queues as part of a task's mailbox. The mailbox itself actually consists of two message queues, one for mail passing up the protocol stack, and the other for messages going down. In addition, a mail message may be given a high priority, which forces it to the front of the message.

MTEX supports numerous software based timers driven from a single hardware timer. It may allocate and de-allocate a software timer from a fixed size pool of timers. The implementation of the timer function within MTEX is as a timed mail message, such that when a task makes a request for a timer with a specific time-out value, it also specifies a mail message and mailbox as part of the same request. At any time before the time-out, the timer may be cancelled. However, if the time-out value is reached, then the mail message is delivered to the mailbox.

MTEX provides memory management using fixed-size pools of memory blocks. The block size and number of same size blocks in each pool can be defined but remain fixed during operation. This method has an advantage in that it is simple and deterministic, and is suited to implementation in an embedded system having a limited amount of memory. One of the main uses of the memory manager is in allocating memory for messages such as primitives that are communicated between tasks.

---

[22]A task control block contains information about the task, such as its start address, priority and so on, which the executive uses to schedule and execute it.

[23]MTEX also supports non-preemptive scheduling and time-slicing for use in other applications.

6.10.2.4 Inter-task communications

Each task has its own mailbox associated with it, to which any task, including itself, may write messages, although due to the layered nature of the protocol stack, the majority of messages tend to be communicated between adjacent layers. Messages are the standard primitives defined in the protocol standards, and non-standard primitives or control information defined for layer coordination and control of hardware devices through the drivers at the top and bottom of the protocol stack.

Two types of message are defined. The first contains a single byte whose value may be used as a command or status indication. Its use is mainly at the driver interfaces of the system. The second type of message contains the primitives and may consist of up to three interlinked parts. The first part is an envelope that contains the type of primitive along with other information that is useful in the processing of the primitive, such as where the message originated from, the SAPI and TEI/CES values, the call ID and so on. If the message also contains protocol data, then a field within the envelope will indicate the address of a packet descriptor. The packet descriptor itself contains the address and length of the header (address and control fields) and information fields of the HDLC frame stored in memory. Addresses are provided for both the header and information parts of the frame to allow them to be stored separately from one another if necessary. This structure is illustrated in Fig. 6.15.

An important aspect of inter-task communications is the way in which messages are transferred between tasks and mailboxes. When a message such as a primitive is defined, an appropriate size buffer is requested from the memory manager and the primitive is created. As the primitive gets processed and is passed between tasks and mailboxes, the physical location of the primitive in memory does not change. Instead, the address of the primitive is the attribute which is communicated, and obviates the need to continually write and read large amounts of data to mailboxes, whose size would, as a consequence, become prohibitively large.

6.10.2.5 Configuration and adaptation

The adaptability of the software is designed through the following features.

- The software is supplied as source code, and is almost entirely written in the C programming language, which is one of the most popular high-level languages for micorprocessor programming. In addition, parts of the software may be compiled and run on a PC, which together with debug tools, allow changes to be made and verified independently of the target hardware platform. Porting the software code to a new microprocessor is also feasible provided a suitable C compiler exists[24].

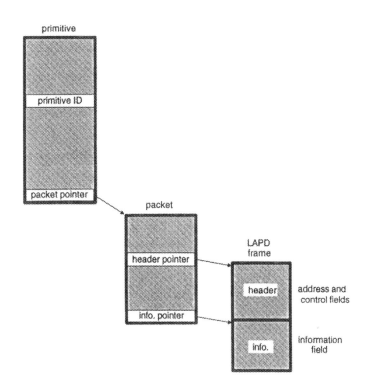

Fig. 6.15 The structure of primitives in the Telenetworks ISDN software.

MTEX has been written to run on a number of different microprocessors. When a task switch occurs under MTEX, the contents of various registers must be preserved in the context of the old task, while the register values of the new task must be loaded before it can run. In particular, each task will possess its own stack, whose current location in memory must be saved whenever its associated task ceases to execute due to a task switch. Each type or family of microprocessors will have its own unique register set, some of which are best supported by a small assembly-language routine to implement

---

[24]The C compiler must be ANSI C compliant.

the low-level stack switching. This is the only processor dependent code required by MTEX.

* The software code implements the country and switch specific variants of each layer that allow the selection of the specific implementation to be made through conditional compilation. Where a particular function is implemented in several ways based on different standards, conditional compilation will only compile the code for the relevant standard by the definition of its corresponding compile time flag. Alternatively, where multiple implementations are selected at compile time, selection of a particular protocol implementation can be made at run-time initialization based on some kind of hardware selection, for example through a switch.

* Two types of application interface are provided at the top of the call control stack. The first is a simple control interface that passes a single byte of information between the call control and the application. Such an interface is suitable for low complexity functions such as on-hook and off-hook, and the transfer of keypad or display information. The second interface is more complex and allows messages to be communicated between the call control and application where the application runs independently of the call control on another microprocessor. Such a situation would exist, for example, on an ISDN PC card that had its own microprocessor to process the call control functions, while the PC runs the applications that controls it.

* The use of the LIF module masks the dependencies of the executive from the tasks, thereby allowing other executives to be used where this is beneficial to the implementation.

6.10.2.6 Development features and tools

The Telenetworks ISDN software contains several features, as well as a software tool, that allow user-written modifications and application tasks to be debugged and verified. As previously mentioned, conditional compilation will allow hardware independent parts of the software to be compiled and more conveniently run under MTEX on a PC. However, such an exercise is only useful if there is also a means of inputting messages into the system and being able to view the resulting output. **Task View** is a tool written for the PC that does just this. Task View is simply another task running under MTEX which interprets commands and data from a text file and translates them into mailbox commands for the tasks under test. The integration of Task View together with the tasks under test is through conditional compilation, and one or more adjacent tasks can be exercised using Task View.

Task View defines a simple language that is used to create a list of commands in a text file on the PC using a simple text editor. For example, when interpreted by Task View, the appearance of the command INFO in the text file followed by a data string will create a mail message primitive with the data as the contents of the information field within its associated packet. A SEND command would then cause the mail message to be sent to a specified mailbox, and a subsequent READ command would capture any mail message output by the task under test and display its contents. In this way, an individual task or group of adjacent tasks can be exercised according to the sequence of input messages, states and output messages that define their operation.

In addition to Task View, the Telenetworks software contains numerous C language debugging statements that when enabled, again through conditional compilation, allow internal variables and states to be monitored during execution. For this feature to be used in the target hardware environment rather than a PC, the target hardware must contain a UART connected to a simple ASCII terminal.

MTEX itself also contains some features which are useful during debugging. Firstly, all MTEX system calls that use system resources return error codes if the resource requested is not available. These messages can be translated into debug messages and displayed on a debugging terminal under conditional compilation as described above. Also under conditional compilation, the memory manager will record the maximum number of memory blocks used during operation. This feature is useful as it is impractical to determine a suitable value for the size of many of the memory blocks prior to running the software under test. Instead, an over-estimated value may be initially assigned to the total number of blocks within a particular memory pool, and the maximum number of blocks actually used is then monitored during operation under all conditions. The final number of blocks assigned to the particular memory pool can then be taken as the maximum monitored value plus a modicum of overhead to provide a degree of safety against unforeseen operating conditions. This procedure avoids wasting RAM storage due to an over cautious assignation of blocks to a memory pool.

Finally, MTEX provides a similar feature to monitor the maximum depth of the stack for each task. Unlike the MTEX memory manager, many microprocessors do not provide any traps to avoid stacks overrunning areas of data when too little memory has been allocated to it. Instead, the data becomes corrupted and the system usually crashes. Again, allocating a more than adequate over-estimate of the stack size will usually solve the problem but will also unnecessarily waste memory. By monitoring the stack size under all operating conditions a more appropriate stack size can be determined.

6.11 APPLICATION PROGRAMMING INTERFACES

It is desirable to make the software communication system independent from any application that may use it such that the application does not need to be conscious of the physical details that govern the operation of the communications system that lies beneath it. To achieve this requires the definition of a common set of abstract functions between the top layer of the communication protocol stack and the application. This set of functions is known generically as an application programming interface (API).

Such independence promotes the development of applications to a common interface, and may allow the same application to be supported by different communication networks according to the communication protocol and hardware that lies beneath it. The API may either be embedded within the top layer of the communication protocol stack for stand-alone applications, or be incorporated within an operating system if it is part of a computer system such as a PC. In either case, the API is considered to be part of the communication system rather than the application.

In reality, it is not possible to define an all encompassing API that interfaces all types of application to all types of communication system. Instead, specific APIs are required according to the type of communication system they are attached to. In the case of ISDN, the lack of standardization in this area until recently has led to the development of several APIs supported in various world regions. The following is a short list of APIs defined for ISDN.

ASI (Application Software Interface) is being standardized through the North American ISDN Users Forum (NIUF).

CAPI (Common ISDN Application Programming Interface) is today the most common ISDN API used in Europe, and is supported primarily by equipment manufacturers in Germany. CAPI supports both call control and B-channel protocols for data applications such as fax and file transfer. CAPI has also been made a standard by ETSI.

COM/APPLI (API between Communications software and Applications software) is supported mainly in France.

FRAPI-A (Foundation for the Promotion of Telecommunication Service API) is currently being implemented in Japan.

ISDN PCI (Programming Communications Interface). This is the API standardized by ETSI for applications interfacing to Euro-ISDN [11].

TAPI (Telephony Applications Programming Interface). An API developed by Microsoft and Intel, and included in Microsoft's Windows 95™ operating system. TAPI is designed to support call control functions for ISDN, PSTN and mobile (wireless) networks.

In response to the need for worldwide standardization, the ITU-T is preparing a Recommendation that describes an architecture framework for APIs within which future interfaces can be defined.

REFERENCES

1. National Semiconductor. MICROWIRE Serial Interface, Application Note 452.
2. Details of SPI can be found on Motorola data sheets for those products that contain the interface. For example, MC68HC11A8 HCMOS Single-chip Microcomputer – Advance Information, Motorola, 1985.
3. Motorola (1989) MC68302 Integrated Multi-protocol Processor User's Manual.
4. Intel (1989) SLD Interface specification, in *Microcommunications Handbook.*
5. Mitel (1987) Implementing an ISDN Architecture using the ST-BUS. Application Note number MSAN-128.
6. Siemens, Alcatel, Italtel, GPT (1989) General Circuit Interface (GCI) Specification, Issue 1.0.
7. Siemens (1990) IOM-2 Interface reference guide, in *ICs for Communications,* edition 1.90.
8. Mietec (1989) MTC-2071 UIC U Interface Circuit Reference Manual, Issue C.
9. Telenetworks (1992) Programmer's Manual for Telenetworks ISDN Software, version 3.3.
10. Telenetworks (1992) Programmer's Manual for Telenetworks Multi-Tasking Executive MTEX, version 3.2.
11. ETSI (1994) Integrated Services Digital Network (ISDN); Programming Communication Interface (PCI) for Euro-ISDN, pr ETS 300 325.

7

Voice and video communications in the ISDN

In the previous chapter, we discussed the implementation of those generic functions of a terminal which allow it to request establishment and control an end-to-end connection between users connected to an ISDN. We now turn our attention to how the user or application information is interfaced and processed at the terminal such that it may be communicated between users once a connection between them has been established.

User, or application, information is transported across the ISDN subscriber loop in B-channels. Within the terminal therefore, user information must be:

- formatted to allow transmission and reception across the network;
- rate adapted, and if necessary compressed, to match the 64 kbps, or multiples thereof, available at the ISDN user–network interface.

In this chapter we deal with the techniques used today to interface, process and transport real-time user information such as voice and video, across the ISDN, while in Chapter 8 we look at the communication of more transactional information, such as remote terminal data and computer file transfer.

Interactive voice and video communications are considered real-time in the sense that people participating in the communication are aware of timing imperfections in the communications process, such as delay, jitter and synchronization, which can cause the quality of the information received to be subjectively degraded. The communications process, which consists of both the processing of user information that takes place within the terminal and the conveyance of the information across the network, must respect the real-time nature of voice and video signals in order for it to provide an acceptable level of service.

7.1 SPEECH AND AUDIO COMMUNICATIONS

Two telephony services have currently been defined in the ISDN for the communication of speech and audio information. The 3.1 kHz telephony service is equivalent to that provided by existing analogue telephony subscriber loops, while a 7 kHz service provides an extended bandwidth for higher quality speech and audio that finds use in applications such as audio- and videoconferencing, and broadcasting where the increased bandwidth improves the intelligibility and natural quality of the speech, and eases speaker recognition.

The PCM codec/filter for analogue–digital conversion in an ISDN telephone was already discussed in the previous chapter. However, using PCM and increasing the analogue bandwidth from 3.1 kHz to 7 kHz results in double the amount of digital data generated, making it necessary to compress the data in order to accommodate it in a single B-channel call. The process defined by the ITU-T to provide the necessary compression is based on a technique known as adaptive differential pulse code modulation, or ADPCM.

7.1.1 ADPCM

As defined in ITU-T Recommendation G.721 [1], the ADPCM technique is used to compress standard 64 kbps PCM equivalent to a 3.1 kHz voice channel, into 32 kbps ADPCM. Although compression from 64 kbps to 32 kbps finds its main application in pairgain and mobile systems, we shall describe the basic operation of ADPCM here before considering its extension to the conversion of a 7kHz voice channel to 64 kbps ADPCM in the provision of high quality speech and music over an ISDN. The G.721 ADPCM Recommendation is designed to provide compression and decompression of both 3.1 kHz speech and voice-band data signals.

Figure 7.1 shows a block diagram of an ADPCM encoder and decoder. At the input to the encoder will be either a linear A/D converter[25] if the input signal is an analogue signal, or a converter to linear PCM if the input signal is already in companded PCM format. In the case of the latter, the ADPCM encoder is usually referred to as a **transcoder** as it is performing a conversion from one digital format to another rather than encoding a signal from its original source.

Two functions within the ADPCM process which contribute to data compression are **adaptive quantization** and **differential encoding**. Differential encoding works on the basis that instead of encoding the absolute value of the signal, only the difference between the current sample and a predicted value of the current sample are encoded. A prediction of the current input sample is generated

---

[25]An A/D resolution of 14 bits is required for ADPCM.

from past sample values of the input signal and a knowledge of the statistics of the signal itself. This function is implemented by a filter whose specification is defined in the G.721 recommendation, and is contained in the signal predictor block in Fig. 7.1. As can be expected, the variance of the difference signal will be less than that of the original signal, and consequently will require a digital value with less resolution, or fewer bits, than PCM when encoded.

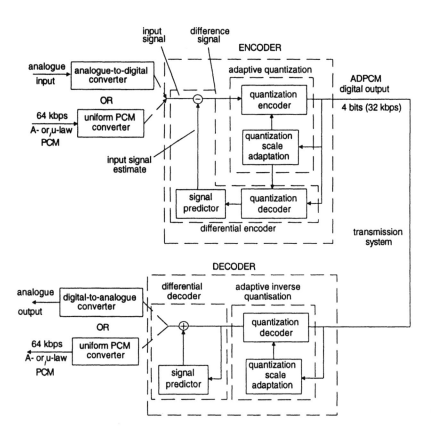

Fig. 7.1 The ADPCM encoder and decoder.

The difference signal is then translated into one of 15 levels represented by a four-bit code (three bits to represent magnitude and one bit to represent the sign) by the quantization encoder. Unlike PCM, where the absolute scale of the eight-bit, 256 level code is constant, the scale of the four-bit code is adjusted according to the magnitude of the difference signal and a speed control factor. This speed control allows the quantization scale adjustment to occur at different rates, and provides rapid adjustment for signals with fast frequency and magnitude fluctuations (such as speech) or slow adjustment for signals with slow fluctuations (such as modem signals). As the difference signal reaches the extremes of the encoding range, the range is increased through scaling it by a factor greater than one. Similarly, when signal is encoded around zero, the scale is diminished through scaling by a factor less than one.

The output of the quantization encoder is the output of the ADPCM encoder, and consists of a four-bit value every 125 µs resulting in a data rate of 32 kbps, and gives a data compression of 2:1 over PCM. At the ADPCM decoder, the encoding process is reversed to yield the signal in its original format. The fact that the prediction process in the encoder involves only past input samples means that the decoder needs have no additional information beyond the four-bits ADPCM to synchronize its operation to the encoder, and the design of the predictor filter and adaptive quantisation is such that the encoder and decoder will inherently re-gain tracking should errors occur during transmission.

ADPCM codecs and transcoders rely on digital signal processors (DSPs) for their implementation. A 16-bit fixed-point arithmetic capability is generally required and may be implemented on general purpose programmable DSP devices such as those from Texas Instruments [2], or as dedicated devices such as those from OKI Semiconductor and Motorola.

As far as the use of ADPCM in ISDN is concerned, there is little incentive to use ADPCM at 32 kbps on a 64 kbps circuit-switched call as the remaining 32 kbps of bandwidth remain unused. Of course, an application may make use of the unused bandwidth in some way to transfer data across the same connection for example, but its usage remains limited. Instead, the main application today for ADPCM codecs is in digital wireless telephony, such as DECT, CT2 and PHP, where a 32 kbps channel is defined for speech communications across the wireless subscriber link between the handset and the basestation. For ISDN terminals, the ADPCM technique is modified as described in the next section to provide a means of communicating higher bandwidth speech and audio signals.

7.1.2 SB-ADPCM

The method used for the encoding of high-quality speech and audio signals with a bandwidth in the range 50 Hz to 7 kHz is known as **sub-band ADPCM** (SB-ADPCM), and its operation is defined in Recommendation G.722 [3–5]. As well as extending the high-frequency cut-off from 3.4 kHz to 7 kHz, the low-frequency cut-off is also extended downwards from 300 Hz to 50 Hz, thus giving the reproduced acoustic signal more depth. The name sub-band ADPCM is derived from the fact that the signal bandwidth is split into two sub-bands, a high band and a low band, and each is then encoded using ADPCM. A terminal that implements G.722 is commonly referred to as a **wideband speech terminal**.

The G.722 SB-ADPCM technique is designed to handle speech and may also accommodate audio signals such as music. However, unlike G.721 it is not intended for voice-band data applications. Instead, it is designed to accommodate low bit rate digital data transfer through an auxiliary data channel which is embedded into the 64 kbps stream. A SB-ADPCM codec may operate in one of three modes, two of which provide for the incorporation of either an 8 kbps or 16 kbps data channel, as shown in Table 7.1.

Table 7.1 SB-ADPCM mixed audio and data modes

Mode	Audio channel	Data channel
1	64 kbps	0
2	56 kbps	8 kbps (6.4 kbps data plus framing)
3	48 kbps	16 kbps (14.4 kbps data plus framing)

In addition to these three modes, mode 0 is also defined as corresponding to 64 kbps PCM (A-law or μ-law) according to G.711, and allows a wideband terminal to interwork on a point-to-point or multipoint conference with normal 3.1 kHz terminals. In addition, under emergency power conditions, a wideband terminal will fall-back into normal 3.1 kHz telephony mode.

Use of a data channel in modes 2 and 3 results in minor degradation of speech quality at the receiver, but provides additional functionality for applications such as document transmittal between correspondents during a conversation. A wideband terminal may operate in more than one mode. Three different terminal types are defined, as listed in Table 7.2.

Table 7.2 Terminal types defined for wideband terminals

Type	Modes	Terminal capability
0	0	3.1 kHz PCM
1	0,1	7 kHz with fall back to 3.1 kHz, no data
2	0,1,2,3	7 kHz with fall back to 3.1 kHz, with data

Switching between the different modes can be performed during a conversation and is controlled according to Recommendation G.725 [6], and discussed later in the next section.

Operation of the SB-ADPCM encoder and decoder is explained in the block diagram shown in Fig. 7.2. The bandlimited audio input signal is sampled at 16 kHz and each sample converted to a 14-bit linear binary code. Splitting the signal into two separate equal sub-bands is then performed by the digital transmit quadrature mirror filters, and results in a lower sub-band of 0 to 4 kHz and an upper sub-band of 4 to 8 kHz, where each sub-band has an effective sample frequency of 8 kHz. Each sub-band is then encoded with a separate ADPCM encoder.

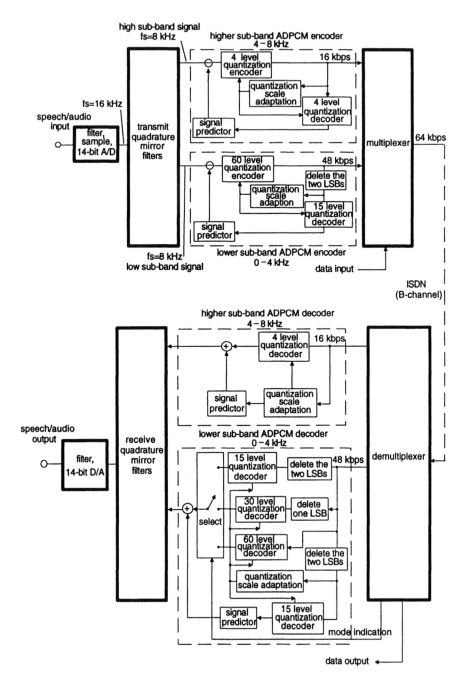

Fig. 7.2 The sub-band ADPCM audio encoder and decoder.

Much of the information contained in the speech signal will be present in the lower frequencies, and makes it necessary to maintain a higher level of signal-to-noise ratio for these frequencies rather than the higher ones. As a consequence, six bits per sample are used by the ADPCM encoder to encode the lower sub-band, thus producing a 48 kbps data stream, while only two bits are used for the upper sub-band, producing a 16 kbps data stream. These two streams are then combined in the multiplexer to produce a 64 kbps stream.

It can be noticed from Fig. 7.2 that the lower sub-band ADPCM encoder uses only the most significant four bits in its feedback loop instead of the entire six bits generated by the quantization encoder. This allows the multiplexer to discard the least significant bit (LSB) or two LSBs from the lower sub-band and substitute them with other data bits, thus creating an 8 kbps or 16 kbps auxiliary data channel. The upper sub-band is always encoded with two bits regardless of the operational mode. A further implication of this substitution is that the decoder must be aware of it such that it can operate with the corresponding quantization decoder. The decoder is informed of the operational mode of the encoder using the in-band handshake signalling described in the next section. Finally, the receive quadrature mirror filters combine the upper and lower sub-bands and a 14-bit D/A and filter reconstitute the analogue signal.

As the G.722 SB-ADPCM encoder and decoder are not intended for voice-band data signals, the quantization scale adaptation process need only have a fast adaptation speed. However, the signal predictors used in each ADPCM encode and decode block are the same as those used in G.721.

7.1.2.1 In-band signalling and framing

Call control, including terminal identification as a 7 kHz wideband terminal, will take place as usual over the D-channel of the ISDN interface between a terminal and the network. However, the D-channel signalling today has no means of communicating the particular mode of operation between the communicating terminals. An in-band signalling scheme is used for this purpose; signalling and frame format are defined in Recommendations G.725 [6] and H.221 [7] respectively. The frame format is shown in Fig. 7.3 and is not specific to just wideband speech terminals, but is used more generally to integrate different media signals, such as voice, video and data, for transmission over synchronous B-channels. H.221 framing is discussed further in this chapter in the context of video communication over the ISDN.

As can be seen in Fig. 7.3, no framing is provided in modes 0 and 1 where the content of each byte is either all PCM or SB-ADPCM. In modes 2 and 3, a frame is constructed from eighty consecutive bytes where the last bit of each byte in mode 2, and the last two bits of each byte in mode 3, constitute an auxiliary channel that contains a frame alignment signal (FAS), a bit rate allocation signal (BAS) and the auxiliary data itself. The purpose of the FAS bits is to provide

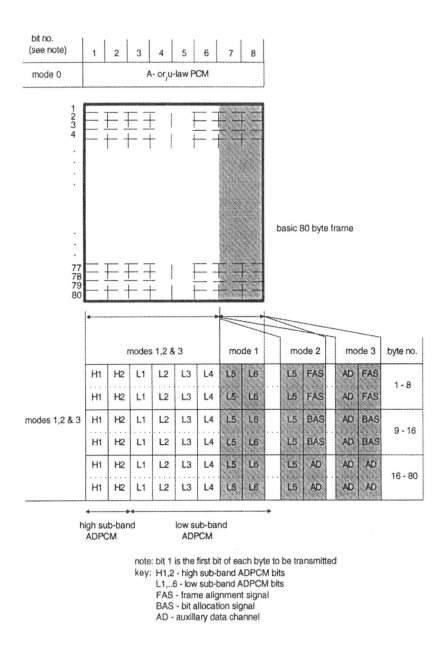

Fig. 7.3 H.221 framing used in SB-ADPCM to accommodate in-band signalling and the auxiliary data channel.

framing indication for the 80 byte frame, while the BAS bits contain a codeword which indicates the mode of operation that defines the use and structure of bits within the frame.

Dynamic switching between modes is indicated by a change in the BAS code. This is simple to achieve between modes 2 and 3 as the framing that includes the BAS codes is present in these modes. However, modes 0 and 1 are unframed such that when a mode switch is required, the transmitting terminal must first initiate the framing in order to allow an exchange of BAS codes to take place, after which the terminals continue operation in the designated mode.

A similar procedure is used during the initialization of terminals. At the beginning of the conversation phase of a call the terminals start to operate in mode 0. Provided that connection is to another wideband terminal (indicated during the call setup sequence), framing is established and used to exchange terminal capabilities. The terminals then switch to their lowest common operating mode.

Most implementations of G.722 codecs today are on programmable digital signal processors from companies such as Texas Instruments, AT&T and Motorola. However, as the market for high-quality conferencing services increases, so too will the demand for equipment and low-cost codecs, and will undoubtedly result in a more widespread availability of dedicated semiconductor codec devices.

7.1.3 Low-bit-rate speech encoders

The aggregate user bandwidth of an ISDN BRA connection is 128 kbps. For videoconferencing applications, audio, video and perhaps data streams must be combined into the 128 kbps aggregate channel. The compression of the video signal represents the most demanding task in terms of the levels of compression that must be achieved, and the more bandwidth of the connection that can be dedicated to the compressed video stream then the better the video quality at the receiver. Higher levels of speech compression beyond that delivered by ADPCM are therefore advantageous in order to make available as much bandwidth for video.

PCM and ADPCM both belong to a class of speech encoding techniques referred to as **waveform coding**. The name is derived from its ability to encode a waveform and then reproduce it as accurately as possible at the decoder, and it produces acceptable quality at data rates of 64 kbps for PCM and 32 kbps for ADPCM. A second class of encoding techniques is known as **source coding**, which unlike waveform coding, attempts to analyse the speech signal to determine its core parameters from which it can then be synthesized, for example voiced or unvoiced sound, pitch, volume and so on. The main advantage of such techniques over waveform coding is that they result in significantly lower data rates, and can today generate coded speech signals at data rates as low as 2.4 kbps.

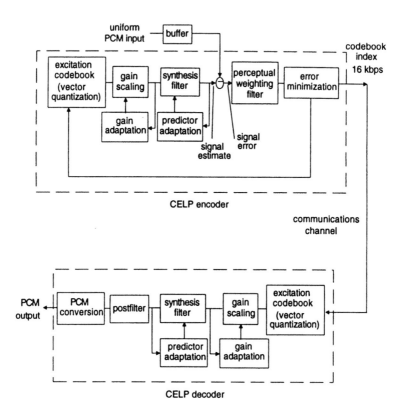

Fig. 7.4 The low-bit-rate CELP speech encoder and decoder.

A particular technique which falls into this category is known as code excited linear prediction (CELP). The ITU-T have standardized a version of this technique in Recommendation G.728 [8] that yields a low-delay CELP (LD-CELP) encoder and decoder for the coding of 3.1 kHz speech at 16 kbps. The principle behind CELP can be explained with reference to Fig. 7.4. As in ADPCM, the encoder loop operation is based on minimizing the difference between the input signal and an estimate of the input signal. However, there the similarities end. The estimate of the input signal is derived from a codeword which when passed through the gain scaling unit and synthesis filter, closely matches the original signal. The codeword is one of a number of such fixed code words stored in a table, referred to as a **codebook**, that represent the underlying characteristics of the signal it attempts to estimate. Once the codeword has been found which generates the lowest error, the index to the codeword is transmitted across the communications network to the decoder where an identical codebook is stored. The fact that the index to the codeword is sent and not the codeword itself allows bit rates of down to and lower than one bit per sample to be achieved with this technique. In G.728, the encoder and decoder operates on groups of five consecutive PCM samples, or 40 bits in 625 µs. The codebook stores a total of 1024 codewords and therefore has an index of 10 bits that results in an output data rate of 16 kbps.

The main design tradeoff in the design of CELP encoders is speed versus accuracy. A codebook that contains a large number of codewords has a higher likelihood of generating a smaller error (resulting in distortion of the original signal) than does a smaller codebook. However, each codeword in the codebook must be tried before the one that generates the minimal error can be selected. Consequently, long codebooks can result in lengthy search procedures unless the code words are organized in some structure that yields a fast search process. The G.728 codebook comprises two smaller gain (three-bit) and shape (seven-bit) codebooks, while other codecs may use tree- or trellis-encoded codebooks to reduce search time. In any case, the codebook search time must be less than the equivalent transmit time of the number of PCM samples operated on at a time, and determines the encoder/decoder delay.

The gain scaling unit and synthesis filter in the encoder use backward adaptation techniques obviating the need to communicate their parameters to the decoder to remain synchronized, while the perceptual weighting filter attempts to include a model of auditory perception into the error minimization loop as traditional methods, such as mean-squared error minimization, are found not to function well at low bit rates.

As with G.722 codecs, the most popular form of implementation today is on programmable DSP devices, although one or two dedicated devices, such as the PSB 7280 JADE (joint audio decoder and encoder) from Siemens Semiconductor, have recently become available.

7.2 VIDEO COMMUNICATIONS

The extension of voice communications to include the real-time images of the correspondents is a natural progression of the features provided by a telecommunications network, as it more closely resembles the way we are used to communicating and interacting with people on a daily basis. The biggest technical challenge is of course that the transmission of an uncompressed video signal requires a channel bandwidth several orders of magnitude higher than that of the PSTN. Today, video compression techniques have been developed that meet this challenge, and the ISDN has played a major role as an enabling technology with which video communications can be effectively delivered to the end user.

Video communication is available today on three broad platforms to meet different applications and users need.

Videotelephone. This is an extension of the standard telephone that allows both visual and audio communications. Videotelephones are of a similar size to a standard voice telephone and typically have a small LCD screen to display the picture. Such devices are today too expensive for general consumer markets, and have remained a niche for business executives.

PC videophone. A videocodec and ISDN terminal adapter card, together with camera, microphone and speakers are added to a personal computer to provide it with videotelephony capability. In addition to desktop videotelephony or videoconferencing, the PC may be used to run software that allows the sharing and joint manipulation of information such as spreadsheets, documents or presentations, across the communications network. These types of applications are sometimes referred to generically as collaborative computing or groupware. Such equipment is expected to become commonplace in the work environment within a few years, given the existing widespread use of PCs in business today.

Videoconference equipment. This equipment provides video communications between two or more conference rooms where there are typically several participants at each location. Hence, large-screen televisions are used and cameras that give a broader field of view. Wideband audio is also used to give an acceptable sound quality in such conference situations.

Room-based videoconferencing systems have been in existence for some time, and before the availability of ISDN, used digital leased lines for communications that made them expensive to run as well as own.

Figure 7.5 shows the core building blocks of a PC based videocommunications for voice, video and data communication applications. Although illustrated here for a PC-based terminal, the basic video and audio elements may be found in any of the types of equipment discussed above. At the heart of such systems is the videocodec whose primary purpose is to provide real-time compression of a moving image from the camera into a data stream that can be transported across the. A camera that outputs a digitally encoded form of television signal will generate a raw data stream of 216 Mbps[26]. Using the standards described below for the framing of the voice and video signals, it means that a compression ratio of up to 1:3140 must be achieved for transmission across the two B-channels of an

---

[26]According to CCIR Recommendation 601 for 4:2:2 sampling.

ISDN BRA interface, assuming G.711 speech is also used. At the receiving end, the videocodec will provide the corresponding decompression into a picture that can be displayed.

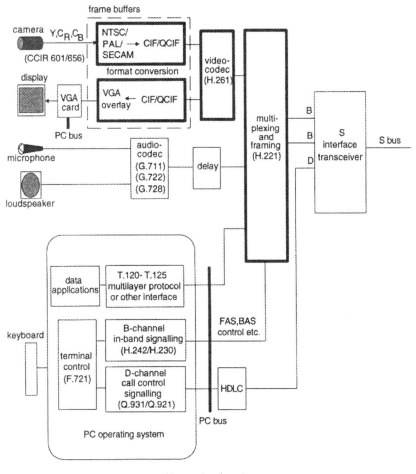

Fig. 7.5 The core building blocks of a PC videophone for voice, video and data communications across the ISDN.

The purpose of the delay block that occurs between the audiocodec and the H.221 framing and multiplexing block is to provide what is known as lip synchronization between the audio and video streams. The encode and decode delays introduced by the videocodec will in general be longer than those introduced by the audiocodec, and direct combination of the audio and video output streams would cause the image of someone talking and their speech to be out of synchronization. The delay at the output of the audiocodec therefore equalizes the delay paths of the video and audio streams in both the transmit and receive directions.

7.2.1 Video communication standards

The technical requirements for terminal equipment that provide visual communications at data rates up to 2 Mbps are outlined in ITU-T Recommendation H.320 [9]. This document describes the basic system architecture together with modes of operation. Further details on various parts of the system are provided by other associated recommendations, in particular H.261 [10] and H.221 [7] which respectively describe the architecture of the videocodec and the way the video and audio streams are combined for transport across the ISDN. Other Recommendations cover in-band signalling aspects (H.242 and H.230) and extensions to the basic call control to handle network connections involving more than one B-channel (F.721).

H.221 may also provide multiplexing and demultiplexing for a data channel in addition to the audio and video. Standard protocols may be used in the channel provided for such data communications. For example, the T.120 series of Recommendations define how data is communicated as part of sessions that may also involve video and audio, and between more than two people in a multiparty conference call. T.120 is discussed in more detail in the next chapter.

A new standard has recently been developed by the ITU-T to allow videocommunications at data rates less than 64 kbps. The new Recommendation, known as H.324 [11] is intended for use on normal PSTN subscriber lines using standard 28.8 kbps V.34 modems. Like H.320, H.324 is an umbrella document that refers to other Recommendations that define the video and voice coding algorithms, and the multiplexing and control. The video coding part, defined in H.263 is an extension of H.261 and defines improvements to the motion estimation algorithm that allows higher video compression ratios to be achieved. In addition, a CELP algorithm defined in G.723 is used to code the speech and achieve low bit rates of 5.3 kbps and 6.3 kbps.

7.2.2 Video signal format

A video signal is generated from a camera in one of the formats used today in the transmission of television signals. Unfortunately, the three different standards that are in use today for the transmission of television signals, PAL, NTSC and SECAM, are each incompatible in terms of frame rate and frame dimension. This has prompted the definition of a **common intermediate format** (CIF), to make the videocommunications using ISDN independent of the video source format generated by a camera or accepted by a display, thus making it inter-operable on a world-wide basis. In Fig. 7.5, the signal from the camera is therefore first converted from its source format to CIF before being compressed by the videocodec. Similarly in the receive direction, the output of the videocodec will be in a CIF format that must then be converted into the correct format for the display device. The CIF format also provides a first stage of data reduction through a sacrifice of picture resolution, it being approximately one quarter of the equivalent television formats.

The source encoding for the transmission of a colour video signal consists of a **luminance** signal and two **chrominance** signals. The luminance signal, Y, represents the brightness, while the two chrominance signals, C_R and C_B, represent the colour information[27]. A fixed relationship exists between Y, C_R and C_B that allows the three primary colour signals, red, green and blue to be derived for display on a colour monitor. As it turns out, the human eye is not as sensitive to colour changes as it is to brightness, and allows it to be sub-sampled without significant loss of picture quality. In the CIF format, the chrominance signals are sub-sampled by a factor of two both in the horizontal and vertical dimensions of the frame, and are spatially oriented between luminance samples as illustrated in Fig. 7.6[28].

Each source frame is built up of a number of horizontal lines which, when operated on in the digital domain, are sampled to give a series of picture elements, or **pels**, each of which are converted to a digital value. According to CCIR Recommendation 601, the horizontal line resolution of a frame is 720 pels for the luminance signal and 360 pels for a chrominance signal. As the CIF format is intended to be approximately one quarter of the overall picture resolution, the CIF luminance resolution should be 360 pels, or half this value. However, as we shall see later, the videocodec operates on 16×16 pel **macroblocks**, so it is convenient if both the horizontal and vertical resolutions of the CIF are multiples of 16 such that a frame contains an exact number of blocks. This determines the final value for the horizontal CIF resolution as 352.

The vertical dimension of a television frame may contain either 625 or 525 lines according to which of the PAL, NTSC or SECAM standards is used. The

---

[27]Y, C_R and C_B are the digital forms of the analogue luminance and chrominance signals. It is assumed here that the output of the camera is in a digital format.

[28]This type of spatial sub-sampling is referred to as 4:2:0 sampling scheme.

CIF vertical dimension may be derived from the 575 active picture lines contained in a 625-line frame. A simple ratio of 2:1 provides the vertical CIF resolution of 288. In a 525-line system, a ratio of 5:3 provides coverage for most of the active area of the frame.

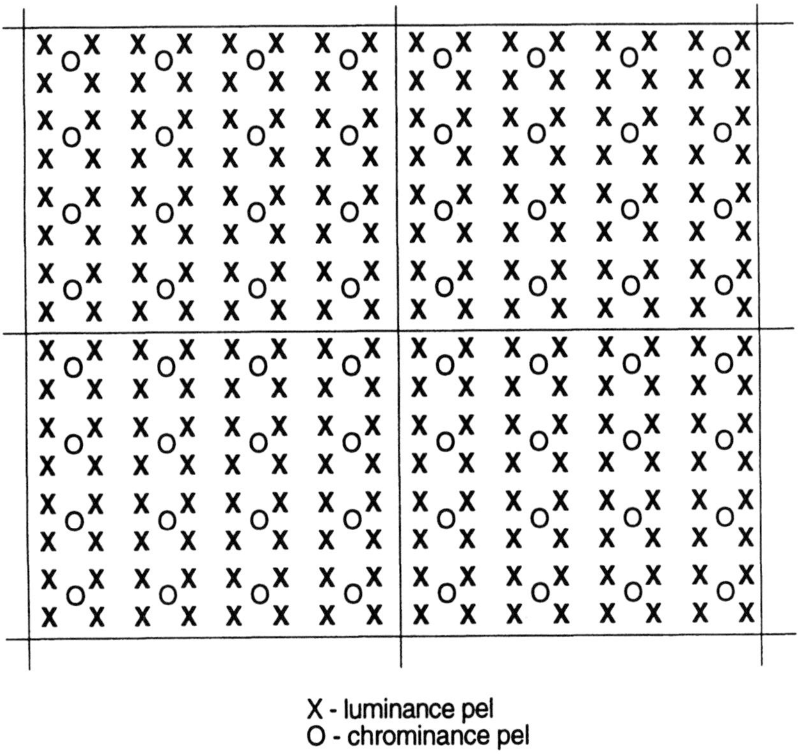

X - luminance pel
O - chrominance pel

Fig. 7.6 Four 8×8 luminance blocks and the relative spatial positioning of the pels in the two chrominance blocks that form a macroblock.

A moving image consists of a sequence frames where the smoothness of the motion depends on the frame rate, the motion content of the image and the rate of change of that motion. In general, where frame rates are lower than 30 fps (frames per second) some motion in the picture may not appear smooth. However, for the so-called talking-heads video communications, such as videotelephony, where there is a minimal amount of motion, a lower frame rate may be acceptable. A CIF frame rate of 29.97 fps at the input to the videocodec is taken directly from the frame rate in the 525-line system.

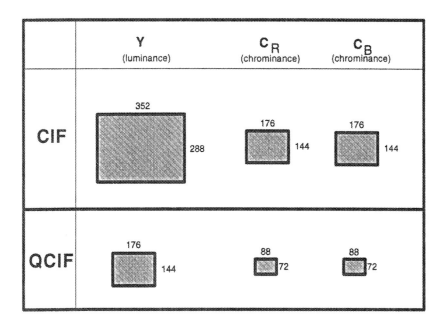

Fig. 7.7 CIF and QCIF frame dimensions.

The dimensions of the CIF frame are shown in Fig. 7.7. Each pel is represented by an eight-bit value, and a quick calculation shows that the channel bandwidth required to transmit video as a series of CIF frames would be $[(352 \times 288) + 2 \times (176 \times 144)] \times 8 \times 29.97 = 36.5$ Mbps. If we remember that the output from a digital camera is 216 Mbps, then it can be seen that conversion to the CIF format has provided a reduction in data of approximately 6:1. However, to reduce this data rate to that available with a 2B-channel call requires a further reduction of 530:1, which in real-time is still a demanding task. To make the task somewhat less demanding, yet a further 2:1 reduction in both horizontal and vertical resolutions is defined in the quarter CIF, or QCIF format, also shown in Fig. 7.7. The equivalent data rate of QCIF is one quarter that of CIF, or 9.1 Mbps, and requires a compression of 132:1 for a 2B-channel call. This remaining reduction in data rate is achieved through compression which takes place in the videocodec.

In H.261, the QCIF format is mandatory for codecs that conform to this recommendation, while the CIF format is optional. Due to the lower performance, size and cost requirements of a QCIF implementation, it is the preferred format for use in videotelephones, while either the QCIF or CIF format is used on

PC-based equipment, and the CIF format with videoconferencing equipment where the higher resolution is required.

7.2.3 The H.261 videocodec

The objective of the videocodec is to compress the raw CIF or QCIF video frame data into a data stream that can be transmitted across the available communications channel. Several standards exist for videocompression, and the one standardized by the ITU-T for videotelephone, videoconference and other audiovisual services available across public networks, is defined in Recommendation H.261 [10].

Block diagrams of the H.261 encoder and decoder are shown in Fig. 7.8. Compression of video data is achieved using a combination of the following techniques:

- differential encoding;
- discrete cosine transformation (DCT);
- variable length encoding;
- quantization;
- motion compensation.

7.2.3.1 Differential encoding

The basic form of the encoder is similar to that of the ADPCM encoder. It operates on the basis of encoding and transmitting the difference between the same portions of the current frame and the previous frame. Rather than process a complete frame in one cycle of the loop, 8 × 8 pel blocks are processed at a time. A frame is actually divided into a number of macroblocks, each of which contain four 8 × 8 blocks of pels arranged in a 2 × 2 macroblock. For a sequence of frames containing no motion the output of the encoder would be zero resulting in no information being transmitted, but as the motion increases so does the difference and hence the amount of data transmitted. The difference information generated at the output of the subtractor can be thought of as a prediction of the current frame based on the previous frame. An encoder loop delay of one frame is achieved by setting the delay of the block buffer in the feedback path such that the combined delay of all functions in the path from the output of the subtractor to its lower input is equal to one frame delay at the output of the codec. In this way it is ensured that the corresponding block of the current and previous frame are compared.

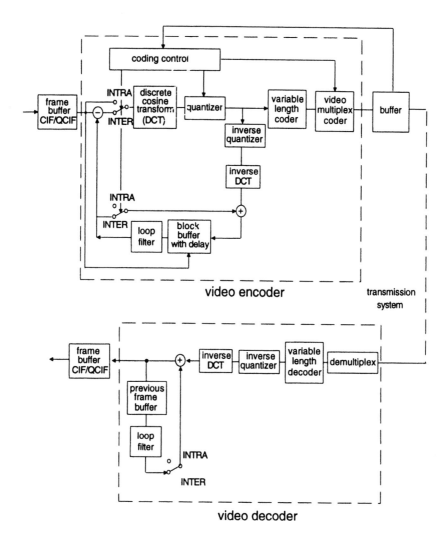

Fig. 7.8 The H.261 videocodec.

Closer examination of Fig. 7.8 shows the existence of two switches that operate in unison to select either an INTER mode or INTRA mode. The INTER-frame mode is that described above where the difference between the same blocks of successive frames is encoded and transmitted. However, an inverse transform (inverse DCT) function is also present in the forward path of the decoder, and in the feedback loop of the encoder such that frame differencing can take place in the pel domain. Owing to result rounding and scale clipping during calculations involved in the transformation of blocks, the result of a transform followed immediately by an inverse transform may not yield exactly the original. These errors can accumulate in the decoder and, because of the recursive nature of the encoder, will eventually cause the reproduced picture to become noticeably corrupted. To avoid the accumulation of these errors to such a point that they become a problem, an INTRA-frame mode is permitted to allow the information in the macroblock itself rather than the difference to be encoded and transmitted. However, encoding the picture information rather than the difference will generate significantly more data and can only be done once every so often. The criteria on which the INTRA mode is invoked are not defined in H.261 apart from specifying the requirement that it should be done at least once every 132 frames. Use of the INTRA mode during startup is typically performed to provide a meaningful picture quickly.

7.2.3.2 Discrete cosine transformation (DCT)

There are typically many instances in a picture where the information in adjacent pels is very similar, or perhaps identical. In other words, there may exist a high degree of correlation between picture elements, which implies that much of the picture information contained within the pel data is redundant. The purpose of a transform is to translate the data into another form which is less correlated, thereby reducing the amount of data required to represent the picture information. The discrete cosine transform (DCT) [12] performs this function in the H.261 videocodec, and transforms the two-dimensional spatial amplitude luminance and chrominance information into corresponding spatial frequency, or picture feature, information.

The DCT is a derivative of the discrete Fourier transform (DFT). The DFT is used to transform a time domain representation of a periodic signal into the complex frequency domain using a transform that involves both cosine and sine functions. But whereas the Fourier transform results in a complex number that represents the frequency and phase information of the signal, only the real component, or the cosine function, is used in the DCT.

The DCT function in the H.261 videocodec operates on blocks of 8 × 8 pels for both the luminance and chrominance parts of the picture, instead of on the whole picture in one go. This speeds up the transformation process and permits real-time operation, and also keeps the size and cost of the videocodec at a reasonable level. To perform the DCT and inverse DCT functions requires a digital signal processor (DSP) with at least 64-bit floating point accuracy. The luminance and chrominance components are spatially related to one another for the purposes of coding and transmission through definition of a macroblock consisting of six blocks made up of four luminance blocks and one of each of the spatially corresponding chrominance blocks (Fig. 7.6). Most of the time, the DCT is actually performing a transformation of an 8 × 8 block which is the difference between the current block and the predicted block, and only occasionally will process the source picture itself depending on whether the loop is operating in INTER or INTRA mode.

The result of a transformation is an 8 × 8 block of 12-bit transform coefficients that represent the amplitudes of the spatial frequencies within the block. If the energy of the block has a low spatial frequency, in other words it contains only slow variations, then the transformation will generate only a small number of significant coefficients that are concentrated toward the top left-hand corner of the coefficient block, as illustrated in Fig. 7.9. The coefficients are then operated on by a quantizer and variable length coder to extract the coefficient information in such a way that allows the encoded picture quality to be varied with the bandwidth available for transmission.

7.2.3.3 Quantizer and the run-length coder

Little data compression takes place as a result of the DCT itself. In fact, the amount of data generated increases because the transformation of eight-bit pel data results in 12-bit frequency coefficients. The first compression step comes from quantization of the output of the DCT into a more restricted set of values that can be represented by fewer bits. Of course, this process is lossy in that once a coefficient has been quantized its original value cannot be recovered, but as this quantization takes place in the frequency domain its consequence is much less severe in the pel domain due to the fact that the inverse transformation in the decoder spreads the effect amongst all pels within the block.

The resulting position of the significant coefficients within the 8 × 8 coefficient block towards the top left of the block allows advantage to be taken of the zero coefficients that occupy the remainder of the block. A second compression step is performed by serializing the coefficients using a zigzag ordering as shown in Fig. 7.9, and subsequent encoding of the serial stream using run-length codes. With this zigzag scan pattern, there is a high probability that long sequences of zero-valued coefficients will be generated that can be encoded as much shorter values. The pairs of most frequently occurring combinations of coefficient value with the preceding number of successive zeros are encoded with a run-length code between two and 14 bits, while remaining combinations are encoded with a 20-bit value.

The severity of the quantization, and hence the degree of compression, can be varied dynamically according to the status of the transmit buffer at the output of the encoder. If a sequence of frames contains a lot of motion that results in coefficient blocks with significantly higher number of frequency coefficients, then the run-length coding process will not yield such high degrees of compression, with the danger that the amount of data generated at that instant may exceed the capacity of the communications channel. To compensate, the quantizer can adjust its step size for groups of blocks to one of 31 values in the range 2–62 in

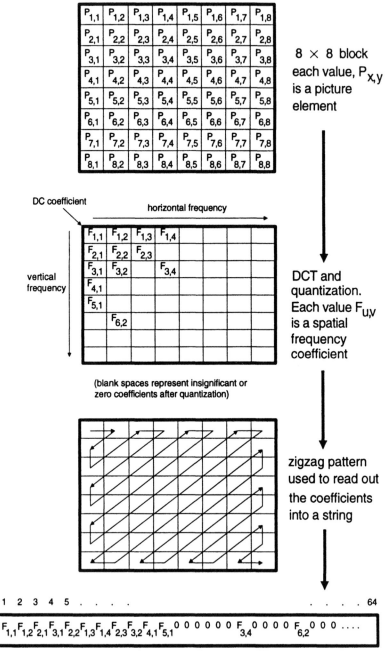

Fig. 7.9 The H.261 picture block transformation and coding process.

increments of two. The result is that as the picture motion increases the quantization becomes coarser in an attempt to balance the increase in the amount of data generated, but at the expense of a reduced picture quality. Further means of controlling the amount of data generated by the codec are realized by discarding complete frames (known as temporal sub-sampling) or discarding blocks within a frame. Given that the frame rate at the input to the codec is approximately 30 fps, these techniques will need to be invoked for low bandwidth channels such as BRA ISDN.

A filter is present in the feedback path of the videocodec and has the purpose of suppressing noise generated by the quantization process. The filter is typically switched into and out of the loop, depending on whether motion compensation is also active, as the presence of the filter during the absence of motion adversely affects the prediction values.

7.2.3.4 Motion compensation

The prediction values that result from the difference between the same block of the current and previous frame may be improved through **motion compensation** that takes place in the encoder, thus providing another mechanism for data compression. For an object which undergoes purely displacement motion within a picture, the motion can be represented as a displacement in both vertical and horizontal dimensions in the difference between one frame and the next. In other words, the size, shape and orientation of the moving object remain the same, but is simply shifted up or down and left or right between frames. This motion can therefore be simply represented as a motion vector.

Motion compensation in the H.261 videocodec takes place on all four luminance blocks within a macroblock, in other words in blocks of 16 × 16. An attempt is made to match the current macroblock with a set of same size blocks in the surrounding area of the previous frame. The motion vector is then computed as the horizontal and vertical displacements between the best matching block in the previous frame and current macroblock. These are used to adjust the delay in the block buffer such that the corresponding points on a moving object in the current and previous block are fed to the subtractor. For purely displacement motion, the output of the subtractor will be a zero matrix and only the motion vector need be transmitted. However, most motion is more complex than a mere displacement, and may involve a change in shape, size and orientation, or may obscure and uncover parts of a background object. In such cases the output of the subtractor will not be zero, but the prediction error will never the less contain less energy as result of the motion compensation and therefore result in a smaller amount of information being output from the videocodec. In this case, both the prediction error and the motion vector are transmitted to the decoder. Rather than transmit the absolute value of the motion vector, the difference between the motion vector in current and previous frames is transmitted.

The method used to determine the motion vector is not defined in H.261, only that its horizontal and vertical components can have values in the range −15 to +15. This means that the macroblock in the previous frame that is being compared with the current macroblock can be displaced by up to 15 pels in any direction from it. In the worst case, the current macroblock would need to be compared with up to 900 macroblocks in the previous frame in order to find the best fit, a task which would be computationally expensive and time consuming. Other methods which are less computationally intensive may be used, and because the method itself has no significance outside the encoder, can be implemented without sacrificing inter-operability.

Motion compensation is optional and can be invoked on a per macroblock basis when the codec is operating in INTER mode.

7.2.3.5 The video multiplex coder

As well as video information, a certain amount of control information must also be communicated in order for the decoder to maintain an equivalent mode of operation. The control and video information is therefore structured in a hierarchical serial multiplex. At the top of the hierarchy is the picture itself which is considered to be made up of a number of **groups of blocks** (GOB), 12 in the case of a CIF frame and three in the case of a QCIF frame. Similarly, each GOB is made up of 33 macroblocks, where each macroblock can be one of four basic types depending on the use of motion compensation, filtering and INTER- or INTRA-frame coding. Finally, each macroblock is made up of four luminance and two overlapping chrominance blocks. The hierarchical structure is illustrated in Fig. 7.10, as well as the information at these different levels.

7.2.3.6 The output buffer

The serial data stream at the output of the multiplexing function has a variable data rate based on the content and motion within a picture. However, the channel available across the ISDN to transmit this data has a fixed bandwidth, so an amount of rate adaptation is necessary. During periods of excessive motion it will be necessary to adjust the quantization step accordingly and perhaps increase the rate at which input frames are discarded in order to bring the data rate down, while during periods of no inter-picture changes the data rate has to be increased through the insertion of **fill patterns** in order to pad out the data rate accordingly. The buffer stage at the output of the codec provides the relevant indication to the coding control section in Fig. 7.8.

Finally, a forward error correcting code known as the BCH (Bose–Chaudhuri–Hocquenghem) code is then added to the serial video data stream before it is passed to the H.221 framer in order to provide a degree of protection against noise and other disturbances during transmission. This is necessary because of the high

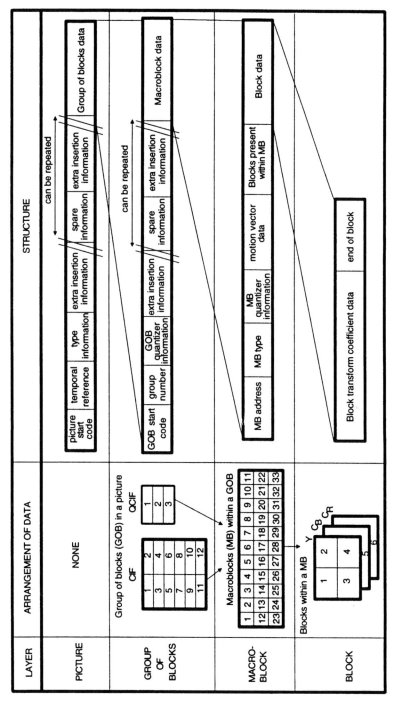

Fig. 7.10 Hierarchical multiplexing of video and control information in H.261.

level of compression and encoding that is used, which means that should an undetected error occur in the received datastream, it may have an impact on a large amount of decoded output data. In addition, because picture information is encoded differentially, an error in the coefficient data may have a lasting effect on the received picture quality.

The video decoder part of the codec essentially provides a reversal of the encoding process, and a comparison with the encoder will show that it has a simpler due to the fact that it operates only on a subset of the information contained in the original picture.

Although the architecture and functions of the videocodec are defined as part of the H.261 Recommendation, there are several implementation issues which are not defined, for example the particular strategy adopted for the choice of quantization step, and how the decision is made for dropping frames based on the output data rate, are all functions which are not defined. While inter-operability is maintained due to the fact that the format of the control information concerning these issues is defined and is transmitted from the encoder to decoder, it does lead to varying degrees of picture quality between different manufacturers' codecs.

7.2.3.7 Pre- and post-processing

Prior to encoding, the video signal may undergo pre-processing in order to reduce the effect of poor lighting that introduces noise into the image background. This noise can cause the encoder to think that there is background motion and can significantly increase the data output of the encoder. The purpose of the pre-processing is to filter the background in such a way that the encoder detects true motion in the image.

Likewise, post-processing is used to compensate image degradation introduced by compression and decompression processes which can lead to the introduction of visible artifacts on the displayed image, particularly if a low-resolution QCIF format is displayed on a large screen. For example, a high quantization level will in general introduce distortions, or quantization noise, into the image. In particular, a common form of artifact is seen, known as a blocking of the picture. This is caused when the encoder is operating with a high quantisation level such that a gradual change in picture colour or brightness can lead to visibility of the 8×8 block boundaries, leading to a blocky image. Filtering of the image around the block boundaries can help to alleviate this problem, but at the expense of reducing the definition of any structure within the image that appears at these edges.

Another common artifact in images that contain small points of high contrast on a plain background is seen as flecks, and is again the result of a high quantization level. The problem here is that the DCT of a small high-contrast point on a blank background generates a matrix whose coefficients are all non-zero, instead of a more usual case where the coefficients are grouped in the top left corner of the matrix with the remaining coefficients being zero. After quantization is applied, some of the coefficients in a full matrix do not get transmitted, such that when an attempt is made to reconstruct the image with the inverse DCT at the receiver, it causes flecks to appear as echoes of the original image.

7.2.4 The H.221 framer

The videocodec makes few assumptions about the type of network it will be connected to, only that it provides a fixed bandwidth channel at data rates that are multiples of 64 kbps. It is likely that in many applications the video will be combined with speech and possibly data for transmission over the same channel. Recommendation H.221 defines a framing structure for such purposes in synchronous channels of 64 kbps, and makes provision for the combination and synchronization of more than one channel up to a maximum of 30, equivalent to a channel bandwidth of 1.920 Mbps.

Already introduced earlier in section 7.1.2.1 and illustrated in Fig. 7.3, the basic H.221 frame consists of 80 consecutive bytes of data, where the most significant bit in each byte is used to create a service channel. The first eight bits of the service channel are defined as the frame alignment signal (FAS), and the second eight bits as a bit-rate allocation signal (BAS). The third eight bits can be optionally occupied by an encryption control signal (ECS) to provide encryption of remaining information in the H.221 frame. The rest of the H.221 frame may be occupied by any combination of voice, video and data. Sixteen frames form a multiframe which is split into eight sub-multiframes of two frames each. The purpose of the multiframing is to allow the significance of the service channel bits in different frames within the multiframe (and sub-multiframe) to take on different meanings. For example, the FAS bits are used not only to convey frame alignment information, but also to perform end-to-end error checking, and allow the insertion of control and alarm information.

When a call establishes a B-channel connection between two terminals, the initial idle data in the B-channel will be all '1's. To establish framing, the terminals will force the frame alignment signal into their transmit data stream such that it attempts to create the H.221 multiframe within the bi-directional channel. Indication that the receiver has detected the framing signal and gained alignment is achieved by it setting an acknowledgment bit in the FAS channel back to the transmitter. Once framing is established, an exchange of terminal capabilities is performed using codes in the BAS channel and an appropriate mode is selected at both terminals which assigns the audio, video and data streams to portions of the available bandwidth for the communications. The frame alignment acknowledgment bit is continually monitored by the receiver to detect a loss of alignment on which a recovery process is invoked.

7.2.4.1 Communications on more than a single B-channel

The higher the bandwidth of the communications channel, the higher the quality of the video picture, at least to a limit governed by the techniques used in the coding and decoding of the video picture. As a BRA ISDN connection has the capability to support two 64 kbps channels, most videocommunications equipment connected to it will attempt to make use of both channels. Although the option exists within Q.931 to establish a single call with a 2B-channel bearer capability, it is not a supported feature in most ISDNs due to the fact that the switching exchanges and routing protocols in the public network are not capable of switching and routing two consecutive 64 kbps channels. As a consequence, the end-to-end path taken through the network by the data in each B-channel may be different, and will result in data that is transmitted in sequence into the two B-channels being received out of sequence by an amount that is equivalent to the differential path delay of the two B-channels.

A method is defined in H.221 that allows the differential path delay to be determined by a receiving terminal thus permitting it to insert the appropriate amount of delay into the channel that experiences the least amount of transmission delay in order to bring the two B-channels back into sequence with one another. The method assumes that the transmission delay remains constant for the duration of a call such that once the differential delay has been established and the amount of compensation determined, no further adjustment will be necessary.

As an example of how the H.221 method operates we consider the setup of a 2B-channel videophone call through the establishment of two separate single B-channel calls to the same destination. The calling videophone requests the establishment of the first B-channel connection using standard Q.931 signalling with a 64 kbps transfer rate and 7 kHz or 'unrestricted digital information' transfer capability. This allows at least speech communications to be established first when the call is answered. When the connection state is reached, H.221 framing is established in the B-channel and an exchange of terminal capabilities takes place using the BAS codes. This procedure will identify to the called terminal that the calling terminal is attempting to establish a call with an aggregate 2B-channel bandwidth capability, and hence a further call is needed to establish the second B-channel. Speech communications will then be established between the two videophones during which the calling videophone then places a second call which is automatically answered[29] by the called terminal. H.221 framing is then also established in the second B-channel.

Alignment at the receivers of both the called and calling terminal between the H.221 streams of the first and second call is performed through use of a four-bit counter embedded within the FAS bits of each H.221 multiframe. For successive

---

[29]On receipt of a SETUP request from the network, the called terminal will respond directly with a CONNECT message rather than an ALERTING message.

H.221 multiframes, the counter is decremented by one. At the transmitters, the channels will be aligned such that each transmitter sends the same numbered multiframe simultaneously. Hence, the differential delay introduced by the network in each channel is measured at the receiver by timing the delay in the receipt of a particular number multiframe in one channel, and reception of the same number multiframe in the other channel. By delaying the first channel by this amount brings the two H.221 streams back into alignment again.

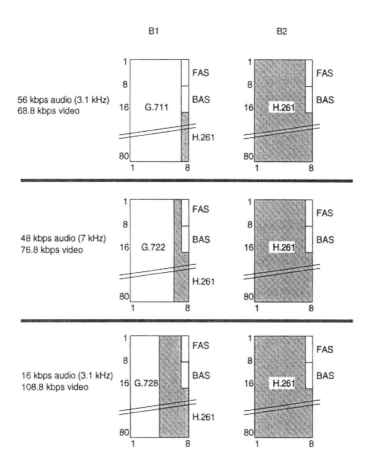

Fig. 7.11 B-channel H.221 formats for three different bandwidth configurations.

Using this method, up to six B-channels can be aligned[30] where each B-channel is numbered using three bits embedded within the FAS bits of a multiframe. In the above case of a 2B-channel call, Fig. 7.11 illustrates the partitioning of the H.221 frame in each B-channel according to the most likely combinations of the H.261 videocodec with different types of audiocodec. As can be seen, use of G.728 in the last combination provides the highest bandwidth for video and would therefore be expected to provide marginally better picture quality over the other two combinations.

H.221 also provides for the synchronization of higher bandwidth H0 channels. An H0 channel has a bandwidth of 384 kbps and is usually provided by leased line or a dedicated switched network service. Using the same technique described above, up to five H0 channels can be synchronized to provide an aggregate channel bandwidth of 1.920 Mbps. Such aggregation of channels allows higher bandwidth, and hence higher quality video, to be communicated as well as the provision of higher bandwidth data channels. H.221 makes the distinction between low-speed data (LSD) at typical rates of 6.4 kbps, 14.4 kbps and 32 kbps, and high-speed data (HSD) at data rates in multiples of 64 kbps.

7.2.5 System considerations

At the beginning of this chapter we briefly mentioned the real-time nature of voice and video signals and the need of the communications process to suitably maintain certain timing parameters such as delay, jitter and synchronization, within bounds that are acceptable to users. The ISDN provides end-to-end connections which are inherently synchronous, and thus preserves the basic timing of the source information, while delay and jitter are controlled by design of the transmission and switching elements within the ISDN and are bounded by limits defined by the network operator.

Within the video terminal, synchronization is controlled through buffering and bit-stuffing in order to match the output rate of the video encoder to that of the ISDN channel. A most visible degradation however is typically seen in the end-to-end delay, and is contributed to significantly by the conversion delays of the encoder and decoder as well as pre- and post-processing functions. One-way delay for a 2B-channel video call can typically be between 0.5 and 1 second and depends largely on the speed of the DSPs used to implement the videocodec.

Although delay is introduced by the video encode and decode, the need to maintain lip synchronization means that the accompanying speech must also be delayed, for which delays between 0.5 and 1 second can be annoying and cause interrupted conversations, particularly in videoconferencing. Second- and

---

[30]The six B-channels can be delivered either from three BRA connections, of from six B-channels within a PRA connection.

third-generation videocodecs will attempt to reduce these delays to under 300 ms which is considered the threshold for maintaining smooth conversation.

7.2.6 Example–the AT&T videocodec chip-set

IC technology is critical to the realization of cost-effective videocodecs and videocommunications equipment. However, IC videocodecs are extremely complex devices, and require high-performance DSP devices whose architecture and instruction set have been designed to meet the types of calculation required in processing real-time video signals. In this section the videocodec chip-set from AT&T[31] is briefly described to provide an example of how an H.320 system can be implemented using VLSI (very large scale integration) devices. Further details can be found in references 13 and 14.

The basic AT&T videocodec chip-set comprises three devices; the AVP-4310E video encoder, the AVP-4220D video decoder, and the AVP-4120C systems controller chip, which amongst other things, implements the H.221 framing. To these devices are added one or more DSP devices to perform the audiocodec function and video pre- and post-processing functions. The use of programmable DSP devices is used here to provide flexibility in the implementation of audiocodec standard and of the video pre- and post-processing functions.

The interconnection of these devices in a video communications terminal is shown in Fig. 7.12. Use is made of several types of bus to provide a communications path between the different devices for video and audio data and control information. The two main buses in the system are referred to as the host bus and the serial bus. The host bus is essentially a microprocessor bus attached to the host CPU system and used to control and monitor the devices by writing commands and reading status information from their registers, while the serial bus carries the compressed video and audio data between the video encoder, decoder, systems controller and audio DSP device in the system. Other buses are used in the system where the nature of the information and its timing is such that it warrants the use of a dedicated bus. For example, the video encoder receives uncompressed CIF or QCIF frame data from the frame pre-processor through a dedicated video bus interface. Similarly, the output of the decoder chip to its frame buffer is through a pixel bus interface. Lastly, the systems controller device interfaces the H.221 formatted channels to the ISDN S transceiver through a TDM serial bus known as CHI (concentration highway interface).

---

[31] AT&T Microelectronics is now kown as Lucent Technologies.

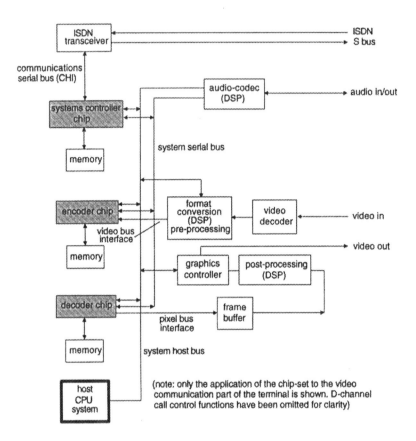

Fig. 7.12 An example of how the AT&T chip-set may be used in an audio/video communications ISDN terminal.

7.2.6.1 The video encoder

The AVP-4310E video encoder is designed to compress video according to the H.261 standard. It is shown in Fig. 7.12 connected to a memory which holds both the current input frame and the previous frame used for INTER frame coding and motion estimation. The encoder contains a number of processors that implement processes specific to the H.261 algorithm. The key processors are as follows.

Signal processor. This function is primarily responsible for performing the DCT calculations on the difference macroblocks. As each macroblock contains six

8 × 8 blocks (four luminance blocks and two chrominance blocks), the signal processor contains six processing elements that can perform the same calculation on each of the six blocks in parallel, an architecture known as SIMD (single instruction, multiple data). In addition, it also performs quantization according to the value supplied by the quantization processor, and inverse quantization and inverse DCT to construct a copy of the decoded macroblock which is then stored in the previous-frame buffer held in memory. The signal processor also performs the run-length encoding of the transformed macroblock.

Quantization processor. The quantization processor is a separate processor embedded within the encoder device. It is based on RISC (reduced instruction set computer) technology and executes a program stored in internal memory. Its primary function is to determine what level of quantization should be applied to each compressed macroblock in order to maintain a constant output bit rate. Control of the quantization is achieved by a signal from the variable-length encoder that indicates how many bits it is generating. When many bits are generated, a coarser quantization scale is chosen to reduce the number of values and thus attempt to maintain a constant output bit rate. Similarly, when the variable-length encoder generates few bits, a finer quantization scale will be chosen. Optimization of the control algorithm for desired output bit rate, encoding delay, frame rate and picture quality is implementation dependent and can be configured through programmable registers accessed via the host bus.

Motion estimator. A separate hardware processor performs motion estimation by comparing the current macroblock with offsets to adjacent macroblocks from the previous frame. An exhaustive search is performed to generate a motion vector from the best match 16 × 16 block in the previous frame to the current macroblock. The address of the best match block is passed to the memory controller, and the motion vector is passed to the quantization processor. Depending on the quality of the match, the motion estimator may decide that a good match cannot be found. In this case it will encode the macroblock as an INTRA-frame macroblock instead of INTER-frame, and inform the quantization processor accordingly as only a single quantization scale is defined for INTRA-frame encoded macroblocks.

Variable-length encoder. The purpose of the variable-length encoder is to combine the run-length encoded macroblocks from the signal processor and the motion vectors from the quantization processor and combine them into the hierarchical bit stream illustrated in Fig. 7.10. Feedback is provided to the quantization processor to maintain a constant bit-rate output.

Memory controller. As the operation of the different functions within the encoder involves intimate interaction with the video frame data stored in memory, the memory controller implements critical functions that assist in locating the relevant data in external memory, reading it into the encoder and writing it back to the memory after it is processed. For example, the memory controller assists the motion estimator by reading from memory the next macroblock of the current frame and the appropriate search area from the previous frame. After computation of the motion vector, it also transfers the best match 16 × 16 block and reference macroblock to the quantization processor. The memory controller provides the necessary hardware support to access and refresh external dynamic RAM (DRAM) memory.

Global controller. The global controller performs a coordination function for all the processors in the encoder to ensure that the video information gets processed according to the desired frame rate.

The AVP-4310E is the most complex of the three devices. Flexibility of most of the programmable processors within the encoder is maintained by allowing them to download their program code across the host bus. This allows adjustments and fine-tuning to be made as standards evolve, and to allows the encoder to implement other similar standards to H.261, such as MPEG. A similar approach is also adopted in the video decoder chip described in the following section.

7.2.6.2 The video decoder

The AVP-4220D video decoder chip is a less complex device than the encoder as it does not contain functions such as motion vector calculation and the control circuitry necessary to maintain a constant bit rate output. In addition, the demultiplexing functions found in the decoder tend to be less complex to implement. The decoder accepts input from the system controller chip via the serial bus and outputs decoded video data either through its interface to the host bus or through the pixel bus to the video output subsystem consisting of the frame-buffer, post-processing and graphics controller. Like the encoder chip, the decoder requires local memory which is used to store the present and previous frame for INTER-frame coded picture information.

The main processing elements of the decoder are a variable-length decoder, a symbol processor, a signal processor, a memory controller and colour converter, which are each briefly described below.

Variable-length decoder. The data stream input to the decoder is first operated on by the variable-length decoder which splits the information into control information such as the quantization level and motion vector data, and the compressed video data represented by the DCT coefficient data. These two sets of information are then passed onto the symbol processor.

Symbol processor. The symbol processor performs most of the control functions of the video decoder chip by operating on the control information passed to it by the variable-length decoder. In addition it generates complete DCT difference blocks from the DCT coefficient data and symbol information which are then passed on to the signal processor. The symbol processor is a RISC processor with additional hardware for processing time critical tasks such as run-length decoding of the DCT coefficients.

Signal processor. Like the signal processor in the encoder, the decoder signal processor is an SIMD processor comprising six processing elements, and performs inverse DCT and inverse quantization functions.

Memory controller. The external memory connected to the video decoder holds the decoded video blocks for the previous frame and the current frame as it is being assembled. The memory controller manages and controls the transfers between the signal processor and external memory as well as reading the picture frame data in a raster-scanned format from memory and passing it to the colour converter. As in the encoder, the memory controller in the decoder provides hardware that supports DRAM devices.

Colour converter. The decoder can output video data either in the original YC_RC_B format or in RGB (red, green, blue) format by means of conversion using a multiplier-accumulator and look-up table.

As the decoder can only generate video information for those frames that were encoded, the buffer in Fig. 7.12 is necessary between the decoder and the display in order to match it to the rate at which the display controller updates the screen. Additional post-processing may be performed by the DSP to remove artifacts introduced into the picture by the encoding process. If the video communication system is designed to be used as an addition to a PC, then the graphics controller shown in Fig. 7.12 will introduce the video as an overlay to the PC video output.

7.2.6.3 The systems controller chip

The main function of the AVP-4120C systems controller device is to provide the multiplexing and demultiplexing of the audio, video and data streams into constant bit rate synchronous channels according to the H.221 recommendation. The device also implements the error-correcting code defined in H.261 and performs buffering and delay for the audio channel to maintain lip synchronization.

At the heart of the systems controller chip is a RISC processor whose instruction set has been designed to optimize its performance in bit processing intensive communications applications. This is also supported by the architecture of the processor which has separate memories and interfaces for its program and data, an architecture more commonly known as a Harvard architecture. The relevance of this is that the bandwidth of the processor–memory interface does not have to be shared by both program instructions and the data it operates on, which is the case in more conventional processors. Instead, the full bandwidth of an interface can be dedicated to the transfer of data between the processor and its memory, giving it a high throughput for the data intensive bit-manipulations necessary in processing the H.221 frame.

The processor is assisted by two hardware functions called the bit server and FIFO server.

The bit server. This function performs efficient processing of the bit-level operations in H.221 and the generation and detection of the error-correcting code defined in H.261.

The FIFO server. This function manages large FIFOs created in the external memory to handle the synchronization between the audio and video streams, prevent underrun and overrun conditions, and provide error recovery.

The systems controller chip interfaces to the communications network through the CHI serial TDM bus to a transceiver that provides the physical connection to the network. CHI is a flexible interface and may be configured to interface with a range of different manufacturers' transceivers, allowing not only an interface to the ISDN, but leased lines and other types of synchronous data networks. In the case of BRA ISDN, the systems controller can provide for the aggregation of its two B-channels.

The host bus interface allows access by the host processor to the control and monitoring of those portions of the in-band signalling implemented by the systems controller. Data for any data channel that may be established in the H.221 frame may also be transferred from the host to the serial controller across the host bus.

The AT&T AVP chip-set is one example of a number of silicon solutions that are available in the market place today. H.261 solutions are also available from LSI logic (L647xx series devices), which are used in many room-based videoconferencing systems, and GEC Plessey Semiconductors (VP2611 and VP2615), while Texas Instruments (TMS320C80) and IIT (VCP3100) have programmable video compression processors capable of implementing a number of different compression algorithms, including H.261.

7.3 CONFERENCING

One of the areas of application for combined audio, video and data communications which is expected to become widely used in the future is **conferencing**. While videoconferencing has been in existence for many years, it is only in the last few years that its use has increased due to the availability of affordable terminal solutions and cheap switched digital subscriber lines in the form of ISDN. Many of the video terminals in operation today are of the room-based type where the equipment in a conference room would host a number of participants communicating with typically a terminal in another conference room. Such systems offer limited capability for data communication, and where it does exist, it is used mainly for facsimile communication between conference participants. The trend in the future will be toward the deployment of personal video communication terminals, typically hosted by a PC that runs business applications, which allow groups of users to work collaboratively through conferencing and the ability to communicate mixed media information. With this trend comes the increasing need to connect more than two locations whenever more than two people need to confer.

The public switched network was designed to interconnect two terminals in a point-to-point configuration. To connect more than two terminals for the purpose of conferencing would require each terminal to establish a separate point-to-point connection to each of the conferenced terminals. For n participants, $n-1$ channels must be provided across the subscriber loop and each terminal must have the ability to process multiple audio and video streams.

An alternative solution is to handle the conference call setup and processing of audio and video streams centrally in a **multipoint control unit** (MCU). The MCU can either be located within the network itself or be part of the customer premises equipment. In either case, the MCU appears to a user's terminal also as terminal equipment, and can be dialled using an allocated telephone number. Instead of every user terminal establishing a point-to-point link with every other terminal, all terminals dial into the MCU which has the responsibility to provide the conferencing service. The advantage of this is that no modification is required to the capabilities of the end-user terminal[32] to allow it to be used for conferencing in conjunction with an MCU, while the connection bandwidth between the terminal and the MCU need only be that of a single point-to-point connection. A disadvantage is that while this is convenient, the single point-to-point bandwidth limits the amount of information that can be conveyed between the MCU and the terminal.

Some features of a typical conference service implemented by an MCU with audio, video and data capabilities, are as follows.

---

[32]It is worth noting that considerations for conferencing using a central MCU are taken into account in the signalling standards associated with H.320.

Call setup procedures: An MCU will typically allow two basic types of call setup. If the MCU has dial-out capability, a user who wishes to establish a conference call can dial in to the MCU using the normal Q.931 call control procedures, and send it, using in-band signalling, the numbers of the additional conference participants for it to dial. These numbers can either be dialled and added to the conference individually by the MCU itself, or at the same time using a **blast dialling** feature. Alternatively, all participants can dial in to the same number of the MCU which then automatically conferences the individual calls and allows participants to join or leave the conference at will. This type of call is otherwise known as a **meet-me** conference call.

H.221 calls: Once point-to-point calls are established between the terminals and MCU, H.221 framing is established. In-band signalling between a terminal and MCU, such as the transfer of additional dialling information above, is achieved with BAS codes within the H.221 frame. More modern MCUs will support the ITU-T T.120 series of protocols for the control of audio, video and data conferencing in the data channel of the H.221 frame.

Reservations: Scheduling the availability of MCU resources is performed either by an automatic or operator-assisted reservation system.

Conferencing: Audio signals can be mixed such that the signal received by a conference participant is the combination of the transmitted speech signals from the remaining participants. A key feature here is that the bandwidth of the mixed signal is the same as its individual components, and can therefore be inserted directly into the receive audio channel within the H.221 frame without modification. In this way, participants at one terminal can simultaneously hear all participants at the other terminals. Combining the speech channels is performed in an audio conference bridge within the MCU.

Video and data cannot be treated in the same way as their simple combination linearly increases the bandwidth of the aggregate signal. As the point-to-point link will typically establish equal bi-directional bandwidth within the H.221 frame, some switching mechanism is typically used in the case of video to select which of the video streams transmitted to the MCU is broadcast back to the terminals. Several modes are possible as described below.

- The MCU provides the video signal from the current speaker to participants through automatic audio switching.
- The decision of whose image to broadcast to other terminals is decided and controlled manually by one of the participants who is assigned the role of **conference conductor**.

- Each participant may have the ability to browse the images from each participant and select which is viewed.
- A **continuous presence** mode may be supported where the CIF image returned to each participant is divided into four QCIF screens in a 2 × 2 format to display the images of the four most recent talkers simultaneously. This can be achieved without the need to decompress individual images and recompress a composite image by manipulation of the picture header and group of blocks (GOB) header in the H.261 stream.

Similarly, data, such as facsimile, that is to be communicated between conference participants must also be switched as the bandwidth constraints allow only a single source of data to occupy the data channel at any one time. Therefore, MCUs typically support a half-duplex broadcast communication mode. The decision of who has access to the channel is usually delegated to the conference conductor.

Figure 7.13 illustrates the network configuration for an audio, video and data conference MCU connected to a public ISDN, and conferencing together a number of H.320-based terminals. The MCU is connected to the ISDN with a PRA or multiple BRA connections, and has a number of ports that become allocated on demand to incoming or outgoing calls. Figure 7.13 shows only three such ports while in reality many more than this will exist according to the capability of the MCU. Each port provides multiplexing and demultiplexing of the H.221 frame such that within the MCU individual audio, video and data streams are available to each of the video and data switches and the audio bridge. Most MCUs have the ability to be cascaded to allow expansion of conferencing capabilities.

For conference calls of five or more participants it is often found that some structure has to be imposed on the flow of communications between participants. In much the same way that a large meeting is conducted through a meeting chairman, so is a conference call conducted through a conference conductor who has control over the MCU functions by means of the BAS control codes. This person may well be the same person or different to the actual meeting chairman, but in either case, receives requests from participants to contribute and then grants the requests sequentially. A similar request/grant mechanism is also used for sending data between participants.

The standard features of an audiovisual MCU are described in ITU-T Recommendation H.231 [15].

7.3.1 The multilayer protocol (MLP)

The requirement for communication of multimedia (voice, video and data) information is expected to dramatically increase during the course of the coming

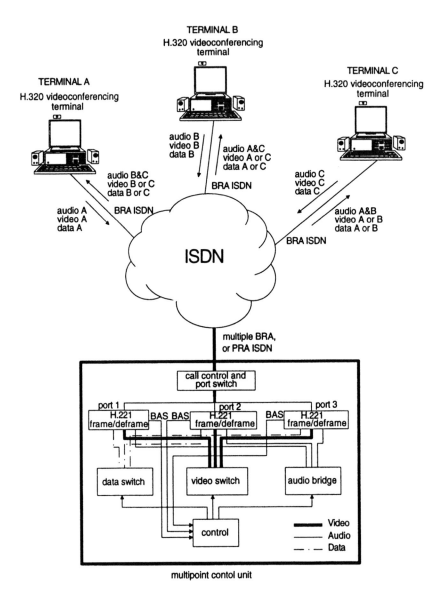

Fig. 7.13 The multipoint control unit (MCU) and multipoint connection for conference call applications.

years, and will be fuelled in part by the use of collaborative workgroup applications that allow not only voice- and videoconferencing, but also data to be shared and worked on in real-time by a number of people simultaneously. However, with such a diverse range of communication networks in existence today it is only reasonable to assume that these tools must be deployed on a variety of platforms and across a number of different networks in both the wide area (for example public networks such as the ISDN) and local area (such as LANs) for them to become pervasive.

With these objectives in mind, the ITU-T is in the process of defining a group of protocols known collectively as T.120, known generically as the multilayer protocol (MLP). Within T.120, protocols are defined that allow data conferencing across local area and wide area networks, and together with H.320 conferencing, will provide the basis on which a range of different information types can be communicated and shared within a conference environment. The concepts and architecture behind T.120 are discussed in the following chapter which deals with data communications in the ISDN.

REFERENCES

1. ITU-T (1993) 32 kbps Adaptive Differential Pulse Code Modulation (ADPCM). Recommendation G.721.
2. Taka M., Murata R. and Ungami S. (1988) DSP Implementations of sophisticated speech codecs. *IEEE Journal on Selected Areas in Communications*, 6(2), 274–282.
3. ITU-T (1993) 7kHz Audio-Coding within 64 kbps. Recommendation G.722.
4. Maitre X. (1988) 7kHz audio coding within 64 kbps. *IEEE Journal on Selected Areas of Communications*, 6(2), 283–298.
5. Mermelstein P. (1988) G.722, A new ITU-T coding standard for digital transmission of wideband audio signals. *IEEE Communications Magazine*, 26(1), 8–15.
6. ITU-T (1993) System Aspects for the Use of the 7 kHz Audio Codec within 64 kbps. Recommendation G.725.
7. ITU-T. (1990) Frame Structure for a 64 to 1920 kbps Channel in Audiovisual Teleservices. Recommendation H.221.
8. ITU-T (1992) Coding of Speech at 16 kbps using Low-Delay Code Excited Linear Prediction. Recommendation G.728.
9. ITU-T (1993) Narrow-Band Visual Telephone Systems and Terminal Equipment. Recommendation H.320.
10. ITU-T (1990) Video Codec for Audiovisual Services at P × 64 kbps. Recommendation H.261.
11. ITU-T (1995) Multimedia Terminal for Low Bitrate Visual Telephone Services over the GSTN. Recommendation H.324.
12. Ahmed, N., Natarajan, T. and Rao, K.R. (1974) Discrete cosine transform. *IEEE Transactions on Computing*, C-23, 90–93.

13. Ackland, B., Aghevli, R., Eldumiati, I., Englander, A. and Scuteri Jr., E. (1993) A videocodec chip-set for multimedia applications. *AT&T Technical Journal*, **27**(1), 50–66.
14. Ackland, B. (1994) The role of VLSI in multimedia. *IEEE Journal of Solid-State Circuits*, **29**(4), 381–389.
15. ITU-T (1993) Multipoint Control Units for Audiovisual Systems using Digital Channels up to 2 Mbps. Recommendation H.231.

8

Data communication in the ISDN

In this chapter we examine the terminal protocols and functions used to process user information for a range of data-oriented applications. Most data terminals do not have ISDN connectivity built-in as a standard function. Instead, adapters are required to achieve the necessary interfacing, whose functions and operation are dependent on the type of terminal which it connects to the ISDN. Two distinct types of adapter can be identified. These are:

- stand-alone terminal adapters (TAs);
- PC or workstation adapters.

The stand-alone TA is an external device that connects to a data terminal through a standard interface, and allows the ISDN to be used as a transparent connection between two pieces of non-ISDN data equipment. For example, a TA could be used to connect a remote data terminal to a centralized computing facility. In the case of a PC or workstation, the ISDN adapter is usually an expansion card that is embedded within the terminal, and essentially transforms the terminal from a TE2 type equipment to a TE1.

Many PCs in use today are not directly connected to the ISDN, but instead are connected to local area networks (LANs) which are themselves interconnected across the ISDN using devices such as bridges and routers. The higher bandwidth and different packet format of the LAN place some special requirements on the protocols implemented, and are also discussed in this chapter.

In addition, the frame relay service is also introduced in this chapter. Frame relay has been defined by the ITU-T as part of the suite of ISDN protocols to provide efficient data communications, and is in use today particularly for LAN-to-LAN interconnection.

Finally in this chapter, the T.120 suite of protocols for data conferencing applications is discussed.

8.1 TERMINAL ADAPTERS

A Terminal adapter provides non-ISDN data processing equipment with an interface to the ISDN. Its function is analogous to a modem that provides data equipment with connectivity to the PSTN over analogue subscriber loops.

An ISDN TA provides the following four basic functions:

- translation and rate adaptation of the terminal data into a format suitable for transfer across the ISDN;
- end-to-end synchronization during data transfer;
- call establishment and disconnection;
- conversion of the electrical and mechanical interfaces between the terminal and the ISDN user–network interface.

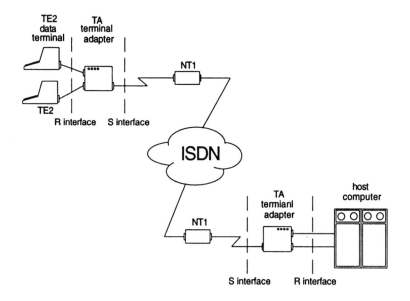

Fig. 8.1 Terminal-to-host connections using ISDN terminal adapters.

A terminal adapter will have two main interfaces. The user–network interface is a BRA or PRA S interface, while the terminal interface, designated the R interface in standards, conforms to one of the standard V-series type interfaces for interconnecting data processing equipment. Figure 8.1 illustrates a typical terminal-to-host application.

The adaptation process used by the TA depends on the type of R interface provided and the type of bearer service employed by the TA to transport the R interface information across the ISDN. The ITU-T have defined three protocols to accommodate the connection of a data terminal to the ISDN. Two of these protocols, defined in Recommendations V.110 [1] and V.120 [2], are end-to-end protocols that rely on a circuit-switched connection through the ISDN to link equipment having V-series interfaces, while Recommendations X.31 [3] and X.32 [4] deal with the support of packet mode terminals by the ISDN.

Prior to the establishment of ISDN, proprietary terminal adaptation protocols have been developed for the interconnection of computers to PBXs using digital circuits. Most notable among these are DMI developed by AT&T, and TLINK developed by Northern Telecom. These protocols helped to shape the definition of ISDN TA protocols, and similar features from both DMI and TLINK can be found in V.110 and V.120.

8.2 V.110

Derived from the European Computer Manufacturers Association standard ECMA102, the V.110 standard is used mainly by computer manufacturers within Europe[33]. V.110 defines an 80-bit frame in which is encapsulated the terminal data from the R interface, and which is transmitted and received in part or all of an ISDN B-channel used in an end-to-end circuit-switched connection. In addition to the terminal data, which we refer to here as the D-bits (not to be confused with the ISDN D-channel bits), the 80-bit frame, depicted in Fig. 8.2, also contains synchronization and control information (indicated by the S-, E- and X-bits).

ECMA102 defines support for both V-series and X-series terminal interfaces, while V.110 defines support only for V-series terminals. ITU-T Recommendation X.30 [5] defines the support for terminals with X-series interfaces, specifically X.21, X.21bis and X.20bis, to the ISDN using the same framing structure as that defined in V.110. ECMA102 can therefore be thought of as a combination of both V.110 and X.30.

---

[33]This stems solely from a geographical preference rather than any technical limitation or suitability.

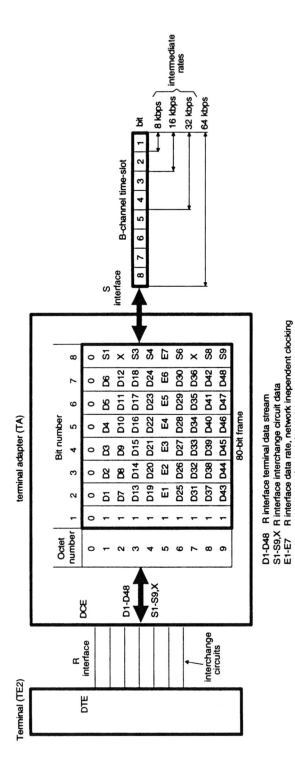

Fig. 8.2 The V.110 frame structure.

8.2.1 Frame synchronization

Once a call is established between two TAs, each must establish frame synchronization of the 80-bit frame before they are able to interpret the contents of the frame and communicate information between their R interfaces. Each receiver in the TAs achieves frame synchronization by detection of the eight '0's at the beginning of the frame followed by the '1' at the beginning of the following nine bytes. Frame synchronization is monitored continuously throughout the duration of a call, and should a framing error be detected in more than three consecutive frames, then the TA enters a recovery phase. During this recovery phase, the D-bits in the 80-bit frame are forced to '1' and the TA searches for the framing pattern, which if not detected after a period of several seconds, disconnects the call.

8.2.2 Rate adaptation

Only part of an ISDN B-channel is required to transport the V.110 frame because the terminal data is rate adapted in the 80-bit frame to one of three intermediate rates, 8, 16 or 32 kbps, that occupy 1, 2 or 4 bits respectively of a B-channel time-slot as illustrated in Fig. 8.2. For example, if the terminal data rate at the R interface is 9.6 kbps, the data is placed into the 80-bit frame and rate adapted to next highest intermediate rate, in this case 16 kbps. Rate adaptation is achieved either through duplication of data bits, such that the value of a data bit at the R interface is assigned to two or more consecutive D-bits in the 80-bit frame, or through the assignation of D-bits as fill bits that have no significance to the R interface data itself and are discarded by the receiving TA. As the 80-bit frame in this case represents a data stream equivalent to 16 kbps, it is then inserted into the first two bits of each B-channel time-slot, allowing the remaining six bits (or 48 kbps bandwidth) to be used to provide rate adaptation for another terminal, or perhaps to support another application. This two-stage rate adaptation process of first adapting to an intermediate rate and then to a 64 kbps B-channel is used for synchronous terminal data rates up to 19.2 kbps. Rate adaptation of higher synchronous data rates at 48 kbps and 56 kbps is performed by a single-step process, and uses modified frame formats to rate adapt directly to a full 64 kbps B-channel.

As the ISDN B-channel is a synchronous channel, adapting an asynchronous terminal data stream to it requires some further consideration as the bit rate at the terminal will not be synchronized to that of the ISDN. The adaptation process therefore requires a third step that provides an asynchronous-to-synchronous conversion at the R interface before the terminal data is framed in the 80-bit frame as described above. This conversion is achieved by the TA continuously sampling the received asynchronous stream at a rate higher than that of the nominal asynchronous rate. In effect, this results in the insertion of additional asynchronous stop bits between characters in the asynchronous stream. The TA sampling rate is synchronized to the ISDN clock and has a value $2^n \times 600$ bps, where n is an integer between 0 and 5 such that the sample rate is the closest rate that is equal or above that of the nominal asynchronous rate. For example, an asynchronous terminal data stream at a rate of 3600 bps is adapted to a synchronous stream by the TA sampling it at a rate of 4800 bps ($n = 3$) before it is then further rate adapted to 8 kbps in the 80-bit frame. As the terminal communication is asynchronous, it is likely that the instantaneous data rate will be marginally higher or lower than the nominal rate, and compensation for so-caused overrun and underrun conditions is achieved through insertion or deletion of fractional stop bits.

Bits E1, E2 and E3 in the 80-bit frame together with the intermediate rate indicate the synchronous terminal data rate. However, the intermediate rate, which determines the number of bits in the B-channel used to carry the 80-bit frame, is communicated in the D-channel between the TAs during call establishment in the call control messages. The E-bits are valid only for synchronous terminal rates, so asynchronous terminal rates must also be communicated between TAs during call setup using call control messages.

8.2.3 Network-independent clocking

Under normal operation, the transfer of synchronous information from a terminal into the ISDN is synchronized to the ISDN clock which is recovered from the S interface signal received at the terminal from the network. However, there may be instances where the synchronous data is generated in a terminal or modem interworking equipment by a clock source that is outside, and therefore independent, of the ISDN. It then becomes necessary to maintain this synchronization across the ISDN to the far-end terminal or interworking equipment to ensure that data is not lost at the interfaces to the ISDN. In effect, the near- and far-end equipment become synchronized independently of the ISDN clock.

A mechanism is provided within V.110 to accommodate this network independent clocking for synchronous data at the R interface at rates up to and including 19.2 kbps. The scheme is illustrated in Fig. 8.3. Bits E4, E5 and E6 in the 80-bit frame are used to encode the phase difference, ϕ_{Diff}, between the same nominal frequency derived from the remote clock source having phase ϕ_r, and ISDN clock source having phase ϕ_{ISDN}. This phase difference is encoded in jumps of 20% relative to the phase of ϕ_{ISDN} and is used at the far end to adjust the phase of the outgoing synchronous clock, $\phi_{r'}$, relative to the ISDN clock, thus allowing it to track the phase of the remote clock source ϕ_r in steps of 20%. The potential problem of phase jitter on $\phi_{r'}$ caused by continuous jumping backwards and forwards between adjacent phase displacements is avoided through the application of hysteresis to the encoding of the phase difference. For persistent slight differences between the two clock frequencies, the phase difference will gradually increase until either an additional data bit must be inserted in the 80-bit frame, or a data bit must be deleted, depending on whether the remote clock frequency is greater than the ISDN clock or the other way around.

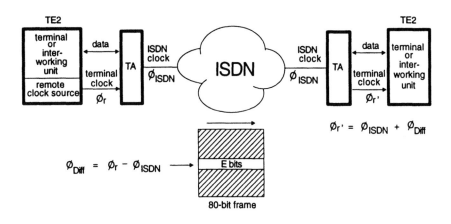

Fig. 8.3 Network independent clocking scheme in V.110.

8.2.4 Operation of interchange circuits

The physical connections across the R interface between the terminal and TA are known as interchange circuits. Using terminology common to the ITU-T Recommendations, the TE2 is known as a data terminal equipment (DTE), which is the originating source and final destination for data transferred across the communications link, and can include all nature of terminals and computing equipment. Similarly, the TA is referred to as data communications equipment (DCE), and contains those functions necessary to establish, maintain and terminate a connection across a communications network, in this case the ISDN. Amongst the interchange circuits which connect the TE2 (DTE) to the TA (DCE) there are the transmit and receive data lines and the clock signal for a synchronous terminal, while the remaining circuits indicate the status of either side of the R interface connection and may be used for call control events and flow control across the R interface. However, there is also a need to provide the TAs at either end of a connection with the status of certain of each other's interchange circuits in order to synchronize the entry to and exit from the data transfer mode by the attached TE2s. Provision is made within V.110 to sample the state of relevant interchange circuits at the transmitting TA and embed this information in the S- and X- bits of the 80-bit frame. At the receiving TA, this information is mapped into corresponding interchange circuits at the R interface.

8.2.5 Flow control

For the connection of asynchronous TE2s to the ISDN, the possibility exists within V.110 for one TE2 to transmit characters into the ISDN at a different rate than it is received by the other TE2. In the case where a TE2 sources data at a rate which is higher than that at which the far-end TE2 is receiving it, flow control procedures are implemented in the TAs that attempt to reduce the rate at which data is generated at the faster TE2 to match that of the slower TE2. Both local flow control and end-to-end flow control are implemented.

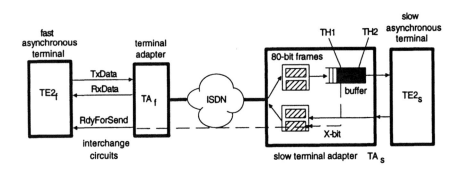

Fig. 8.4 V.110 flow control for asynchronous terminals.

The flow control scheme is illustrated in Fig. 8.4. Reducing the rate of a faster terminal to that of a slower terminal requires the presence of a local buffer in the TA connected to the slower terminal, TA$_S$. Data received from the ISDN is unpacked from the 80-bit frame and stored in a buffer before being passed on to the slower terminal, TE2$_S$. The buffer has a two pre-defined thresholds. The first threshold, TH1, represents an upper buffer level which if exceeded causes the single X-bit in the 80-bit frame, sent in the reverse direction toward the TA of the faster TE2, TA$_f$, to be turned off. When received at the TA$_f$, it causes it to use its local flow control to prevent TE2$_f$ from sending more information through control of local interchange circuits, thus preventing any new information from being received at the TA$_S$ and entering its buffer. As the TA$_S$ continues to send characters from its buffer to the TE2$_S$, the level in the buffer decreases until it falls below TH1, and eventually below the second lower threshold level, TH2. At this point, the TA_S turns the X-bit on again, causing $TE2_f$ to resume its transmission. The X-bit is also used in other instances where it is necessary for a TA to inhibit transmissions from a far-end terminal, for example when a TA has lost its frame synchronization.

Local flow control by the TA over the TE2 can either use the interchange circuits to prevent the TE2 from sending more data, or use in-band control characters called XOFF and XON within the data stream from the TA to the TE2 to stop and start data from the TE2.

8.2.6 Mode initialization

In an idle state, when no connection exists between TAs, a terminal is prevented from sending information by the state of the interchange circuits connecting it to the TA. A connection between two TAs will be established using the normal D-channel call control procedures, and during which the operational modes to be used for communications are specified by the calling TA using the low-layer compatibility (LLC) information elements within the D-channel Q.931 messages. For V.110 the following parameters may be defined:

- TE2 as a synchronous or asynchronous terminal;
- ability to perform in-band negotiation;
- terminal rate (not necessary for synchronous terminals);
- intermediate rate;
- ability to process network independent clocking information (only for synchronous terminals);
- ability to process flow control.

A TA which is unable to operate in the mode specified will reject the incoming call, otherwise a call is established and a 64 kbps channel exists between the TAs. Frame synchronization is then established, after which the TAs use the interchange circuits to bring the terminals into a data transfer state. Terminal Adaptation ceases when the call is disconnected or an error is detected, for example loss of frame synchronization.

The procedure described above uses D-channel call control signalling to establish a common operational mode between the two TAs. This method is fine provided that all communications exist within the ISDN, but is inadequate when the ISDN must interwork with the PSTN in order to complete an end-to-end connection between two terminals. The inadequacy is due to the fact that the PSTN has no out-of-band signalling channel that would allow the LLC information to be communicated across the two networks. V.110 therefore specifies an additional in-band parameter exchange (IPE) mechanism that may take place between the two TAs in the absence of out-of-band signalling.

IPE parameters are exchanged in a block that consists of groups of a command or parameter identifier followed by the parameters themselves. As no error correction capabilities are provided for in V.110, each command or parameter identifier is a byte that is repeated at least 32 times, thereby allowing the receiver to deduce the correct command or parameter identifier from the persistency of the byte value received. Similarly, the parameter data itself is contained in pairs of bytes following the parameter identifier, and which are transmitted three times such that the receiver may use a majority-vote principle to select the received value. For IPE, the operational mode parameters also include the possibility to assert loopback test modes in TAs that support such maintenance features.

The communication of IPE information can take place at a rate of 64 kbps, 56 kbps, 19.2 kbps, 9.6 kbps or 4.8 kbps. At the 64 kbps and 56 kbps rates the information bytes are inserted directly into the B-channel. The bytes are aligned with the B-channel time-slots and rely on the network maintaining this alignment across the end-to-end connection. The 56 kbps rate is included for operation on North American T1 type transmission systems that steal the eighth bit of the time-slot for signalling. At the lower data rates, bytes are transferred using a start–stop asynchronous character format, but must first be rate adapted to the 64 kbps B-channel. Hence, the V.110 80-bit frame format is used at the appropriate intermediate rate.

8.2.7 V.110 TA design

The intensive bit-level processing required to implement V.110 framing and terminal adaptation results in a mostly hardware-oriented design for V.110 terminal adapters, and therefore not surprisingly, there exist dedicated devices from a number of semiconductor vendors that implement the functions as defined in V.110. Specific devices are discussed later in section 8.4 as many of them provide both V.110 and V.120 capabilities.

Figure 8.5a illustrates the main components within a V.110 terminal adapter. Framing, terminal adaptation and network independent clocking are implemented in the V.110 controller which provides a direct connection between the R interface and a B-channel on the TDM system interface bus connecting it to the S transceiver. Processing of terminal data is therefore handled entirely by hardware elements within the TA, leaving software to handle the usual D-channel call control signalling, mode control and status monitoring of the V.110 controller, and optionally the IPE, although framing for the IPE information would still need to be provided by the V.110 controller. The split between hardware and software functions is illustrated in the protocol diagram of Fig. 8.5b. The hardware processing of the terminal data in V.110 has the benefit of speed resulting in low delays, but requires dedicated hardware to provide the specific frame synchronization and 80-bit framing which is not found on general purpose serial communications devices.

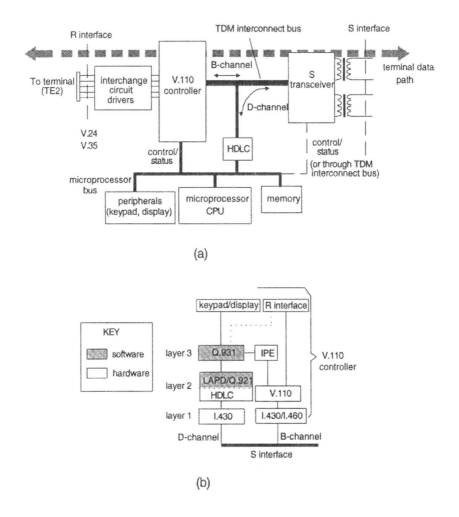

(a)

(b)

Fig. 8.5 The V.110 TA showing: (a) the main functional blocks and their interconnection; (b) the hardware and software partitioning.

Call control commands such as on-hook, off-hook, dial number and so on, can originate from a number of sources. The stand-alone TA itself might provide a keypad and display, or alternatively the terminal can generate these signals and communicate them to the TA across the R interface using modem control signals such as those defined in V.25bis, or an in-band command protocol such as the Hayes AT extended command set for ISDN.

8.3 V.120

Unlike V.110's dependence on a unique frame structure, V.120 uses the popular HDLC frame. Much of V.120 is based on LAPD which, with the advantage that many of V.120's functions are similar to those already present in the layer 2 of the call control signalling. However, unlike call control messages that use the D-channel, V.120 frames are communicated in B-channels. Like LAPD, V.120 can support multiple logical links, which means that the connection between TAs across the ISDN can support multiple point-to-point links between pairs of terminals connected to the same TAs. In-band signalling using messages similar to Q.931 messages are used to establish and terminate the links within the connection. The V.120 terminal adaptation protocol is used primarily in North America.

8.3.1 Multiple logical links

The V.120 frame is shown in Fig. 8.6, from which the similarities with HDLC and LAPD can be seen. Following the opening flag are two address bytes, 13 bits of which are used to identify the logical link, known as the LLI (logical link identifier). Three links are automatically present when a call is setup between two TAs. One of these links has LLI number 256 and is the default link that assumes that there will be at least one terminal connected to the TA at each end of the connection which uses this default link for communications. LLI number 0 is also defined, and is the in-band signalling link between the TAs used to set up further links as required with numbers between 257 and 2047 for use by additional terminals. The third link, LLI number 8191, is used for in-channel V.120 layer management. All other LLI numbers are reserved for future use.

V.120 frames that contain terminal data for a specific logical link are statistically multiplexed with those of other links into the 64 kbps channel established between TAs. Statistical multiplexing takes advantage of the fact that channel will not be occupied all the time by the frames transmitted on a particular link, and can thus be shared with other links, provided the aggregate link data rate remains lower than the channel bandwidth.

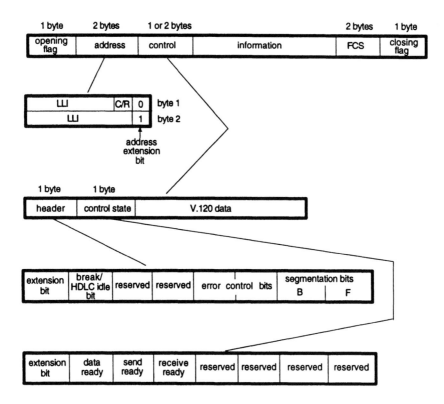

Fig. 8.6 The V.120 frame format.

8.3.2 Connection control for logical links

Once a connection is established between two TAs, the control of logical links between TAs within this connection is performed with Q.931 like messages in LLI 0. Assignment of logical links must be assumed by one of the TAs on either side of the connection. In some cases, a TA may have pre-assigned LLIs in which case it must take on the role of assigning LLIs for the connection. A bit is provided in the LLC information elements during call setup to indicate if the TA wishes to assume this role. If not, then it is not obvious which of the two TAs should perform this function and some procedures are required to resolve it, as described below.

The first TA to request the establishment of a link, other than the default link, will take on the role of assigning other LLIs. This it does by the exchange of four Q.931-like messages with the other TA. SETUP and CONNECT messages are used for link establishment and may contain low-layer compatibility and LLI information elements, while RELEASE and RELEASE COMPLETE messages are used for clearing the link and may contain the cause information element. These messages have the same structure as Q.931 messages and are distinguished from them by containing a different protocol discriminator, 0x09 instead of 0x08 for Q.931. Like Q.931 messages, they use a layer 3 call reference value to reference the request and subsequent establishment of a link to its layer 2 LLI value.

The role of the TAs as either LLI assignor or assignee during these transactions is reflected by the status of a bit in the LLC information elements. Should a conflict occur at setup time such that both TAs send SETUP messages simultaneously, the request is granted first to the TA whose message contains the higher call reference. In the unlikely event that both TAs contain the same call reference, both requests are terminated with RELEASE COMPLETE messages and the TAs attempt to establish the links once more with different call reference values.

8.3.3 V.120 terminal adaptation

V.120 supports the adaptation of both asynchronous and synchronous HDLC format terminal data streams to the ISDN. Protocol information within the terminal data stream, such as start and stop bits for asynchronous terminals, or flags, bit-stuffing bits and FCS bytes for synchronous terminals, are stripped prior to its encapsulation in a V.120 HDLC frame in the TA for communication in an ISDN B-channel. At the receiving TA, the information is reinserted into the data before being sent to the terminal across the R interface. This process of stripping and reinserting the protocol information within the terminal data streams requires the TA to be equipped with appropriate framing and buffering hardware.

The length of a V.120 HDLC frame, shown in Fig. 8.6, is identical to that of a LAPD HDLC frame. For asynchronous terminal characters, groups of stripped characters up to the maximum allowed in the information field can be sent per V.120 frame. Synchronous terminal data, however, may arrive at the TA in frames that, after stripping of the protocol information, are longer than the maximum permitted V.120 information field. In this case the frame from the terminal must first be segmented into smaller portions before being encapsulated in V.120 frames, while at the receiving TA, the portions of frame must be reassembled before the correct protocol information can be reinserted and sent to the terminal. The segmentation and reassembly process is controlled by two bits in the header byte (Fig. 8.6) that indicate if the V.120 frame contains either the beginning of the terminal frame (the B-bit) or the end of the terminal frame (the F, or final, bit). Of course, for terminal data frames that are shorter than the maximum V.120 information field length, these bits will reflect that it contains both the beginning and end. The header byte also contains error control bits and a bit that indicates the presence of HDLC idle condition at the R interface for synchronous terminals, or the break condition for asynchronous terminals.

In addition to the **protocol sensitive** mode of operation described above, V.120 also supports a **bit-transparent** mode similar to the way V.110 operates. Bits are taken from the terminal interface and simply encapsulated in the V.120 HDLC frame without processing.

8.3.4 V.120 rate adaptation

V.120 frames are transmitted and received in a B-channel at 64kbps. Rate adaptation to the terminal rate, which is less than 64 kbps, is achieved by varying the rate at which V.120 frames are communicated (not the bit rate) according to the rate at which information arrives from the terminal, and filling the inter-frame period in the B-channel with HDLC flags. This method is inherent to the way that V.120 works. For example, during the adaptation of asynchronous terminal data, terminal characters arriving at the TA will be stripped of start and stop bits and buffered in a memory store. One mechanism that may be used to determine when to send a V.120 frame is to use a buffer limit[34]. When the buffer reaches a defined limit, its contents will be transferred to a V.120 frame and transmitted in a B-channel. For a fixed-size buffer, the rate at which V.120 frames are sent will depend on how fast the buffer fills up, which is determined by the rate at which characters are communicated across the R interface towards the TA. In fact, this process rate adapts to the terminal character rate and not the terminal data rate which is negotiated between the TAs during call setup.

A similar scheme would be used in the case of the bit-transparent mode, but because the TA is insensitive to any protocol information in the data stream, the rate at which the buffer fills will be dependent on the rate at which the TA samples the terminal's transmit data line (related to the terminal data rate) regardless of whether information is being communicated by the terminal or not. In both cases, the size of the buffer is implementation dependent, and involves a trade-off between efficiency and delay. A large buffer size will result in a large V.120 frame information field such that the frame overhead, consisting of the flags, address, control field, header and so on, consumes only a small portion of the overall B-channel bandwidth in comparison to the information. The disadvantage is that a large buffer size will inevitably cause longer end-to-end

---

[34]A buffer limit is not the only mechanism that can be used to trigger the generation of a V.120 frame. Others include use of a timer, or receipt of a special character in the terminal data stream such as a carriage return.

delays. Likewise, a small buffer would produce small end-to-end delays but poor B-channel efficiency.

In the case of synchronous data terminals, each terminal frame with a length up to the maximum information field allowed in the V.120 frame, will be sent in a V.120 frame. Those that are larger will be segmented into several V.120 frames at the transmitting TA and then reassembled at the receiving TA as described above. V.120 frames are sent as required by the TA according to the rate at which terminal frames arrive at the TA.

8.3.5 Flow control and the control of interchange circuits

Control state information follows the header byte in the V.120 frame and reflects both the status of the sending TA entity, and of the appropriate interchange circuits at its R interface. The control state information is included in a V.120 frame whenever a change of state occurs. Three bits are used to reflect Data Ready that indicates that the R interface is active, Send Ready that indicates that the terminal is ready to send data, and Receive Ready that indicates that the terminal is ready to receive data. Through coordination of the Send Ready and Receive Ready variables, flow control can be implemented on an end-to-end basis between terminals. These three states can be mapped to the interchange circuits of the R interface standard in use on the TA.

As mentioned earlier, V.120 is based on LAPD, and is capable of both numbered I frame and unnumbered UI frame communications. Both I frames and UI frames may be used in the communication of R interface data for which the TA has stripped its protocol sensitive information, while UI frames only are used in bit-transparent mode. Although the TAs will do their best to match the flow of information at both sides of the link, the buffers within the TAs that are used to temporarily store terminal data as it is transferred between the R interface and ISDN S interface, may both experience underrun and overrun conditions. In the asynchronous mode, buffer underrun in not a problem as the character byte is the smallest unit of information communicated, and the TA simply waits until the terminal sends more characters and meanwhile sends HDLC flags in the V.120 link. However, a buffer overrun condition can occur at either sending or receiving TA, and may be relieved using either local flow control, the control state flow control mechanism described above, or by using the LAPD message flow control mechanisms (RNR and RR messages) between TAs.

For TAs operating in synchronous mode, both overrun and underrun conditions can occur as buffers may become full or empty during the transmission or reception of a frame. An underrun condition is only of significance from the TA towards the R interface, in which case the frame is aborted or sent with a faulty FCS. An overrun condition in the same direction can be treated in the same way, although sufficient buffer memory should be provided within the TA for this not to happen, while an overrun condition toward the ISDN S interface is indicated in the error control bits contained in the V.120 header byte.

As the TA has no control over the contents of the data stream in the bit-transparent mode, it is capable only of discarding the contents of a buffer on overflow and starting afresh. For a buffer underflow toward the R interface, the data line is held in the high state until more data arrives, and will be interpreted as an abort signal by the terminal.

Network independent clocking, as supported in V.110, is not directly supported in V.120, although can be indirectly supported through monitoring the rate at which the receive buffer within the TA fills with information. Variations in this fill rate can be made to adjust the clock toward the terminal in order to normalize the fill rate.

8.3.6 Differences between V.120 and LAPD

V.120 procedures are based on those of LAPD. However, as LAPD is intended to support the communication of call control signalling messages between the terminal and the network, it has some features which require modification such that they are better suited for full-duplex data communication in V.120. Below is a brief description of the major differences between V.120 and LAPD (Q.921).

- **I frame response.** In LAPD, I frames are sent as commands only, whereas in V.120 I frames can be transmitted either as commands or responses. Reception of an I frame response simply causes V.120 to treat it as a valid I frame command with the poll bit not set.

- **Symmetrical C/R bit.** In LAPD the polarity of the command/response bit is inverted depending on whether the terminal or the network originated the command/response sequence. This allows the originator to be identified while also allowing the two address bytes in the HDLC frame to be identical for a command and its associated response frame, thus complying with true HDLC principles. In V.120, total symmetry of the C/R bit is maintained as both TAs have equal functionality and it is not necessary to distinguish between terminal and network sides of a connection.

- **Frame reject (FRMR) response.** In LAPD, a data-link entity is permitted only to receive a FRMR response and not send one, while in V.120 a FRMR response can be transmitted for an error that cannot be recovered from by resending an identical frame, and results in a data link being reset.

- **Address management procedures.** In LAPD, assignment and removal of the terminal's TEI value must be managed by the network through procedures that reside both within the terminal and the network. Likewise, in V.120 the LLI values must also be managed, but because of the equal capabilities of both TAs, the role of LLI assignor must first be negotiated or assumed.

8.3.7 V.120 TA design

Unlike the predominantly hardware solution of V.110 TAs, a significant part of the protocol processing in V.120 is typically handled in software, and owing to the similarities with LAPD, it is often found that available software solutions are capable of both LAPD and V.120. Figure 8.7a illustrates the main components of a V.120 TA.

A consequence of software processing is that terminal data at the R interface must have its bit-level protocol terminated such that relevant information content can be stored in memory and processed. For operation in protocol sensitive mode, the TA must remove the flags, bit-stuffing bits, and FCS from a synchronous (HDLC oriented) terminal data stream, while the start and stop bits are removed from asynchronous (UART) terminal characters. In effect, the TA must contain the appropriate serial communications controller at the R interface to perform HDLC and UART framing functions. Such devices are usually designed to meet one or a number of specific serial protocols and may also be able to operate in the protocol insensitive mode that allows just the simple serial-to-parallel conversion required for bit-transparent mode. A separate HDLC controller will provide the framing support for V.120 frames that interface to a B-channel on the systems interface bus.

Moving information between the TA's R interface and memory, and between memory and the B-channel HDLC controller, is usually automated through DMA (direct memory access) to relieve the burden on the microprocessor and provide it with more time to process V.120 link information rather than simply move it between memory to I/O devices. Devices supporting lower throughput rates such as UARTs may alternatively use a FIFO (first-in-first-out) buffer and rely on an interrupt service routine to transfer blocks of information in one go between the device and memory. The boundary between hardware and software functions is illustrated in the protocol diagram of Fig. 8.7b.

8.3.8 Multiprotocol TAs

TAs that are designed to operate in both V.110 and V.120 modes, as well as other proprietary modes, must have the ability to determine which mode to operate in when receiving a call request. For connections between TAs that are contained only within ISDNs, the out-of-band signalling provided between TAs allows the

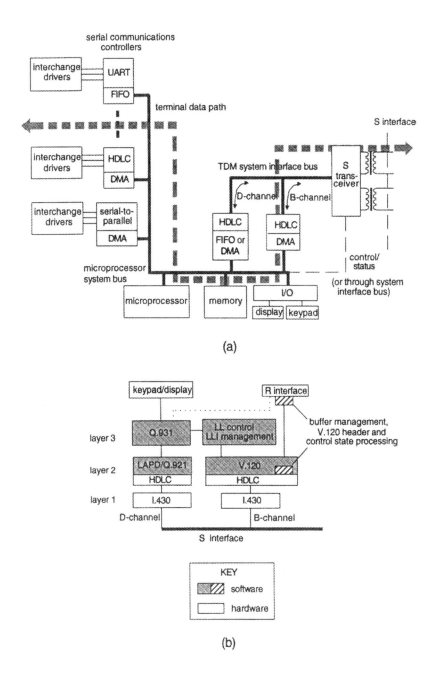

Fig. 8.7 The V.120 TA showing: (a) the main functional blocks and their interconnection; (b) the hardware and software partitioning.

lower layer compatibility information element within the Q.931 SETUP message to be used to identify whether V.110 or V.120 is to be used. However, in cases where the connection must traverse a PSTN in which no out-of-band signalling is possible, the type of TA must be pre-determined or an in-band signalling method must be employed to identify the protocol. This may be as simple as sequentially attempting to recognize the framing pattern of each supported protocol until the correct one is found, or an interworking protocol may be used (for example those recommended in ITU-T series I.500 Recommendations).

8.4 TA DEVICES

Several devices exist that allow TA solutions to be realized with varying degrees integration. V.110 provides the best opportunity to achieve a high level of integration due to its hardware intensive processing requirements that require only a minimum of software support beyond call control. However, the use of a specific framing protocol results in functions that are unique to V.110. This contrasts with a V.120 solution, which can use serial communications controllers that provide HDLC framing at the B-channel interface, and UART and HDLC framing at the R interface, both of which are used widely in other data communication applications and therefore more widely available. In addition, many serial controllers also include a bit-transparent mode to allow proprietary framing formats to be implemented, and may be used to support the bit-transparent mode in V.120 which is a basic requirement of V.120 TAs.

Table 8.1 lists some of the devices that could be used in the implementation of ISDN TAs. All the devices listed are hardware engines that support V.110 and may also provide framing support for V.120 and other protocols. In this respect, the Fujitsu device goes further and provides a state-machine that also executes the LAP procedures for a single V.120 logical link. Not all V.110 devices implement network independent clocking (NIC) and in-band parameter exchange (IPE) as they are features only required for interworking between an ISDN and PSTN. The more general purpose devices, such as HDLC controllers and UARTs, that can be used in V.120 TAs, have not been included in the list.

The Motorola part, MC68302, which was also mentioned in Chapter 6, is unique in that the device is a microprocessor system with a MC68000 microprocessor core. The V.110 and V.120 protocols are supported by a separate communications processor that is integrated on the same chip and consists of its own RISC (reduced instruction set computer) processor with three serial communication controllers that can support a range of framing protocols, including HDLC, UART and V.110. The three serial channels could be configured with one of them operating as the D-channel HDLC controller for call control signalling, another providing V.110 or V.120 (HDLC) interfacing to a B-channel, while the third one provides the R interface connection to the terminal. In the case of V.110, the formatting of terminal data between the R interface and the 80-bit frame is achieved in software instead of in hardware as in the other devices listed. However, the features of the MC68302 allow a highly integrated multiprotocol solution to be achieved.

Table 8.1 A selection of terminal adapter devices

Device	Protocol	TDM system i/f bus	Remarks
Fujitsu RIU rate adaptation interface unit MB86442	V.110 V.120	GCI SLD	IPE support with microprocessor/software. Provides V.120 LAP for 1 LL.
Mietec MSRA multi-standard rate adapter MTC 2073	V.110 V.120 +others	V* ST-bus	Supports NIC. DMA used for V.120 data transfer between MSRA and memory. Other protoccols supported are DMI, SIS and DT80.
Mitel RIM R-interface module MT 89500	V.110	ST-bus	
Motorola IMP integrated multi-protocol processor MC68302	V.110 V.120	IDL GCI SLD	Microprocessor with integrated serial controllers that implement framing and other protocol functions.
Philips DRA data rate adapter PCB 2320	V.110	2 Mbps terminal highway interface (THW)	
Siemens ITAC ISDN terminal adapter circuit PSB2110	V.110 V.120	IOM-2 (GCI)	IPE support with microprocessor/software

8.5 CONNECTING PACKET MODE TERMINALS TO THE ISDN

Packet-switching networks have been in existence since the 1970s and provide a service that is more suited to the communications of bursty traffic that is typical of computer terminal-to-host type applications and the transfer of small files over connections that are established over long periods of time. In most cases, the packet services within ISDN will be provided by an existing packet-switched network, with access provided by the ISDN. The ITU-T Recommendation X.31 [3] identifies two basic ways that a terminal can gain access to a packet-switched network from an ISDN subscriber loop connection. The difference between them is based on whether the access point to the packet-switched network is considered from an architectural standpoint to be inside the ISDN or is outside it.

If the access point lies outside the ISDN, then the ISDN is simply used to establish a 64 kbps channel between the terminal and the port of an access unit into the packet-switched network as illustrated in Fig. 8.8a. In X.31 this is referred to as Case A, and represents a minimal integration scenario for packet services in the ISDN. The port of the access unit itself will be assigned a number that allows it to be dialled up by the TA using normal ISDN D-channel signalling protocols. When a connection is established, the terminal may then use normal X.25 packets directly to the access unit across this connection in order to establish the X.25 call and then transfer data. The called X.25 terminal may either be connected directly to the packet network, or like the calling terminal, be interfaced through some other access network such as the ISDN or the PSTN.

For TAs that provide ISDN call control independently of the terminal, it is sufficient that the terminal packets are communicated transparently to the TA which, apart from establishing a connection from the terminal to an access unit, serves only to provide a physical layer interface between the terminal and ISDN B-channel. This is illustrated in Fig. 8.8a which also indicates the relevant protocol stacks for the switched B-channel mode. Instead of a dialled-up switched B-channel connection, a TA may provide a semi-permanent B-channel service in which the logical connection between the terminal equipment and the access unit always exists from the time at which subscription to such a service is taken out. Under such circumstances, the D-channel signalling is no longer needed.

In broad terms, two types of TA are possible. The type described above which provides no protocol processing for the B-channel represents one type of TA. When such a TA is in use, the terminal is considered the DTE while the access unit provides X.25 DCE functionality. However, for TAs that rely on the terminal to originate call requests, and to interface different implementations of X.25 in the terminal, it may be necessary for the TA to also provide X.25 protocol processing as illustrated in Fig. 8.8b and Fig. 8.8c. When using this second type of TA, the DCE functionality is now either wholly or partially contained within the TA.

The alternative network configuration is where the access point to the packet network lies inside the ISDN, and a packet handler interworking function is

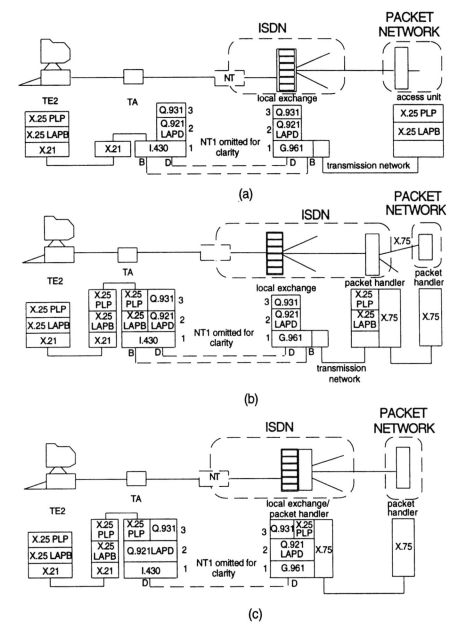

Fig. 8.8 Access methods to a packet network using the ISDN. (a) Switched access; (b) using the ISDN virtual circuit service in a B-channel; (c) using the ISDN virtual circuit service in the D-channel.

provided between the ISDN and the packet network. In X.31 this is known as Case B. A terminal and its TA may gain access to the packet handler through what X.31 refers to as **ISDN virtual circuit bearer services** using either an ISDN B-channel as before, or the D-channel. In both cases, connection to the packet handler is implied by the packet mode bearer service requested by the TA in its Q.931 call setup request to the ISDN, thus obviating the need to supply an address or number for the packet handler. Figure 8.8b illustrates the configuration and the relevant protocol stacks. Note that as stated previously, the requirement for X.25 protocol processing within the TA does not depend on its use of the ISDN virtual circuit bearer service, but is instead dependent on the X.25 capabilities of the terminal.

Access to the packet handler through the D-channel is significantly different in a number of ways. Firstly, the X.25 packets are no longer communicated in a dedicated B-channel, but must now share the D-channel with Q.931 signalling packets. As X.25 packet communication is a secondary function of the D-channel, the call control signalling packets are assigned a higher priority at the physical layer (Chapter 3). To be able to share the same channel, the X.25 packets must use LAPD at the data-link layer, which allows them to be distinguished from signalling packets through the use of a separate SAPI (SAPI=16) in the LAPD address. Secondly, as the D-channel only exists between the terminal and the local exchange, some processing must be performed in the local exchange to filter-off the X.25 packets arriving in the D-channel and route them to the nearest packet handler.

An early implementation example of D-channel access to a packet network is provided by the French ISDN 'Numeris' and French packet service 'Transpac'. D-channel access to Transpac was provided to subscribers at the end of 1992 as part of the ISDN upgrade known as VN3. The access between Numeris and Transpac is between a selected ISDN exchange and a local packet access point using n × 64 kbps channels on 2 Mbps semi-permanent connections. As the local exchange of the subscriber may be different from that to which connection into the packet network is made, processing of D-channel X.25 packets at the subscriber's local exchange involves routing the packets along the correct n × 64 channels. To facilitate this routing function, the LAPD addressing format is extended inside the network by replacing the normal TEI byte with three bytes that contain a link identification code (LIC) that uniquely identifies the channels in the connection between exchange and the packet access point into Transpac.

The connection configuration and relevant protocol stack for access via the D-channel are shown in Fig. 8.8c. The packet handler is shown here as part of the local exchange as an illustration only, and may also be located remotely.

There are several advantages of using the D-channel to access a packet network.

- Once the physical layer of the subscriber loop has been activated, the D-channel is always available for communications. As the D-channel is not switched, it is essentially non-blocking.
- Up to eight terminals can be connected to a passive S-bus, each being able to use the D-channel for access to the packet network, and each being uniquely addressed by the network according to its TEI.
- The same HDLC controller used for call control signalling is used to access the packet handler.

Although use of the D-channel for X.25 packet communications makes use of the bandwidth not wholly consumed by call control signalling, the entire D-channel is limited to 16 kbps and restricts effective bandwidth available for packet communications to a typical value of 9.6 kbps.

During the establishment of a call by an X.25 terminal it may be necessary for the terminal to identify itself to the packet-switched network for the purpose of billing and accounting. Additionally, in some cases when a call is being received by a terminal, there may also be a need for the packet network to identify itself to the terminal. X.32 [4] defines the functions and procedures at the R interface reference point between the TA and the terminal that supports both terminal and network identification during incoming and outgoing call establishment.

8.6 TELEMATIC SERVICES

Data terminals that access the public networks for a defined data communications service are known as telematic terminals, while the services they access are known as telematic services. The term telematics is often used in a broad sense to encompass all non-voice applications, although in reality it covers mainly data-oriented applications. Examples of such applications are teletex, Group 4 facsimile, videotex, and mixed mode which supports the communication of documents that have both character and facsimile elements.

Telematic applications are supported by a seven-layer protocol stack, the idea being that a common set of network independent layer protocols (layers 4 to 6) can be used to support the different application layers (layer 7). In reality, there exists a base-line set of standards, as for example those referred to in [6], to which extensions may be required to satisfy a particular application.

The network dependent protocols (layers 1 to 3) will depend on the network being used to connect the terminals. In the case of ISDN, a terminal may connect to another terminal either through the packet network as described above, or it may employ a circuit-switched connection through the ISDN. As in the case of access to a packet network, a terminal using a circuit-switched connection will use HDLC/X.25-based protocols with some modifications. Unlike the DTE-DCE relationship between terminal and network found in an X.25 or X.31 connection to a packet network, direct communication between terminals over a circuit-switched connection does not involve any termination of the user plane network dependent protocols by the network. Each side of the communication is therefore treated as equal rather than the computer (DCE) and terminal (DTE) type relationship found in X.25. The modifications required for such DTE-to-DTE communications are defined in Recommendation X.75 [7]. This Recommendation is very similar to X.25, the main difference being that it deals with communications between two entities which may be considered as having an equal status, rather than that which exists between a terminal and a network.

Figure 8.9 shows the protocols used in a telematic terminal that accesses a circuit-switched ISDN connection.

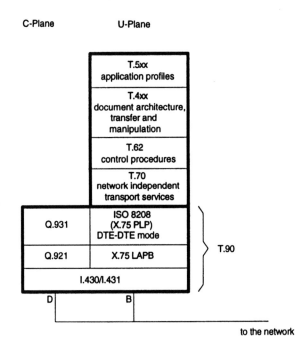

Fig. 8.9   Example of telematic terminal protocols for circuit-switched communications across the ISDN.

8.7 THE PC TERMINAL ADAPTER

The 1980s and 1990s have seen the emergence of the personal computer (PC) as a widely used tool for processing business and engineering applications, and is now becoming commonplace in the home, either as part of a home office or for personal and leisure applications. Data communication applications for PCs today include browsing the World Wide Web on the Internet, remote database access and file transfer, electronic mail, facsimile and remote access to LANs.

To provide interconnectivity to the ISDN, a PC will be equipped with an ISDN adapter card that plugs into the PC's expansion bus. As a result, the PC terminal adapter (PCTA) is significantly more integrated with the applications that use the ISDN than the stand-alone TA.

A PC adapter card can be designed either as a simple peripheral to the main microprocessor of the PC, also known as a **passive** ISDN card, or as an intelligent peripheral with its own local microprocessor. In the former case, the control over all functions of the adapter card is exercised by the PC, including all protocol and application processing that is necessary. Although such cards are inexpensive due to their minimal hardware requirements, they impose protocol processing requirements on the PC whose resources could otherwise be used for processing applications. This is in most cases not a limiting factor for a modern single user PC. In the latter case, the addition of a local microprocessor to the adapter card, while increasing its cost, will off-load protocol processing functions from the PC and allow it to devote more of its attention to application processing. This option is frequently chosen for server applications where the PC and its applications act as a central resource for multiple users.

8.7.1 PC adapter card

The following description of a generic PC ISDN adapter card is intended as an illustration of the features and functions to be found on a passive PCTA card .

Figure 8.10 shows an example of a passive PCTA card with a BRA ISDN interface. The majority of data communication protocols used in ISDN employ HDLC framing, and many applications use protocol stacks that build on top of the basic network dependent TA protocols (layers 1 to 3) to implement communication systems for specific applications such as facsimile, videotex and file transfer (for example, see Fig. 8.9). Three HDLC controllers are provided, one for the D-channel and one each for the two B-channels for simultaneous independent data calls. Each HDLC controller is interfaced with DMA control to local memory that acts as FIFO buffers for the delivery and recovery of HDLC frames between the PC and the PCTA. An interrupt request is used by the PCTA to instruct the PC to refresh the transmit and receive buffers for the B- and D-channels.

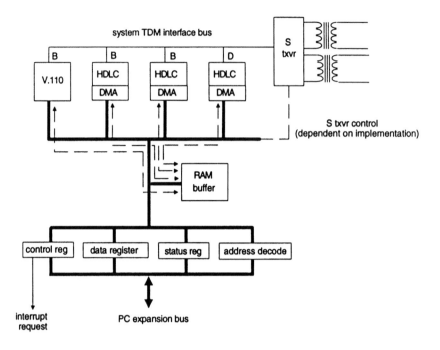

Fig. 8.10 Implementation example of a passive ISDN PCTA card.

A prime consideration in the design of an ISDN adapter card is the efficiency of B-channel data communications. A circuit-switched B-channel connection is tariffed according to the time duration of the call, so efficient use of the B-channel will help to realize a low cost per unit of information transferred. It is therefore important that the data path through the PCTA does not limit the effective bandwidth that can be used across the ISDN. For example, during the continuous exchange of data across the ISDN, the PCTA should be capable of processing packets fast enough to maintain back-to-back transmission and reception of frames in the assigned B-channel, for simultaneous data calls on both B-channels. Key to this ability is sufficient local buffer memory such that the PC is not burdened with processing frequent interrupt requests to refresh the buffers.

A V.110 controller is included in Fig. 8.11 to allow interconnectivity with other data processing equipment connected through V.110 TAs to the ISDN, while the hardware requirements for V.120 are satisfied through either of the existing B-channel HDLC controllers. However, for both V.110 and V.120 in a PC adapter environment, no standard R interface exists. Instead, user information from an application will be transferred between the PC and adapter card across the expansion bus of the PC rather than through an external interface. In effect, the PC expansion bus becomes the R interface. This is particularly important for V.110 as the architecture of the V.110 controller must be capable of supporting an interface to an external memory buffer in which the R interface data is stored.

In general, the type of PC expansion bus has little bearing on the performance of the ISDN adapter card. The majority of PCs in existence today use the ISA (industry standard architecture) bus which, while considered slow for many of today's PC applications, is sufficient to provide the information bandwidth between the PC and adapter card necessary to efficiently utilize the two 64 kbps B-channels of a BRA ISDN interface. ISDN adapter cards may also be found with EISA (extended ISA) bus interfaces and are typically intended for applications in PC servers where this bus type is often used. Future cards designs will in the long term include the faster bus standards, such as PCI (peripheral component interconnect), although many PCI-based PCs also include an ISA expansion bus for backward compatibility.

8.7.2 Multi-functional terminal adapters.

Most telephone wiring inside the residence is unsuitable for use as the ISDN S bus, and re-wiring can be a costly undertaking. In addition, a user who upgrades from analogue subscriber lines to the ISDN is likely to already have a range of analogue-based terminal equipment that cannot be directly connected to an ISDN S bus. A solution to these two problems has been a multi-functional terminal adapter that, like other terminal adapters, connects to the NT1 through an S interface connection. At the R interface, a range of interfaces are provided that allow both interconnection of standard analogue equipment to the ISDN, such as analogue telephones and Group 3 facsimile machines, and standard digital interfaces, such as the RS-232 serial interface, for the interconnection of PCs to the ISDN. The multi-functional terminal adapter provides all necessary interfacing, signalling and rate adaptation functions, and may also provide an additional S interface for a passive S bus extension, which is particularly useful for the connection of a PC equipped with an ISDN PCTA card.

The multi-functional terminal adapter can be placed in close proximity to the NT1, thereby minimizing the amount of S bus cable required, and analogue connections to the terminal adapter can use the existing wiring. Due to the status of the NT1 in America, it may be found integrated into the terminal adapter. Although the NT1 remains a separate entity in Europe, multi-functional terminal adapters are popular as they provide an economical way to provide home offices with ISDN.

8.8 PC COMMUNICATIONS IN THE ISDN: ACCESSING THE INTERNET

Although PCs are used today in numerous communications-based applications, one application that has recently been credited with creating a significant growth in the demand for ISDN is the World Wide Web (WWW) and access to the Internet. Apart from the popularity of the WWW itself, this demand has been created by a need for high-speed connections to the Internet that provide the user with fast download times for WWW pages. This is an issue that has become even more prominent with the increasing use of graphics and other media such as sound and video, that have significantly increased the size of Web pages and made Internet access using modems slow and uncomfortable.

The Internet is a world-wide network of computers and computer networks linked by virtually any means, but mostly by leased lines and packet data networks. The Internet has its origins in research and military networks established in America, but has recently experienced unprecedented growth world-wide that has been stimulated by a graphical user interface and hypertext document retrieval system known as the World Wide Web.

The protocol which governs the way information is communicated in the Internet is known as the **Internet protocol**. This is a suite of protocols whose core is known as TCP/IP (transmission control protocol/Internet protocol) which are roughly equivalent to the transport layer and upper part of the network layer in the OSI-RM. The IP part of TCP/IP is responsible for routing individual messages, called IP datagrams, between computers in the Internet no matter where they are located and what combination of communication networks provides a path between them. In order to achieve this, each computer in the Internet is identified by a unique 32-bit number known as its IP address. (A good introduction to the Internet protocols can be found in reference 8.)

Access to the Internet is provided by an Internet access provider (IAP) who has purchased connections into the Internet backbone network. These links are terminated at the IAP in an Internet router that provides a number of ports to public network connections such as the ISDN or PSTN. A user wishing to have access to the Internet then subscribes to a service offered by the IAP who provides the user with an account and a password. To use the service, the user simply dials in to one of the ports on the router of the IAP, validates his or her identity, is assigned a temporary IP address, and then proceeds to access the Internet. Any applications that run on the PC will expect to transmit and receive IP datagrams, and the underlying data connection must therefore support the transfer of IP datagrams between the PC and the router.

Where the underlying connection is established with ISDN, IP datagrams are encapsulated within HDLC frames and transported between the PC and router. Two protocols are commonly used that define how the encapsulation takes place and how the process is controlled. The first protocol uses the framing and procedures of X.75 in a circuit-switched B-channel. As noted earlier, X.75 is essentially X.25 between two users rather than between a user and a packet-switched network. The use of IP over X.75 is found mainly in Europe due to the higher usage of public packet-switched data networks. The second protocol is known as the **point-to-point protocol** (PPP) and has found widespread use in both Europe and America. Like IP over X.75, it is used for both access to the Internet and for LAN interconnection, and uses a similar data link based on HDLC framing and X.25 procedures, but a different network layer. PPP is described later after a brief introduction to LAN-to-LAN communications.

8.9 LAN-TO-LAN COMMUNICATIONS

Within a localized area such as an office, building or campus, communications between PCs, Workstations and computers is by means of a local area network (LAN). Unlike ISDN, whose services employ a connection-oriented mode of communications, a LAN uses **connectionless** packet communication techniques. This means that for the network dependent protocols of a LAN (usually the equivalent of the physical and data-link layers), instead of the three phases of call setup, connection and disconnection that are normally associated with a connection-oriented protocol, only the connection phase ever exists. Hence, any device can send information to any other device connected to the LAN at any time, and is a direct consequence of the shared transmission medium used in LAN communication systems that effectively connects every device to every other device. As no logical connection exists, every LAN packet contains not only a destination address but a source address as well, such that the recipient of the packet can identify its originator.

A further distinction is that LANs operate at relatively high bit rates in comparison to the ISDN. For example, a bit rate of 10 Mbps is common for a LAN based on the Ethernet protocol, although the actual information throughput achieved may be significantly less than this depending on the LAN configuration and network traffic load.

Communication between remote LANs across the public networks may use various services, such as leased lines, X.25, ISDN and analogue modems connected to the PSTN. Of these, ISDN has become a popular choice because it offers both digital circuit-switched and packet mode services from a single network connection, and the speed of ISDN digital call signalling allows on-demand connections to be established quickly and with increased flexibility and reliability.

8.9.1 Bridges and routers

Interconnecting LANs across the ISDN involves an amount of protocol translation in order to interface the information contained within the packets between the LAN and the ISDN. If we think of the LAN and the ISDN in terms of their OSI-RM seven-layer protocol stacks, then the point of interface normally occurs either at the equivalent of the data-link layer or the network layer. A connection between a LAN and the ISDN at the data-link layer is performed by a piece of equipment known as a **bridge** whose operation is illustrated in Fig. 8.11a. In simplistic terms, the bridge processes frames from the LAN up to the top of the data-link layer, which includes the partial removal of the data-link framing from the LAN packet. What remains of the LAN packet is then passed across to the top of the data-link entity of the ISDN protocol stack which processes the frame downwards through the physical layer and into the ISDN. This involves encapsulating the LAN packet in an HDLC frame, and use of X.75 procedures on circuit-switched connections, or X.25 procedures on connections to packet-switched networks.

Bridging therefore provides a translation of only the physical and data-link layers between the LAN and ISDN, and assumes that the higher layers of the protocol stacks in the interconnected LANs are compatible with one another. As the bridge provides interconnection at the data-link level only, it is typically used in conjunction with fixed links such as leased lines or semi-permanent connections between two LANs at pre-determined locations. As such, no call control is involved.

A more flexible solution that can provide connections to numerous locations can be achieved if the LAN to ISDN interfacing includes the equivalent of the network layer. Such equipment is known as a **router**, as shown in Fig. 8.11b. The network layer includes knowledge of network addresses and allows a LAN packet with a particular destination address to be mapped into a corresponding ISDN connection to the appropriate remote LAN to which the equipment with that address is connected. The router will therefore contain a table which provides the LAN address to ISDN connection number mapping, and allows switched ISDN connections to be established by the router to interconnect the LANs as required. However, owing to the fact that LAN protocols are connectionless in nature, once an ISDN connection is established between two LANs, the router has no way of knowing when a dialogue between PCs on two remote LANs has ended. The ISDN connections are therefore provided with a time-out such that if no activity is detected before a time-out value, then the call is disconnected. Further dialogue between the PCs will then cause the connection to be re-established, and due to the speed of the digital signalling in ISDN, can be made to appear virtually transparent to the application and the user. Such techniques are known as **spoofing** as the application and higher layers of protocol are spoofed into thinking the ISDN connection is permanent rather than intermittent.

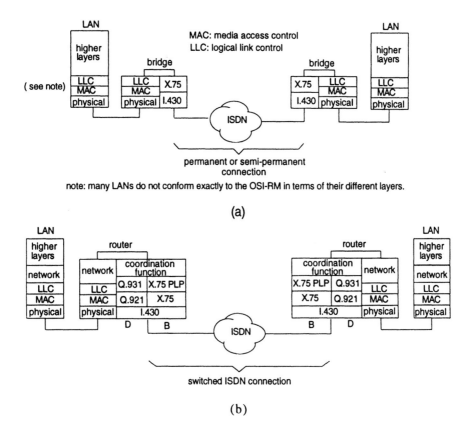

Fig. 8.11 LAN-to-LAN interconnection using: (a) bridges and ISDN permanent or semi-permanent connections; (b) routers and switched ISDN connections.

A router may have access to a number of public and private network services, including ISDN, and will be charged with the responsibility of choosing the most economical service and route for the LAN interconnection based on traffic profile, time of day and quality of service requirements. In many cases, leased lines are used between LANs for corporate networks as they are more secure and are usually justified in terms of the volume of traffic communicated. However, additional ISDN lines are used in order to provide a backup service should the leased line connection become unavailable or drop below a certain quality of service level. Again, the speed with which digital connections can be established in the ISDN means that the switch-over from leased line to ISDN service can be automatic with little impact to the overall service level.

8.9.2 Communicating LAN traffic across the ISDN

Many of the techniques in use today for the communication of LAN traffic across the ISDN using routers are based on proposals by the Network Working Group within the Internet Engineering Task Force (IETF). They involve a technique of encapsulating the equivalent of a LAN network layer protocol data unit, also known as a datagram, in an HDLC-based frame that is sent across either an ISDN circuit-switched connection or into the packet network using ISDN as the access network. However, there are numerous LAN network layer protocols in use, which creates a need for additional compatibility checking to ensure LAN-to-LAN inter-operability. Some of the common LAN network layer protocols are the IP (Internet protocol), the OSI CLNP (connectionless network protocol), SNAP (IEEE sub-network access protocol) and Novell IPX.

Many router products today have to provide connections for LANs that may use different network layer protocols. Identification of which protocol is in use on each of the LANs to be interconnected is therefore essential to ensure connection only between LANs of the same type. These routers are known as multi-protocol routers.

8.9.3 LAN interconnection using packet mode ISDN

X.25 public packet data networks are commonly used today for the inter-connection of LANs, and due to the commonality between X.25 and the ISDN packet mode services, the techniques used can be equally applied to ISDN. Figure 8.12 shows the X.25 frame formats for call request and data packets that encapsulate a LAN network layer datagram. If an X.25 virtual circuit connection between two LANs is being established to transport a single type of datagram, then the LAN network layer protocol is uniquely identified in the call request packet using the network layer protocol identifier (NLPID), as shown in Fig. 8.12a. Once the virtual circuit is established, datagrams such as those illustrated in Fig. 8.12b can then be exchanged without the need for further identification. Identification of the local LAN network layer protocol in the call request packet is necessary for the called router to be able to determine if it is compatible with that of its own LAN. If not, then the call may be rejected.

If a virtual circuit connection is required to support multiple types of datagram, then the virtual circuit is initiated using a call request packet with the null identifier. Each subsequent X.25 data packet is then required to also contain the NLPID (Fig. 8.12c), which identifies the datagram type, and allows it to be routed to the correct network at its destination. However, it is not necessary for a router to support all possible protocols for it to be able to accept a call request with a null NLPID. Instead, the router must only be capable of processing the null encapsulation for those LAN network layer protocols that it can support, and is allowed to discard any datagrams it receives for protocols it does not support. The encapsulation scheme described above is defined in RFC 1356 [9].

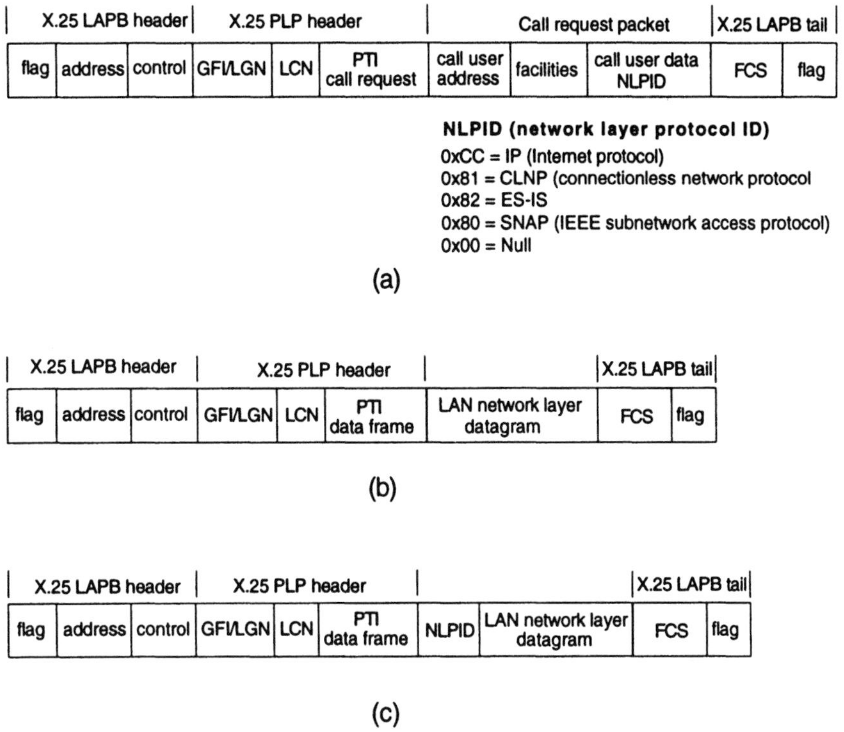

X.25 LAPB header	X.25 PLP header				Call request packet			X.25 LAPB tail
flag	address	control	GFI/LGN	LCN	PTI call request	call user address	facilities	call user data NLPID	FCS	flag

NLPID (network layer protocol ID)
0xCC = IP (Internet protocol)
0x81 = CLNP (connectionless network protocol
0x82 = ES-IS
0x80 = SNAP (IEEE subnetwork access protocol)
0x00 = Null

(a)

X.25 LAPB header	X.25 PLP header			X.25 LAPB tail
flag	address	control	GFI/LGN	LCN	PTI data frame	LAN network layer datagram	FCS	flag

(b)

X.25 LAPB header	X.25 PLP header		X.25 LAPB tail
flag	address	control	GFI/LGN	LCN	PTI data frame	NLPID	LAN network layer datagram	FCS	flag

(c)

Fig. 8.12 Packet formats for conveying LAN network layer PDUs across an ISDN/X.25 network. (a) The call request packet format; (b) data packet for a single protocol encapsulation; (c) format for the encapsulation of multiple protocols.

Invariably, the length of a LAN datagram will be larger than the maximum which can be accommodated by an X.25 data packet. If this is the case, then the packet segmentation and reassembly functions built into the network layer of X.25 are used to split the LAN datagram into more appropriate sizes for communication across the packet network. Routers based on RFC 1356 must be capable of handling LAN datagrams up to a size of at least 1600 bytes.

8.9.4 LAN interconnection using circuit-switched ISDN connections: the point-to-point protocol (PPP)

The point-to-point protocol (PPP) provides a method for communicating LAN datagrams across serial point-to-point links such as ISDN circuit-switched B-channel connections, and as a method for remote access to IP based networks such as the Internet. PPP is defined in RFC 1661 [10] and includes definition of a data-link protocol and a family of network layer protocols in addition to a LAN datagram encapsulation scheme. Figure 8.13 illustrates the protocol stack and the frame encapsulation used in PPP.

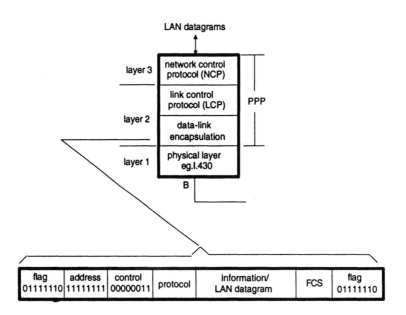

Fig. 8.13 The protocol stack and frame structure for the point-to-point protocol.

The data-link layer is provided by the link control protocol (LCP) whose purpose it is to establish, configure and test the data-link connection for a variety of different types of serial link and networks. Configuration of the link parameters is handled through an extensive capabilities exchange protocol that allows the link to be configured for both standard and non-standard networks. If required, the LCP also allows authentication of the peer data link to take place, using for example an exchange of passwords. Once a data link is established, the network control protocol (NCP) exchanges packets with its peer to select and configure the link to handle one or more LAN network layer protocols. A family of NCPs is defined as each is defined to handle the specific requirements of the respective network layer protocol in use. When this has been achieved, datagrams can be transferred between LANs across the serial link.

As can be seen from the frame structure shown in Fig. 8.13, LAN datagrams are encapsulated in an HDLC formatted frame. The frame shown represents the default frame structure which may be modified through negotiation by the LCP to suit the particular requirements of the serial link. The address field contains a broadcast address (0xFF) in both directions as there is no DCE/DTE distinction as in X.25. Instead, the PPP link connects equipment of an equal status. Only UI command frames are used on a PPP link, and other types of frame that are received are discarded. The control field is extended by two bytes into a protocol field that contains a value that identifies either the LAN network protocol when the information field of the frame contains a LAN datagram, or the particular NCP in use when the information field contains an NCP related datagram. A default maximum information field length is set at 1500 bytes. Finally, the end of a frame is terminated with an X.25 type frame check sequence (FCS) and closing flag.

8.9.5 Multilink PPP

The ISDN BRA interface offers two independent B-channels that can be used to establish two separate 64 kbps connections, or links, between the same end-points, thus delivering an effective bandwidth of 128 kbps. Unfortunately, as already mentioned, the different end-to-end delay on each link, caused by the independent routing of each connection through the network, will cause information that starts synchronized in both channels to be received with a time displacement between the two links. Unlike the channel aggregation technique used in H.221 that equalizes the delay of the fastest link to that of the slowest, an extension to PPP is used that simply orders individual packets between the two links.

In addition to a unique protocol identifier for multilink PPP, a sequence number is added to each packet before the packets are assigned to the links. In the case of BRA ISDN, each connection, or link, has an equal bandwidth and so packets can be divided equally between them. At the receiving end, the sequence number is used to reassemble the stream of IP datagrams before delivering them to higher protocol layers. The sequence number is a 24- or 12-bit number, the size being defined during LCP negotiation, that is continually incremented as packets are created from the stream of IP datagrams to be transmitted. Loss of a packet is simply identified at the receiver by a missing sequence number.

Multilink PPP provides an option that large datagrams can be segmented and suitably encapsulated and framed fragments apportioned to the different links according to the bandwidth capability of each link, the idea being to share the load as evenly as possible amongst the links in order to keep each link occupied and to make optimal use of the aggregate bandwidth available. Of course, this applies to any interface with multiple connection capabilities, and not just to BRA ISDN. Two bits are used in the multilink PPP header, one that indicates that the IP fragment contained in the packet is the beginning of an IP datagram (the B-bit), while the other indicates that the fragment is the end of a datagram (the E-bit). Packets that have both B- and E-bits contain an entire IP datagram.

Figure 8.14 illustrates the multilink PPP frame structure. If an IP datagram has been segmented, then this structure is applied to the first fragment, while subsequent fragments of the same datagram contain just the protocol ID and the multilink PPP header. Multilink PPP is described in RFC 1717 [11].

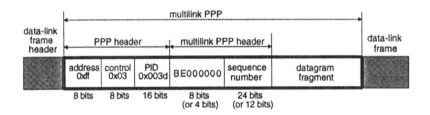

Fig. 8.14 The multilink PPP frame format.

8.10 PROVIDING MORE CONNECTION BANDWIDTH

A single circuit-switched B-channel connection through the ISDN will provide a dedicated 64 kbps bandwidth. However, the line bit rate of a LAN is typically between 2 and 16 Mbps. Although the difference between these two rates appears alarmingly high, the rate at which LAN packets are communicated is significantly less than the equivalent of this bit rate, added to which the ISDN may not be required to communicate every packet that appears on one LAN to the other. Typically, a router will provide some form of address filtering such that only LAN packets that are addressed to terminals on the remote LAN are transmitted across the ISDN, while those packets addressed to local terminals are not. A more important measure of the ISDN connection bandwidth required is therefore given by the aggregate rate of messages and their size that are addressed to a remote LAN. This in turn will depend on the application and network configuration. For example, packets containing simple terminal commands will typically be short and infrequent such that a single 64 kbps connection may be sufficient to provide an adequate quality of service. On the other hand, the transfer of large files using just a single 64 kbps connection may lead to slow response times due to delays and long access times. Under such circumstances, the provision of more connection bandwidth will improve the quality of service.

Obtaining more connection bandwidth can be achieved in two ways. The first would be to subscribe to a switched wideband service provided by the network operator. However, such services are typically expensive, so much so that significant volumes of LAN traffic would need to be communicated in order for it to be cost effective. In addition, switched wideband services are not as widespread as the PSTN and ISDN, and are mostly available only in large metropolitan areas. The international availability of these services is also low. A second approach would be to provide multiple 64 kbps ISDN connections between the same points, and equipment at each end of the connections that aggregates the individual bandwidths such that from the LAN's point of view, it sees a single large bandwidth connection. The equipment that performs this function is known as an **inverse multiplexer** because instead of multiplexing lots of small bandwidth channels into a single large bandwidth channel for transmission between two points, it performs the inverse operation of taking a large bandwidth channel and splitting it up into lots of smaller bandwidth channels for transmission. This process is shown illustrated in Fig. 8.15. Of course, inverse multiplexing can be used for numerous applications besides LAN-to-LAN communications, and was already discussed briefly in Chapter 7 in the context of videoconferencing.

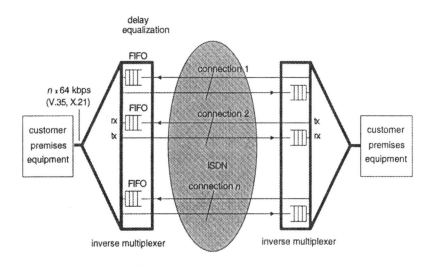

Fig. 8.15 The principle of inverse multiplexing.

8.10.1 Inverse multiplexing

Circuit-switched connections across an ISDN are established in units of 64 kbps[35]. More bandwidth can be added to a single 64 kbps connection through the establishment of further 64 kbps connections, and the aggregation of the individual channels into a single larger bandwidth channel. The difference in end-to-end delay on each connection must be compensated for through adjustable delays, in the form of variable length buffers, for each 64 kbps path, as shown in Fig. 8.15, such that the end-to-end delay for each connection can be equalized. The speed and relative ease with which bandwidth can thus be adjusted to meet the immediate requirements of an application has inspired the generic term **bandwidth on demand** to be applied to such a capability.

Most inverse multiplexing equipment today comprises standalone devices that provide a connection between a customer's terminal, such as a videoconferencing unit, or other equipment such as bridges and routers. A number of standard interfaces are usually provided at the customer equipment interface side, such as V.35 or X.21, while multiple BRA connections, or a PRA connection, are used to connect to the ISDN.

Two inverse multiplexing standards exist today. The first is the method defined in H.221 for audiovisual communications, and is described in more detail in Chapter 7. The other standard is defined by an industry consortium known as the Bandwidth on Demand Inter-operability group, and is known as the BONDING™ protocol [12][36]. The BONDING protocol is intended for use in a wide range of both audiovisual and data applications, including LAN-to-LAN interconnection. More recently, the ITU-T has attempted to consolidate these two methods under a single recommendation, H.244 [13].

8.10.2 BONDING

The BONDING protocol was born out of the need to provide inter-operability between different manufacturers' inverse multiplexers, and today nearly all inverse multiplexing equipment provides at least a mandatory subset of the BONDING protocol.

To establish an $n \times 64$ kbps call, an inverse multiplexer must make n individual calls to another inverse multiplexer, where each call adds a 64 kbps channel to the overall available bandwidth. However, each inverse multiplexer will typically be known by only a single number to which the first call is made. The first call establishes what is known as the **master** channel, and is used to negotiate the operational mode between the two inverse multiplexers. Each inverse multiplexer will store a list of the numbers that represent its additional lines to the public

---

[35]In North America, connections may be based on 56 kbps channels instead of 64 kbps.
[36]BONDING is a trademark of the BONDING Corporation.

network, either through BRA or PRA connections. During the negotiation phase, the called inverse multiplexer has the opportunity to communicate the numbers of its remaining lines back to the calling inverse multiplexer which can then undertake establishment of the remaining channels. Consequently, an inverse multiplexer need only be identifiable with a single ISDN number rather than a group of numbers. Information that is communicated during negotiation is conveyed in a unique frame structure called the information channel, shown in Fig. 8.16a.

Once the negotiation process is complete, the master channel reverts to the frame structure shown in Fig. 8.16b of which the information channel then becomes a part. The basic BONDING frame structure is 256 bytes in length. Of these 256 bytes, four are framing and control overhead bytes that are inserted by the transmitting inverse multiplexer and stripped out by the receiving inverse multiplexer, and are therefore transparent to the user. The frame alignment word (FAW) allows the frame boundaries to be detected while the CRC byte contains a four-bit CRC as well as status bits that indicate a loss of frame alignment and occurrence of a CRC error. The information channel continues to provide an in-band exchange of information between the inverse multiplexers, but obviously has less bandwidth available to it now that is part of a larger frame structure.

A multiframe is created from 64 basic frames. Each of the 64 frames is sequentially numbered with a six-bit value contained in the frame count (FC) byte, and provides the basis by which relative frame delay can be measured between the master channel and further channels added to the call. Given that frames in each channel at the transmitter are synchronized and have the same FC value, then the relative delay between two channels at the receiver is provided by the difference in their FC values and whatever offset exists between their FAWs. This information is then used to adjust the delay of each channel through adjustable FIFOs such that they are brought back in to alignment with one another.

The maximum delay that can be compensated is determined by the time duration of the multiframe because there is no way to distinguish between the same frames in subsequent multiframes. At a channel rate of 64 kbps, the multiframe duration is 2.048 seconds. If we consider the absolute delay of a particular channel to be a reference point, then the adjacent channel which must be equalized to it can have either taken a quicker route through the network or a longer route. The relative maximum delay that can be accommodated between channels is therefore one half of the multiframe duration, or 1.024 seconds.

Two parameters in the information channel also play an important role during the aggregation process. Firstly, the group ID (GID) is used to associate a channel with a particular $n \times 64$ kbps call. As an inverse multiplexer may be capable of supporting multiple simultaneous calls, it becomes necessary to identify each channel with a particular call process. The GID is a six-bit value and allows a maximum of 64 simultaneous calls to be established. Secondly, within an $n \times 64$ kbps call, if the individual calls are made by the calling inverse multiplexer

0	1	1	1	1	1	1	1
1	Channel ID (CID)						1
1	Group ID (GID)						1
1	operating mode			res			1
1	RMULT						1
1	SUBMULT			BCR	RES	MFG	1
1	RI	req	ind	REV			
1	subaddress						1
1	transfer flag						1
1	1	1	digit - 1				1
1	1	1	digit - 2				1
1	1	1	digit - 3				1
1	1	1	digit - 4				1
1	1	1	digit - 5				1
1	1	1	digit - 6				1
1	1	1	digit - 7				1

(a)

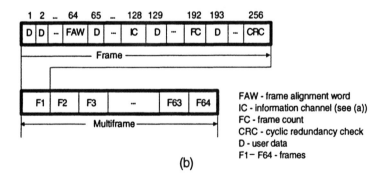

(b)

Fig. 8.16 Frame structures used in the BONDING protocol.

simultaneously or in rapid succession, then due to differences in call routing, the calls may arrive at the receiving inverse multiplexer out of sequence. This would result in a mixing of the channel order and hence a mixing of the time-slot order in the combined data stream toward the customer equipment. The channel ID (CID) is therefore included in the information channel to identify the correct ordering of the received channels. Like the GID, the CID is also a six-bit value which allows a maximum of 64 channels to be combined in a single $n \times 64$ kbps call ($\max(n) = 64$). Following channel aggregation and delay equalization, user data is communicated between the connected inverse multiplexers.

Four modes of operation, numbered mode 0 to mode 3, are defined in the BONDING protocol. The mode used on an $n \times 64$ kbps call is identified during negotiation in the information channel, and where the capabilities of the two inverse multiplexers are different, the negotiation process selects the lowest common operating mode. In mode 0, after the initial negotiation phase, user data is transmitted with no channel delay equalization. This allows the inverse multiplexer to operate with terminals and other customer premises equipment that perform their own delay equalization, for example videoconferencing equipment using H.221.

In mode 1, after negotiation and initial delay equalization, the framing and control overhead bytes are removed in each 64 kbps channel, which allows the full channel bandwidth to be employed to communicate user information. However, once the framing and control bytes are removed, there is no means of detecting errors in a particular channel or recovering from a loss of synchronization. In such cases, the entire $n \times 64$ kbps call must be terminated and re-established. Mode 1 is the common mode of operation for inverse multiplexers that implement the BONDING protocol.

Mode 2 employs the framing and control overhead bytes during the transfer of user information. While this slightly reduces the available bandwidth for user data (one byte in 64, or 1.56%) it allows dynamic channel alignment and synchronization to take place as well as channel error detection. This capability permits the inverse multiplexer to provide a very flexible and reliable service. For example, if persistent errors or loss of synchronization occur in a particular 64 kbps channel, the connection associated with the channel can be terminated and another more reliable connection established without termination of the entire $n \times 64$ kbps call. In some applications such as LAN-to-LAN interconnection, the required connection bandwidth may vary according to the time of day, in which case additional channels can be established to meet the demand[37] (in other words, bandwidth on demand).

---

[37]If the data-link connection between the inverse multiplexer and the customer equipment is based on HDLC frames, then a measure of the bandwidth demand can be derived from the occurrence of interframe-fill flags. The fewer flags detected, the more bandwidth is being used to communicate frames of information.

Mode 3 is a mixture of modes 1 and 2. In some applications, it may be necessary to exactly match a fixed synchronous bandwidth that is a multiple of 64 kbps from the customer equipment with the same bandwidth $n \times 64$ kbps connection across the ISDN. The establishment of n channels in this case would leave no bandwidth to accommodate the framing and control bytes during transmission of user information. This is the case in mode 1. However, in mode 3, the framing and control bytes are added to each channel, and the displaced user bandwidth is made up with the establishment of an $(n + 1)$th channel including framing and control. Spare user bandwidth in this additional channel is filled with null data.

8.11 FRAME RELAY

X.25 and packet-based data networks were developed during the 1970s when subscriber and transmission network technology was not as reliable as today's all digital networks. As a result, packet networks were designed to have fairly comprehensive error detection and correction protocols at each packet switch within the network, and at the subscriber access point to the network. However, this reliability is achieved at the price of a slow end-to-end throughput due to the protocol processing required at each packet switch to process individual packets. In effect, every information packet that arrives at a packet switch must typically be loaded in to memory through an HDLC controller, processed according to data-link and packet layer procedures, routed to the appropriate outgoing link queue, again processed according to packet and data-link procedures for the outgoing link, and finally transmitted through another HDLC controller. Much of the protocol processing will be performed by software running on a microprocessor system that is required to share its available processing bandwidth between several incoming and outgoing ports, and may result in an input-to-output delay through a packet switch in excess of several hundreds of milliseconds.

With the arrival of reliable digital transmission networks and digital subscriber loops, the requirement on the network to provide such a high level of error detection and correction no longer exists, and has prompted the definition and deployment of a new service known as **frame relay**. In fact, frame relay was defined as one of the new bearer services to be provided as part of the ISDN, but has been deployed in many countries as a separate network service aimed at providing fast (64 kbps up to 2 Mbps) packet services for data applications, in particular LAN-to-LAN interconnection. Although the service is known somewhat generically as frame relay, there are two techniques employed, one known by the same name, frame relay, and the other as **frame switching**. The difference between them lies in the functions performed by the data-link layer in the user plane and are best described in conjunction with the protocol stack diagrams shown in Fig. 8.17.

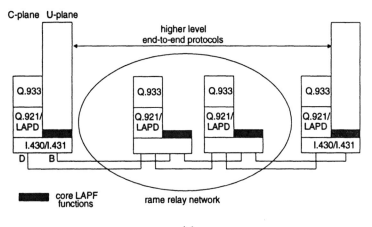

(a)

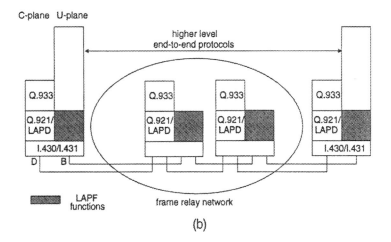

(b)

Fig. 8.17 Protocols used in: (a) frame relay; (b) frame switching.

One of the first things that is noticed from Fig. 8.17 is that ISDN signalling is used in both frame relay and frame switching networks rather than the in-channel signalling that is employed in X.25. Extensions to the basic Q.931 call control functions are defined in Q.933 [14] to allow it to handle signalling for the frame mode, while Q.921 is used at the data-link layer. However, in the user plane at the network side of the interface, the requirement to provide a faster throughput than an X.25 packet switch has led to the exclusion of a packet, or network layer protocol, and use of either part or all of just a data-link layer based on a version of LAPD known as LAPF (link access protocol for the frame mode bearer services) [15]. Frame relay, shown in Fig. 8.17a makes use of only a part of the data-link layer known as the **core functions**, and includes only the framing functions defined in LAPD such as HDLC framing, bit-stuffing, and CRC and frame length error checking. Hence the name frame 'relay' as frames are simply relayed between the nodes in a frame relay network. In addition to the framing functions, the frame addressing supports the use of multiple logical links in the same transmission channel in the same way as that described for V.120 and LAPD. In keeping with the terminology used in LAPD, the address which identifies a particular logical link is known as the data-link connection identifier (DLCI).

Like X.25, frame relay is connection oriented, and during the call establishment phase, a logical route will be established through the network. Each link between nodes in the network and across the subscriber loops to the customer premises equipment will be identified by an individual DLCI. A user frame that arrives at a node will have its framing and length verified and CRC checked for errors, while a look-up table is used to translate the DLCI of the incoming link into the DLCI of the outgoing link. The frame then gets queued at the relevant output port ready for transmission. The processing of only basic framing functions in the user channel of a frame relay network can therefore be achieved mostly in hardware, and allows a significantly higher throughput, typically ten times or more, to be achieved in comparison to X.25 packet networks.

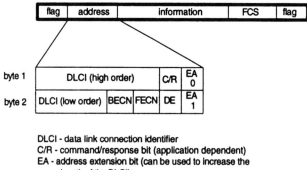

DLCI - data link connection identifier
C/R - command/response bit (application dependent)
EA - address extension bit (can be used to increase the
      length of the DLCI)
DE - discard eligibility indicator
BECN - backward explicit congestion notification
FECN - forward explicit congestion notification

Fig. 8.18 The core LAPF frame format.

Figure 8.18 illustrates the core frame structure used for frame relay. Only unnumbered frames are used which obviates the need for a control field. As the default frame size is intended to be the same as LAPD, the two control field bytes are added to the information field giving it a maximum default size of 262. However, as no packet layer exists, packet fragmentation and reassembly of PDUs (protocol data units) greater then this size that are to be communicated across the frame relay network, must either be performed by whatever protocols are placed above the core LAPF layer within the end equipment, or a larger information field size must be negotiated at call setup.

Another function not present within a frame relay network is flow control. This necessary feature is now provided only on an end-to-end basis through the FECN (forward explicit congestion notification) and BECN (backward explicit congestion notification) bits in the address field. Frames that are lost or duplicated within the network are therefore undetected and rely on higher level protocols operating within the customer equipment for detection and correction action.

The second type of frame relay service is frame switching. As illustrated in Fig. 8.17b, frame switching employs a fully featured data-link layer, which in addition to the features described above, also provides acknowledged frame transfer between network nodes and between the customer equipment and the network, and is therefore capable of detecting and correcting for lost and duplicated frames. While this provides a more reliable service than frame relay, it does require additional processing within the node resulting in longer delays than frame relay, but still significantly less than X.25.

8.12 DATA CONFERENCING

Sharing data, as well as voice and video, is often an essential requirement in teleconferencing activities. In the past, videoconferencing rooms may have been equipped with a separate facsimile machine or document camera as the only means of communicating such information. However, with an increasing trend towards personal conferencing using suitably equipped PCs, a requirement exists for the communication and presentation of information derived from commonly used applications, such as word and document processors and graphical tools. Applications may then be used by conference participants to interactively manipulate the information, thereby allowing collaborative work activities to take place when participants are geographically separated from one another.

The ITU-T has recently produced a core set of recommendations for data collaboration known as T.120. These Recommendations define the protocols that control and manage the connections between conference participants, control shared applications, and transfer different types of data, such as still images, binary files and facsimile, between the conference participants. Figure 8.19 illustrates the building blocks that go to make up the T.120 suite of protocols for use with the ISDN. The core of T.120 is made up of the multipoint communications service (MCS), the generic conference control (GCC), and a set of underlying network dependent protocols that interfaces them to a particular type of network. The diagram also illustrates how T.120 combines with the audio and video communication protocols to create an audio, video and data conferencing system. An objective of T.120 is to provide operation across a number of different network types, so other transport stacks below the MCS suitable for other networks besides that of ISDN are also possible, for example the PSTN (using modems), X.25 and LANs.

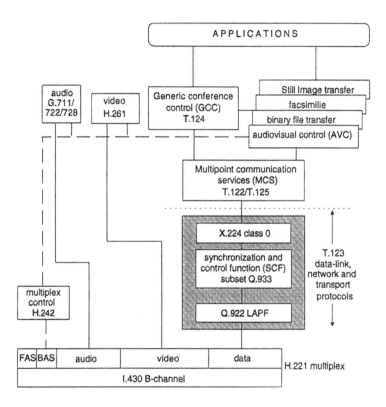

Fig. 8.19 The T.120 data conferencing protocols on BRA ISDN B-channels together with a videoconferencing application.

T.120 also attempts to provide network independence in the way in which conference connections are configured, either centrally controlled through an MCU, or distributed amongst the participating terminals. In the case of conferences established using public networks, the use of an MCU is envisaged where participants dial into an MCU. The MCU would typically provide voice bridging and video switching as well as T.120 data conferencing services.

T.120 information is communicated across the network between terminals in dedicated channels which we refer to here collectively as the **T.120 channel**. For a data only conference call, the T.120 may occupy a 64 kbps B-channel established between a terminal and MCU, or alternatively, for a conference call involving voice and video as well as data, the T.120 channel may occupy the auxiliary data channel in the H.221 frame (Chapter 7). These channels convey both application data and conference control and management data, and are established using the usual ISDN call control protocols.

At the heart of the T.120 lies the multipoint communication service and the generic conference control which are discussed in the following sections.

8.12.1 The generic conference control

The purpose of the Generic Conference Control (GCC) is to establish and manage conference sessions between the conference participants. It includes administration functions that allow a potential participant to view a list of conferences currently in progress, and request to join a particular conference. The addition of a new participant to the conference is then indicated to other participants. The GCC also provides security measures, and may require the new participant to enter a password before being allowed to join a conference.

Structure within the conference activities is provided by one of the participants acting as the conference chairman who can terminate the conference and control participation. If necessary, chairmanship can be passed to another participant.

The GCC operates independently of any underlying network, and is masked from it by the functions of the multipoint communication service and the network specific transport protocols.

8.12.2 The multipoint communication service

The multipoint communication service (MCS) is responsible for establishment and management of specific channels within the T.120 channel that create the multipoint connections between conference participants across different types of network. MCS nodes from different networks may connect together to form a conference, known as a **domain** in T.120 terminology. Users may participate in more than one domain, or subgroups within a domain, at the same time.

The MCS can create three types of channel: a multicast channel which communicates data from a single source to multiple destinations, a single-member channel which creates point-to-point links between participants, and a private channel which is used to convey private correspondence to an individual or group of participants. In the case of multicast channels, the MCS also manages the routing of data between source and destinations, the most simple approach being along the shortest path between them. This approach is optimal if there is only a single source transmitting data at any one time, but in the case of simultaneous transmissions from more than one source, then the difference in paths between participants may lead to the data being received out of sequence to that in which it was originally transmitted from the sources. This is overcome by data from all sources being sent to a single **top provider**, such as an MCU, that is designated as the coordinator of such multicast transmissions. The data is then dispatched from the top provider to all receiving participants in the correct order.

Where channel resources are shared, access to them can be controlled through a token mechanism. The token is sent to a particular terminal to grant it exclusive access to the channel after which the token is relinquished.

The MCS sits on top of a network dependent stack that provides the point-to-point connection to an MCU or other participants. The MCS will reside in both the terminals and MCUs involved in a T.120 conference.

8.12.3 Application interface modules

The way in which application information can be shared and manipulated during a conference is determined by a range of optional modules included as part of T.120, each of which is associated with a particular type or types of information. For example, four such modules are shown in Fig. 8.19: audiovisual control, binary file transfer, facsimile and still image transfer. The range of modules is not limited to this set and is likely to include others for handling sound files, moving picture files, private information that has to be encrypted, and so on. The functions performed by these modules is different depending on the information they process, but provide a standard interface to the applications that sit above them. As an example of the type of function these modules perform, we can consider an application that generates a still image, say from a camera whose image is then encoded according to the JPEG standard. The JPEG image is then passed down to the still image transfer module which initiates its communication to those conference participants who are capable of receiving it. This depends on them also having the still image transfer module and is determined through a capabilities exchange during establishment of the conference. Once the image has been received and displayed, a local operation on the image can be indicated at the remote terminals through movement of a pointing device at the local terminal that moves a pointer icon on the screens of the remote terminals. If several participants are active in the discussion, then several pointers may be active on the screen at any one time, each with an indication as to whom they belong to. Management and control of the pointers is performed within the still image transfer module.

The wide range of different options available within T.120, not only of the modules that are present but also the capabilities of those modules, require definition of system **profiles** to ensure that a terminal with a specific profile will interwork with another terminal having the same profile. Definition of these profiles will be included within T.120 and range from simple data-only conference terminals to fully featured multimedia terminals capable of audio, video and data.

8.12.4 Network dependent protocols

The lower layers of a T.120 protocol stack provide basic network dependent services to support the communication of the protocol data between peer MCSs on a point-to-point connection. Different networks may require different lower layers in order to mask different network dependencies from the MCS. The lower layers consist of the data-link, network and transport layers and are defined in Recommendation T.123. In the case of ISDN, the data-link layer, LAPF, and network layer, SCF, are those used in the frame relay service and are similar to those found in the ISDN D-channel.

Like LAPD, LAPF is also an HDLC-based protocol, but is also capable of supporting user-to-user as well as user-to-network connections. LAPF is capable of establishing multiple logical links which are used for the different information streams generated by the layers above. One of these links is reserved for control and management of the conference, and passes protocol messages up through the SCF and is eventually terminated by the GCC. Each data link may be assigned one of four priorities based on the type of information it carries.

Above the data-link layer, the network layer SCF is responsible for the establishment and release of point-to-point network connections. It is only used for control and management messages, and is null for the communication of application information. The top layer in the T.123 stack is the transport layer according to Recommendation X.224 class 0. As both high-priority interactive data such as mouse and pointer movements, and low-priority bulk data such as image files, must share the services of the data-link layer, the main function of the transport layer is to provide segmentation of large amounts of lower priority data in order to provide an acceptable frame latency for the higher priority data.

REFERENCES

1. ITU-T (1992) Support of Data Terminal Equipment with V-Series Type Interfaces by an Integrated Services Digital Network. Recommendation V.110 (I.463).
2. ITU-T (1992) Support by an ISDN of Data Terminal Equipment with V-Series Type Interfaces with Provision for Statistical Multiplexing. Recommendation V.120 (I.465).

---

3. ITU-T (1992) Support of Packet Mode Terminal Equipment by an ISDN. Recommendation X.31.
4. ITU-T (1992) Interface Between DTE and DCE for Terminals Operating in the Packet Mode and Accessing a Packet Switched PDN through a Public Switched Telephone Network or an ISDN or a Circuit Switched PDN. Recommendation X.32.
5. ITU-T (1993) Support of X.21, X.21bis and X.20bis based data terminal equipment (DTEs) by an Integrated Services Digital Network (ISDN). Recommendation X.30.
6. ITU-T (1988) Applicability of Telematic Protocols and Terminal Characteristics to Computerised Communication Terminals (CCTs). **VII.5**, (Blue Book). Recommendation T.65.
7. ITU-T (1993) Packet Switched Signalling System between Public Networks providing Data Transmission Services. Recommendation X.75.
8. Hendrick, C. Introduction to the Internet Protocols. Can be found on the WWW at http://www.cis.ohio-state.edu/htbin/rfc/hedrick-intro.html.
9. Malis, A. (1992) Multiprotocol Interconnect on X.25 and ISDN in the Packet Mode. RFC 1356, IETF.
10. Simpson, W. (1994) The Point-to-Point Protocol (PPP). RFC 1661, IETF.
11. Slower, K., Lloyd, B., McGregor, G. and Carr, D. (1994) The PPP Multilink Protocol (MP). RFC 1717, IETF.
12. ISO/IEC (1994) Digital Channel Aggregation. CD 13871.
13. ITU-T (1995) Synchronised aggregation of multiple 64 or 56kbps channels. Recommendation H.244.
14. ITU-T (1994) Digital Subscriber Signalling System No.1 (DSS.1) – Signalling specification for frame mode basic call control. Recommendation Q.933.
15. ITU-T (1992) ISDN Data-Link Layer Specification for Frame Mode Bearer Services. Recommendation Q.922.

9

The private branch exchange (PBX) and primary rate access (PRA) ISDN

Voice communication within a large organization is often provided by a private branch exchange (PBX). Similar in function to a local exchange switch found in the public network, the PBX will provide call switching between user's telephones within the customer premises, and also provide them with access to public network connections for external calls. However, unlike a public network switch, the connection between the PBX and a user's telephone need not be based on standards as the public user–network interface now lies between the PBX and the public network and not the PBX and the user's telephone. This has led to the use of alternative protocols that, while offering BRA ISDN-like bearer capabilities, are implemented using different techniques aimed at reducing the cost of the subscriber loop and the user's handset. One of the topics in this chapter is the alternative digital technologies used for PBX user connections.

For a large PBX installation, the connections to the public ISDN should support the number of expected simultaneous external calls, both incoming and outgoing, that can be in existence at any one time. It is therefore desirable to use a physical connection that is capable of delivering a large number of channels between the PBX and the ISDN. Primary rate access (PRA) connections to the public ISDN can deliver 30 individual B-channels in Europe, or 23 B-channels in America, and represent a more cost-effective connection than multiple BRA subscriber. Although PRA ISDN may be used wherever there is a need to deliver many channels between the network and customer premises, it is described in this chapter in the context of the PBX as this is one of its main applications.

9.1 AN INTRODUCTION TO THE PBX

With reference to the ISDN reference model, the PBX is the point at which the public network is terminated, and is therefore a network termination (NT). The NT marks the boundary between the customer premises and the public network, and terminates the customer end of the physical connection to the public network local exchange. However, unlike the simple NT1 discussed in Chapter 4, the PBX provides local switching or concentration between the terminals connected to it and the public network. The PBX therefore belongs to the NT2 type of network termination, and is perhaps one of the most common examples of an NT2 in use today. The PBX connects to the public network through an NT1, typically with PRA interface, that may be found integrated into the PBX.

The PBX is a private switch that provides a range of voice services to a number of users within the customer premises, and provides connections for those users to the public network. It is typically used by organizations with anywhere between 30 users up to several thousand[38].

A major justification for the use of a PBX is cost. A PBX is owned by the customer and resides at the customer premises. Therefore, a call made between two users within the customer premises can be locally switched by the PBX without any need for a connection to the public network, thereby avoiding any call charges from the public network operator. In addition, depending on the demand for external calls, the local users connected to the PBX can share on a demand basis a smaller number of connections to the public network. While this means that a probability exists that a user may attempt an outside call when all external connections are already occupied, the ratio of the number of external connections to internal users can be adjusted according to call traffic patterns such that this probability is kept at an acceptably low level. The number of external connections, and hence cost, is therefore optimized according to usage characteristics.

In many ways, the PBX is similar in functionality to a small local exchange switch used in the public network. However, as the PBX is owned by the customer and resides at the customer premises, it is not part of the public network. As such, many of the internal interfaces of the PBX within the customer premises, for example the connection between the PBX and the telephone handset, are not bound by the ITU-T standards to which public network switches must conform. This results in the following two important PBX features.

- The equivalent of the S interface connection between the terminal and the PBX may use proprietary physical and signalling protocols. The primary reason for such a deviation from ISDN standards is that the requirements of I.430, Q.921 and Q.931 result in a comparatively expensive terminal and linecard solution. Equivalent 2B+D channel capability can be provided over a single-pair cable rather than the two-pair cable required by I.430 by use of a low-cost time compression multiplexing (TCM) technique, thereby allowing wiring costs to be reduced. Furthermore, use of alternative, less complex, data-link and call control protocols can result in significantly reduced

---

[38]For use by small numbers of users, a simplified version of a PBX known as a keysystem may be used, although today the distinction between PBXs and keysystems is becoming increasingly blurred.

processing and memory requirements that permit the design of lower cost terminals and linecards.

- A PBX is not constrained by the services provided by the public network for calls that are local to the PBX, and allows a feature rich set of services to be supported for internal calls. Many of the services offered by public ISDNs have been available on PBXs for many years, although these services have been restricted to users connected directly to the PBX. Now, with the digital signalling capabilities of ISDN, these services can be extended to users connected to public networks. In addition, standardized digital inter-PBX signalling protocols, such as Q.Sig, allow the service features to be provided to users across a group of PBXs interconnected by public ISDNs.

Despite these benefits, there is a penalty in adopting non-standard interfaces and protocols. Both the above features result in PBXs which are mainly proprietary in nature, with manufacturers having designed different implementations that allow only terminals and other peripherals from the same manufacturer to be connected to it. In a large organization, several sites each having their own PBX could only collectively benefit from the services and features provided by the individual PBXs if they were also from the same manufacturer, and interconnected using leased lines through the public network.

In reality, many PBXs are designed with sufficient flexibility as to allow a range of both standard and non-standard interfaces to be offered within the customer premises. The telephone interface is in most cases proprietary as this usually represents the largest number of connections to a PBX, and is therefore the most cost sensitive. However, many businesses already possess analogue Group 3 fax machines and perhaps other data terminal equipment with standard modem connections, and these may be provided for with standard analogue connections. Similarly, standard ISDN connections may also be provided to the user.

A range of different connections may also be found between the PBX and the public network. While the bulk of connections may be provided by high-capacity lines such as PRA ISDN, additional BRA ISDN and analogue connections may also be used to provide additional capacity and backup capability.

9.2 PBX ARCHITECTURE AND OPERATION

Figure 9.1 illustrates the main functional blocks within a PBX and how they are interconnected. Calls enter and exit the PBX either through the **user linecards** from user's telephones, or through the **trunk linecards** that interface to the public network. At the centre of the PBX is a switch matrix which is responsible for making the connections between the appropriate channels of the user linecards and trunk linecards. Inside the PBX, information is conveyed in the usual 64 kbps PCM channels which are time-division multiplexed into higher order serial streams referred to as PCM **highways**. These are similar to the TDM interconnect buses described in Chapter 6, and are used to interconnect the different linecards within the PBX to the central switch. The PCM highways are synchronized through frame synchronization signals to the public network clock.

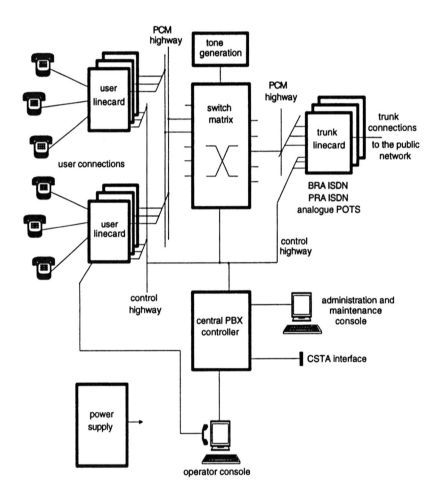

Fig. 9.1 The main functions within a PBX and their interconnection.

A central controller[39] within the PBX provides processing for signalling and switching control functions as well as administration and maintenance functions. The signalling and control information in Fig. 9.1 is conveyed on a separate control highway which links the linecards and switch matrix with the controller. In some cases, highways may combine both the voice PCM channels and control by allocating the control information to specific time-slots within the highway. In large PBX systems, the control functions will typically be implemented by a number of controllers, each allocated to specific tasks within the PBX.

Many call state indications are provided to a user by means of audible tones. With digital signalling, these tones may be generated locally at the user's handset in response to messages received from the PBX. However, an alternative approach is generate the tones centrally within the PBX and distribute them by switching the tone into the relevant channel within the switch. In addition, DTMF is still used extensively as a means of remotely controlling various types of CPE, such as voicemail and voice response systems, from a user's handset. In-band DTMF generation is therefore provided as a standard feature on many PBX systems.

9.2.1 Linecards

A PBX contains two types of linecard: a user linecard that connects a user's telephone or other terminal to the PBX, and a trunk linecard which connects the PBX to the public network. Many of today's PBXs are fully digital and have the capability to provide digital connections similar to ISDN. In many cases a physical layer transceiver is used that delivers full-duplex 2B+D channels similar to a BRA ISDN transceiver, but across a single-pair cable rather than a two pair cable required by I.430. A trunk linecard however, must use standard interfaces to the public network, of which a PBX may offer a number of alternatives, such as POTS, BRA and PRA ISDN, and leased line connections.

The architecture of a PBX user linecard and trunk linecard are similar in many respects. However, an important difference between them lies in the fact that the trunk linecard must act as a regenerator of the public network clock for use within the PBX. This synchronization signal is distributed within the PBX by the PCM highways to the switch and user linecards.

Figure 9.2 illustrates the basic building-block functions of a digital linecard and their interconnection. The physical layer of each subscriber line is terminated by a transceiver with its associated line protection and power feeding circuitry. At the systems interface side, each transceiver will provide a number of B-channels that carry user information, and a D-channel that carries the signalling information.

---

[39]The central controller within the PBX is essentially a computer system. As many of the innovative features in a PBX are implemented in software, the PBX may be considered as a computer with telecommunication specific peripherals such as the switch and linecards.

The B- and D-channels are interfaced to a serial TDM bus on the linecard which also interconnects channels from other transceivers. Programmable control circuitry, either on the transceiver itself or on the linecard, will allocate a transmit and receive time-slot from the TDM bus to each of the B-channels at the transceiver. In many cases, this allocation is performed on a demand basis as calls are established. A backplane interface device then maps the B-channels between the linecard TDM bus and the channels in one of the PCM highways which interconnect other linecards to the main PBX switch.

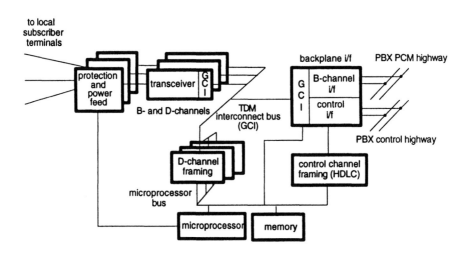

Fig. 9.2 A PBX linecard with on-board D-channel processing.

Call control signalling messages may be processed on the linecard itself, in which case the D-channel of each user connection is terminated on the linecard in a D-channel framing device, such as an HDLC controller. A local microprocessor system is then used to process the call control messages. Further call processing must be coordinated between the linecard and the central PBX controller in order to route the call through the switch to another linecard, initiate a call to the called user, and control any additional services used during the call. These are functions of the central controller in the PBX, with which the linecard must communicate across connections provided by the control highway. It is also typical that HDLC based protocols are used as the basis for this intra-PBX signalling, although the higher level protocols and messages may be proprietary in nature. As well as local call control, the microprocessor system on the linecard is responsible for overall linecard control and supervision functions such as allocation of TDM bus time-slots to transceivers, and monitoring and control of subscriber loop power-feeding functions.

An alternative to the processing of call control functions on the linecard, is to adopt a centralized approach with an additional facility in the PBX that is shared between all linecards. The rationale for doing so is that for a PBX serving a large number of users, a central shared facility is cheaper to implement than the increased power and complexity required of the microprocessor system on each linecard. The impact this has on the linecard is to remove the need for D-channel framing controllers, and to reduce the processing requirements of the microprocessor system to a level that can be met with a significantly lower cost device[40].

In Fig. 9.2, the general circuit interface (GCI) TDM bus (Chapter 6), is used as it integrates the B- and D-channels, and provides a means of monitoring and controlling the transceivers by the microprocessor system on the linecard through the GCI interface, shown here integrated into the backplane controller. The motivation for this is that a single standard bus can be used to interconnect transceivers and other components from different manufacturers, and as a consequence it simplifies the wiring layout of a linecard design.

9.3 PROPRIETARY DIGITAL SUBSCRIBER LOOPS

The variable costs of a PBX system are determined by the number of users to be connected to the PBX, as this determines the number of telephone handsets and linecards that are required. Reducing the cost of the telephone and linecard is therefore of great concern to PBX manufacturers. While it is logical to consider the use of BRA ISDN as a technology for PBX user connections, it is in many ways burdened with complexities that render it an expensive solution. In addition, the need for a standardized subscriber loop interface is not as compelling in the case of a PBX. In most cases a customer will purchase the telephone handsets from the PBX manufacturer as they will be designed specifically to support the features and services provided by the PBX.

Contributing toward the cost reduction of the telephone handset and linecard is the use of proprietary signalling and alternative physical layer transceivers. As was discussed in Chapter 6, implementation of the ISDN D-channel call control protocol requires a microprocessor system with external memory due to the size of software code. A size in excess of 64 kbyte just for basic telephony is typical, while this can double when supplementary services are also included. External ROM and RAM devices represent a significant contribution to the cost of a telephone handset and can only be avoided if the call control protocol is simplified to such

---

[40]Yet another approach that is sometimes used is to combine the two methods such that the D-channel signalling is partially processed on the linecard, for example up to layer 2, with the remaining layer 3 functions processed centrally within the PBX. Each linecard therefore terminates the layer 2 framing, but requires a significantly smaller and lower performance microprocessor system.

an extent that its size will allow it to fit into the memory that can be integrated into a single-chip microcontroller. Although sizes of 32 kbyte ROM and 1 kbyte RAM are found today, these represent the top end of the market for such devices, with sizes of 8 kbyte ROM and 128 byte RAM being more normal. Also, while integrated microprocessor devices that contain HDLC controllers exist today, they tend to be priced at a premium above more general purpose devices that have more common and less complex serial interfaces such as a UART. Use of simple UART framing for example, and a proprietary message format and protocol in the D-channel, will allow all signalling and control functions for a telephone handset to be implemented using a low-cost general purpose single-chip microcontroller.

At the linecard, these changes also result in lower call control processing requirements, but where the processing is performed locally on the linecard, the need to service multiple lines and provide a control channel to the PBX central controller inevitably requires memory sizes and hardware interfaces that dictate the use of a microprocessor system rather than a single-chip microcontroller.

Further cost reductions can be made using an alternative to the I.430 S transceiver. In most cases within an office environment, a simple point-to-point connection is sufficient. In addition, full-duplex digital communications can be achieved without expensive echo-cancellation technology on a single-pair cable through the use of a time compression multiplexing techniques. The following section discusses proprietary transmission schemes for PBX subscriber loop applications that provide similar services to BRA ISDN.

9.4 TIME COMPRESSION MULTIPLEXING

An alternative technique that achieves full duplex transmission without the need for echo cancellation is known as time compression multiplexing (TCM). The idea is that the time period taken to transmit a frame containing the interface channels is split into two halves. During the first half the entire frame is transmitted in one direction, say from the PBX to the telephone, while in the second half the direction of the line is reversed and the frame is transmitted in the opposite direction from the telephone to the PBX. If we assume that the frame contains 2B+D channels, which is equivalent to a full duplex data rate of 144 kbps, then in order to transmit the frame in just half the normal time period requires a line transmission rate of 2 × 144 kbps, or 288 kbps. In reality an even higher bit rate than this must be used to allow time for the transceivers at each end of the subscriber loop to change from being a transmitter to a receiver at one end, and from a receiver to a transmitter at the other end. Therefore, as the transmit and receive signals at a transceiver are uniquely separated in time, there is no need for a hybrid or echo cancellation circuit to provide separation. TCM is also known as **burst-mode** transmission, or **ping-pong** transmission in reference to the continuous mode change from transmit to receive employed by a transceiver.

A disadvantage with TCM is that a line rate of more than twice the channel bit rate must be used, which decreases the effective transmission range for the same line code, cable type and error rate. Despite this, distances in excess of 1 km can typically be achieved using TCM, although the majority of PBX subscriber loops lie within a range of 500 m. Furthermore, in comparison with the BRA ISDN S interface, TCM transceivers can only support point-to-point wiring configurations due to timing complexities at the line interface, and not the point-to-multipoint configurations possible with S transceivers. However, the TCM technique was used in early ISDN U interface transceivers before echo cancellation transceivers were developed for this application and before U interface standards existed.

9.4.1 Example: The TP3401 DASL

The National Semiconductor TP3401 digital adapter for subscriber loops (DASL™)[1], is a TCM transceiver designed for PBX applications. It provides a 2B+D channel interface between the telephone and PBX using a line rate of 384 kbaud and an AMI (alternate mark inversion) line code similar to the BRA ISDN S interface. On 24 AWG cable the DASL has a range around 1.8 km.

9.4.1.1 TCM and framing

The TCM transmission method is illustrated in Fig. 9.3. To coordinate the transmissions from either side of the subscriber loop, the PBX transceiver is designated as the master and is responsible for initiating the transmit–receive cycle, while the terminal end acts as the slave, transmitting its information only after it has received information from the master. Making the PBX transceiver the master also simplifies timing synchronization with the public network clock.

Each transmission burst from the master or slave transceiver consists of two consecutive 2B+D frames combined together with a start bit. As two frames of 2B+D information are transferred each transmit and receive cycle, the cycle repetition rate is once every 250 μs as opposed to 125 μs. The diagram in Fig. 9.3 shows the line data as seen at the PBX (master) end of the subscriber loop, and at the terminal (slave) end. During a master transmit burst, the master transceiver will have its receiver input disabled such that the transmit signal does not upset any of the receiver's adaptive circuitry. Due to the finite time taken for a signal to propagate down a cable, the master burst starts to arrive at the receiver at some time t_d later, depending on the length of the connection between the PBX and the terminal. The receiver at the slave monitors the line for the start bit which allows it to synchronize its operation to receive the following 36 bits, after which it waits for a period of time, known as the **guard time**, before disabling its receiver and transmitting its burst back to the master. Meanwhile, having finished its transmission, the master disables its transmitter and waits for a period of time, equal to the guard time, to allow the line to settle before enabling its receiver. When the master has received the burst from the slave, the direction of the line is again reversed and the cycle starts again, and is repeated every 250 μs. Therefore, although the transmissions are only unidirectional at a point in time, the line reversal method allows full-duplex communications to be supported. At a line rate of 384 kbaud (with AMI, one bit is transmitted per line symbol), 193 μs is required to transmit the master and slave frames, leaving a budget of 57 μs for the guard time and signal propagation time.

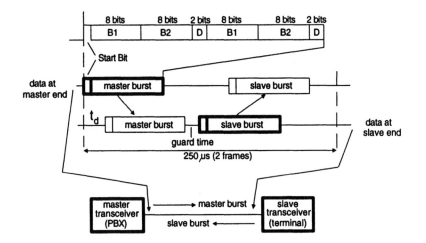

Fig. 9.3 Burst mode (ping-pong) transmissions as used in the DASL.

9.4.1.2 Transceiver operation

A block diagram of the DASL transceiver is shown in Fig. 9.4. In the transmit path, B- and D-channel data is taken from the system interface and formatted according to the frame shown in Fig. 9.3. The frame less the start bit is then scrambled before being encoded using an AMI scheme, which is the same encoding scheme used in the I.430 S transceiver. Scrambling is performed to ensure that an adequate number of '0' symbols are present within the output data stream. When encoded as an AMI symbol, the line signal will contain a transition where the '0' occurs, thus making it easier for the receiver to recover clocking information from it. In addition, AMI encoding results in zero DC component in the line signal which improves its transmission characteristics. Rather than

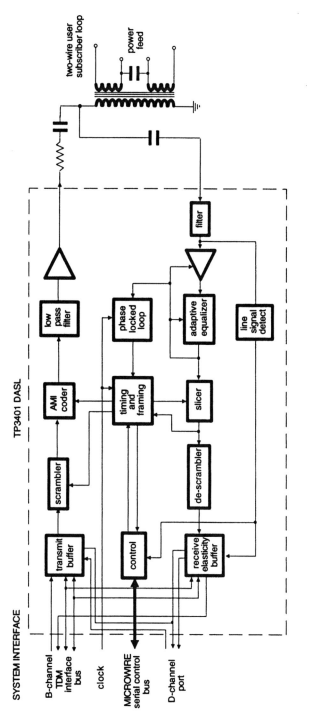

Fig. 9.4 Block diagram of the TP3401 digital adapter for subscriber loops (DASL).

present block-shaped pulses to the line, the AMI pulse is shaped through a filter that has the effect of reducing the higher order frequency components in the line signal. This reduces the amount of radio frequency energy emitted, reduces the amount of crosstalk, and reduces inter-symbol interference (ISI). The transmit signal is coupled to the subscriber line through a transformer as shown.

In the receive path, the line signal is bandlimited by a filter to minimize noise. Pulse distortion is corrected for by an equalizer which adapts itself to a range of cable characteristics. The corrected receive signal is used as an input to a phase-locked loop (PLL) which then recovers the timing information from the signal. This timing information is used to derive the optimum sampling instant for the received symbols which are then converted from AMI to a normal NRZ data stream and unscrambled before being taken to the system interface.

9.4.1.3 Synchronization

As in the ISDN, it is necessary that synchronization with the public network $125\,\mu s$ timing be maintained. At the master transceiver, a master clock synchronized to the network timing is input at the system interface as part of the serial TDM bus, and the frame synchronization signal which is synchronized to the $125\,\mu s$ public network clock is used to initiate the transmit bursts. At the slave transceiver, the PLL will recover timing from the incoming signal and generate a clock signal which is used to recover the received symbols as well as for transmitting frames back to the master. In addition, the start bit in the received frame is detected and used to generate a local frame synchronization signal that it synchronizes to the locally generated clock. The slave transceiver outputs the frame synchronization and clock signals at its system interface to allow other devices connected to it to also maintain synchronization with the public network clock.

Back at the master, the PLL is also used to recover timing from the received signal. However, it will be delayed relative to the master transmissions owing to the round-trip propagation delay of data down the connection from the PBX to the terminal and back again, and is compensated for by the addition of further delay to the next frame synchronization pulse at the system interface by an elasticity buffer in the receiver.

9.4.1.4 System and control interface

The system interface provides a simple TDM interface port for the B-channels, and a separate port for the D-channel data that allows direct connection of a framing device, such as an HDLC controller or UART, without the need for time-slot control. An input signal to the master transceiver is provided that allows the transmissions of a number of DASLs to be synchronized such that they occur at the same time. This feature is used to reduce the crosstalk on subscriber loops that may be bundled together in a single cable. By synchronizing the master transmissions of the transceivers that drive these user connections it is ensured that all their receivers are turned off during transmissions, thus minimizing the crosstalk effect where the transmission of one transceiver is coupled into the receiver of another.

Access to the programmable functions of the DASL is provided through a separate MICROWIRE serial control port which allows a microcontroller equipped with a similar interface to write to and read from its internal registers to provide control and status monitoring of the transceiver.

Table 9.1   A selection of burst mode transceivers

Feature	National Semiconductor DASL TP3401	Motorola UDLT I MC145422/6	Motorola UDLT II MC145421/5	Siemens Up0 PEB2095	Toshiba TC35321P/F
line coding	AMI	modified DPSK	modified DPSK	half-baud AMI (50%pulse width)	AMI
line rate	384 kbaud	256 kbaud	512 kbaud	384 kbaud	512 kbaud
channel number/ type	2B+D	1B+2D (D = 8 kbps)	2B+2D	2B+D	2B+D
system interface	TDM separate B- and D- channel ports	TDM separate B- and D- channel ports	TDM separate B- and D- channel ports	IOM-2 (GCI)	TDM separate B- and D -channel ports
control interface	MICROWIRE serial port	hardwire I/O	hardwire I/O	IOM-2 (GCI)	None
range/cable	1.8 km/24 AWG	2.0 km/26 AWG	1.0 km/26 AWG	3.0 km/0.6 mm	600m

9.4.1.5 Activation

When the user connection is inactive, both master and slave transceivers can be placed into a power-down mode in which much of the internal circuitry is in a static mode consuming minimal power. Activation of the subscriber loop consists of bringing the activating transceiver into normal power mode and instructing the transceiver to activate. Both of these actions are performed by controlling bits in the control register of the DASL and cause the activating transceiver to send bursts of scrambled '1's to the line. At the opposite end of the subscriber loop, these transmissions are detected by the Line Signal Detect function which, even though the transceiver may be in low-power mode, will cause a hardware interrupt signal to become active. This will alert an associated microcontroller device to the activation attempt, which will return the transceiver to normal power mode and then activate the transceiver. When synchronization across the subscriber loop is achieved, each transceiver sets an appropriate status bit and alerts the local microcontroller or microprocessor with an interrupt. Total subscriber loop activation time lasts approximately 50 ms.

9.4.1.6 TCM transceiver devices

The DASL is just one of a number of transceivers aimed at providing low-cost digital subscriber loops for PBX applications. Table 9.1 lists a selection of other devices and compares some of their features.

9.5 PRA ISDN

Primary rate access (PRA) subscriber loop connections are capable of delivering either 30 or 23 B-channels between the network and the customer premises equipment depending on whether the PRA connection conforms to either the European 2.048 Mbps digital hierarchy or the 1.544 Mbps North American system. In both cases, the higher data rate means that such connections must be provisioned by specially conditioned four-wire subscriber loops, and unlike BRA ISDN cannot be delivered using the same two-wire PSTN subscriber loop cables for telephony. However, four-wire 1.5 Mbps and 2 Mbps connections have been available for some time within the transmission network, and also as leased line connections to users. Most of the physical layer and framing used by PRA ISDN is therefore based on these types of connections, and because of their availability prior to ISDN, meant that PRA ISDN was often made available by network operators in advance of BRA as ISDN services were being developed and deployed in the public networks.

The reference model for PRA ISDN is identical to that of BRA ISDN (refer to Fig. 2.4) although the standards which govern the physical layers are different. Within the customer premises at the S/T interface, the ITU-T Recommendation is I.431 [2], with the equivalent ETSI standard being ETS 300 011 [3] and ANSI standard being ANSI T1.408 [4]. However, implementation of the U transmission system between the NT and the local exchange, which is considered as part of the network, is dependent on the proximity of the user's premises to the nearest local switch, and the technology available to the network operator. In most cases, copper cables, either shielded twisted-pairs or coaxial cables, provide the U connection which conforms to ITU-T G.703 [5] and G.704 [6] (Fig. 9.5). These are also used as the standards for leased line connections at the corresponding data rates.

The installation of specially conditioned cable is expensive and time consuming because of the need for frequent repeater stages at every 1 to 2 km that must be placed either in manholes or mounted above ground in cabinets or on overhead cable poles. Also, as a four-wire transmission system is used, it must be ensured that all pairs in a cable bundle have the same direction of transmission so as to reduce cross-talk. The cable pairs must also be free from bridge taps. Alternative technologies that are capable of delivering these types of bandwidth to a customer premises without the need for intermediate repeaters or specially conditioned cables are therefore attractive. These are just beginning to be introduced, and may be based on advanced digital subscriber loop technologies such as HDSL (high-bit-rate digital subscriber loop), or optical-fibre-based technologies which are now widely regarded as the most economic means of distribution within the future local network. Some of these newer technologies are discussed in the following final chapter.

Apart from the differences in channel capabilities and wiring requirements, a PRA subscriber loop differs fundamentally from a BRA subscriber loop in the following ways.

- A PRA connection is only point-to-point. The point-to-multipoint connection allowed on a BRA subscriber loop is not provided for.
- The PRA subscriber loop remains permanently activated and cannot be activated/deactivated according to the presence/absence of call activity.
- Power feeding across the PRA user–network interface can only be provided using an additional cable pair. Where it is provided, the direction of the power feed is from the terminal to the NT.

A PRA interface provides a 64 kbps D-channel which is used for the same purpose as for a BRA interface, in other words call control signalling and data communications. However, being four times as fast, it possesses sufficient additional bandwidth to carry the expected higher volume of call control messages that result from controlling up to 23 or 30 simultaneous calls rather than just 2. The same layer 2 and layer 3 D-channel protocols are used, although operation of the layer 2 TEI management can be replaced with a fixed TEI assignment for the point-to-point link.

9.5.1 PRA ISDN channels

A PRA ISDN connection carries either 30 or 23 individual 64 kbps B-channels, each of which can be used in independent simultaneous calls. As many public ISDNs route each B-channel independently through the network, a single call can be used to establish only a 64 kbps end-to-end connection. This has led to the aggregation technique described in Chapter 8 that equalizes the transmission delays through the network on each connection so as to create a single synchronized $n \times 64$ kbps connection.

Today, most of the channel usage on PRA connections are as single B-channels. However, the individual B-channels remain synchronized across a PRA connection, and can therefore be simply combined to give higher bandwidth channels such as H_0 (384 kbps) and H_{11} (1920 kbps or 1536 kbps) between the customer premises and the network. The benefit of this is realized when networks are more widely available that are capable of supporting switched connections at these bandwidths instead of just 64 kbps. Such network services, known as Multi-Rate ISDN, are just becoming available, most notably in America, and are described in more detail later.

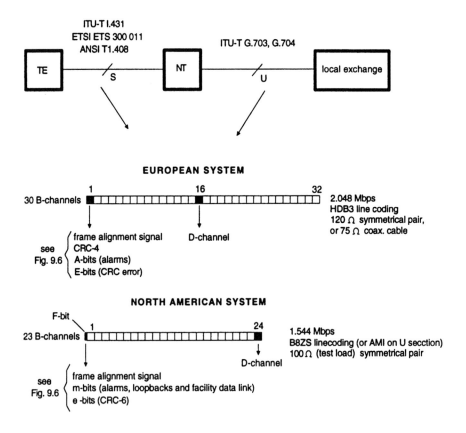

Fig. 9.5 European and North American PRA ISDN systems.

Besides the bearer channels, a 64 kbps D-channel is provided for signalling. Should a D-channel exist on another similar connection, then signalling for all associated subscriber loops may be performed on that D-channel, leaving the redundant D-channel to be used instead as a B-channel. In addition, maintenance information relevant to the operation of the physical link, such as loss of signal alarms and CRC information, are embedded within the bit stream using different techniques according to the 2 Mbps or 1.5 Mbps system framing rules.

Framing of the channels within a PRA multiplex is the same on both sides of the NT for the same system (North American or European). In Europe, the 30 B-channels are transported in a 32 time-slot frame at a rate of 2.048 Mbps. Maintenance information related to the physical link is multiplexed into the first time-slot of a frame in the following way. A multiframe structure of 16 frames is defined in which time-slot 0 of alternate frames carries a frame alignment signal (0011011) that is allocated to the last seven of the bits. The first bit in the same time-slots is assigned to carry one of four bits associated with a four-bit CRC whose value therefore spans eight successive frames within the multiframe. In time-slot 0 of the remaining frames, the third bit is known as the A-bit and used to transfer remote alarm information, while the first bit in the last two alternate frames within the multiframe is used to define two E-bits that convey CRC error indications. This multiplexing arrangement is depicted in Fig. 9.5. Time-slot 16 in every frame may be used as a 64 kbps D-channel.

In North America, the 23 B-channels are carried in a 1.544 kbps 24 time-slot frame that has an additional single F-bit used for framing. The F-bit position is also used to convey additional information known as the m-bits and the e-bits. To do this, a 24 frame multiframe is defined in which the first bit in every fourth frame provides the frame alignment pattern, the first bit in every odd-numbered frame contains an m-bit, while the first bit in the remaining six frames contains an e-bit. Collectively, the m-bits of every multiframe can be used as a 4 kbps channel known as the facility data link to carry messages between the ends of the communications link, and the e-bits are used to carry a six-bit CRC value. Figure 9.5 illustrates the multiplexing arrangement. The 24th time-slot is allocated as a 64 kbps D-channel.

9.5.2 Maintenance functions

Both European and North American systems detect a number of basic alarm conditions, such as loss of signal and loss of layer 1 capability, and where possible communicate the condition back to either the network or the terminal using the status of the A-bits and E-bits (European system) or a repeating sequence of bits in the m-bits (North American system) for as long as the condition prevails. The m-bits may also be used to transfer loopback commands. However, the North American system offers the scope for far greater flexibility in the provision of maintenance features through use of the m-bits as a **facility data link** channel. In this channel, messages are communicated using LAPD UI frames that report the performance of the communication link using parameters that indicate the instances of CRC errors, errored framing, frame synchronization error, line code violations and slip events. The address information within the LAPD frame, that is the SAPI, TEI and C/R bit, is used to indicate the monitored locations, and to which location in the subscriber loop the performance report has to be sent. A similar 4 kbps data-link channel can be established in the European frame using some of the multiframe spare bits, indicated with an 's' in Fig. 9.6a, although their use is dependent on individual national specifications.

frame no.	bits 1 to 8 in the frame							
	1	2	3	4	5	6	7	8
1	C1	0	0	1	1	0	1	1
2	0	1	A	s	s	s	s	s
3	C2	0	0	1	1	0	1	1
4	0	1	A	s	s	s	s	s
5	C3	0	0	1	1	0	1	1
6	1	1	A	s	s	s	s	s
7	C4	0	0	1	1	0	1	1
8	0	1	A	s	s	s	s	s
9	C1	0	0	1	1	0	1	1
10	1	1	A	s	s	s	s	s
11	C2	0	0	1	1	0	1	1
12	1	1	A	s	s	s	s	s
13	C3	0	0	1	1	0	1	1
14	E	1	A	s	s	s	s	s
15	C4	0	0	1	1	0	1	1
16	E	1	A	s	s	s	s	s

A = Remote Alarm bit
C = CRC-4 bits
E = CRC-4 error indcation bits
s = spare bits

(a)

frame number	frame alignment signal	m-bits	e-bits
1	-	m	-
2	-	-	e1
3	-	m	-
4	0	-	-
5	-	m	-
6	-	-	e2
7	-	m	-
8	0	-	-
9	-	m	-
10	-	-	e3
11	-	m	-
12	1	-	-
13	-	m	-
14	-	-	e4
15	-	m	-
16	0	-	-
17	-	m	-
18	-	-	e5
19	-	m	-
20	1	-	-
21	-	m	-
22	-	-	e6
23	-	m	-
24	1	-	-

(b)

Fig. 9.6   The allocation of maintenance bits within PRA multiframes. (a) Allocation of bits within the first time-slot of each frame in the European 2.048 Mbps multiframe; (b) the allocation of F-bits in the 1.544 Mbps North American multiframe.

Error indications at the physical layer are passed on to a management entity protocol through the usual MPH-XX primitives where they can be logged and if necessary corrective action taken. Although under normal operating conditions the physical layer remains permanently activated, a serious failure causing loss of communications will cause a PH-DI primitive to be issued toward the data-link layer, thus allowing operation of the D-channel call control layers to stop communications in a controlled fashion.

9.5.3 Power feeding

It is anticipated that terminal equipment with PRA ISDN interfaces will require a power supply greater than that which can be feasibly delivered across the subscriber loop interfaces between the local exchange and the NT, and will therefore be equipped with their own local power supply. Consequently, power is not made available from the network. However, an NT1 may be simple enough that it can be supplied with power from a remote source. In such cases, a power of up to 7 W may be delivered across the user–network S interface from the terminal to the NT on a separate pair of wires from those used to carry the S interface signal transmission. A voltage of between −20 and −57V may be used.

9.5.4 PRA ISDN devices

Due to the similarities between I.431 and G.703/704, physical layer devices are often designed that are capable of meeting both standards. Traditionally, the solution for these interfaces has been made available in two separate parts consisting of a framing device, and a transceiver for the line interface containing the line drivers, clock recovery and equalization. However, more recently, fully integrated parts that perform both framing and transceiver functions have been developed and brought to market [7], as well as transceivers that perform framing according to both European and North American standards [8].

The I.431 and G.703 standards do not stipulate minimum transmission distances. Instead, an attenuation range of 0 to 6 dB is given for the 2 Mbps system, and 0 to 18 dB for the 1.5 Mbps system. In practice, higher attenuations can be tolerated by most transceivers. Devices from some manufacturers therefore incorporate a variable gain equalizer that can also automatically adjust for short connections (6 dB) that may typically be found at the user–network interface between the terminal and NT1, and for longer connections (43 dB) that will be found between the NT1 and local exchange.

Primary rate framing devices interface to a transceiver on one side (if not integrated within the same device), and a TDM bus on the other that allows connection to other linecard components. A D-channel HDLC controller will be required as well as some microprocessor system to handle alarm indications and other maintenance functions, and call control signalling if this is to be performed on the linecard. An additional serial controller would be required to handle the facility data link in the 1.5 Mbps system. An example of such a device is the PRISM (primary rate interface signalling and maintenance controller – PEB 3035) [9] from Siemens which integrates two fully independent serial communications controllers, one of which could be configured as an HDLC controller for D-channel processing, while the other can be configured to handle the facility data link. On a linecard, the 24 or 30 channels from the framing device will interface to a PCM highway through a backplane controller.

9.5.5 Multi-rate ISDN

Multi-rate ISDN is a service that allows calls to be made that use any number of B-channels up to the maximum provided by the subscriber loop. The availability of this service relies on the network being able to maintain the correct channel ordering and synchronization of the 64 kbps channels that make up the connection. Many networks, having evolved from those designed to carry telephony, are capable of switching and routing only a single 64 kbps channel. However, mainly through changes made to software that controls the allocation of channels within the network, most networks can be upgraded to support multi-rate ISDN services, albeit with a degradation to the service availability. This relieves the customer premises equipment from the burden of making and aggregating multiple calls. Multi-rate ISDN is also known as $n \times 64$ for obvious reasons.

At the subscriber interface, three types of channel allocation scheme can be identified. The first scheme defines a **fixed** allocation of channels. In addition to the normal allocation of a single B-channel for low-bandwidth calls such as telephony, it allows support of multiple H_0 (364 kbps) wideband[41] calls, as well as H_{11} (1536 kbps) and H_{12} (1920 kbps) calls. In the case of an H_0 call, the H_0 channel is allowed only in certain fixed positions within the time-slot multiplex as illustrated in Fig. 9.7a. Should a single 64 kbps channel within any of the defined groups be already allocated as a B-channel, then the remaining 64 kbps channels cannot be allocated to a wideband call. This compromise restricts the flexibility intended for true $n \times 64$ bandwidth selection, but is defined in order to provide backwards compatibility for pre-standardized services.

The second method of channel allocation is known as **floating** allocation. Here, any number of 64 kbps channels may be selected for a call, but the channels must be contiguous as shown in Fig. 9.7b. Of course, for an $n \times 64$ call to be accepted, n

---

[41]Wideband in this context means a bandwidth equivalent to a number of channels n that lies in the range $1 < n <= 32$. Bandwidths greater than 32 channels, or 2.048 Mbps are usually referred to as Broadband, while bandwidths of 64 kbps or less are referred to as Narrowband.

contiguous channels must be available. The last method is known as **flexible** allocation and again allows any number of channels to be selected, but in this case need not be contiguously ordered within the multiplex. As can be seen from Fig. 9.7c, the position of the individual 64 kbps channels that make up the wideband channel can be chosen according to where these channels are available.

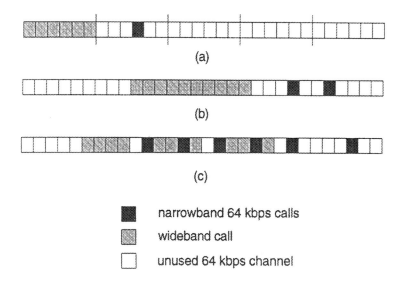

(a)

(b)

(c)

■ narrowband 64 kbps calls

▨ wideband call

☐ unused 64 kbps channel

Fig. 9.7 Different channel allocation schemes in multi-rate ISDN.

In terms of signalling, the Q.931 channel ID information element is defined to allow the position of the individual channels that make up the wideband channel to be identified. During call establishment, a SETUP message containing the channel ID is sent to the network. The channel ID now contains three or four bytes where each bit within these bytes represent the position of one of the 64 kbps channels. If the bit is set to a '1' it is being requested for use as part of the wideband channel. Unused channels, or those currently in use are indicated by a '0'.

9.6 INTER-PBX SIGNALLING

It is often the case that a large organization has several offices or facilities at different locations, and a PBX may be provided at each location to serve the local workforce. Traditionally, each PBX would provide only local access to the features and services it provided. For example, a user making a call to another user connected to the same PBX would need to dial only the PBX extension number, typically a three- or four-digit number. However, when dialling a colleague at another location, first an external trunk connection to the public network must be accessed followed by dialling the full telephone number of the PBX and the colleague's extension. Similarly, services such as call forwarding can only be used locally to forward calls to local extensions rather than to an extension on a remote PBX. Working practices could be significantly improved and made more efficient if, to the user, the PBXs at different locations could be used as if they were a single large PBX with services uniformly available to all users regardless of location. This can only be achieved if signalling between PBXs is supported that allows the services to be controlled amongst the connected PBXs.

Such signalling has been available between PBXs using leased line connections for some time, although the use of proprietary signalling protocols has typically restricted its use to PBXs from the same manufacturer. In 1985, an open standard developed in the UK was introduced to overcome the vendor specific dependencies of inter-PBX signalling. Known as the digital private network signalling system (DPNSS), it was derived from the signalling used in the early pre-standard ISDN service deployed in the UK in the mid 1980s. DPNSS was adopted by several PBX manufacturers thus providing multi-vendor inter-operability. However, due to its lack of conformance with standardized ISDN protocols, it is now being superseded by an international standard known as QSig.

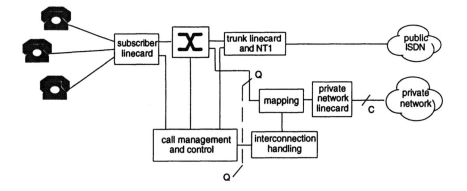

Fig. 9.8 Interface of a PBX to a private telecommunications network.

The definition of QSig standards started in mid 1990 by the ISDN PBX Networking Specification Forum (IPNS) consisting of major European PBX manufacturers such as Alcatel, Siemens, GPT, SAT and Telenorma. The first specifications were produced by the beginning of 1991 and forwarded to the European Computer Manufacturers Association (ECMA) for standardization. Subsequently, the standards have moved under the umbrella of ETSI where they are in the process of becoming full European standards, and then world-wide standards through ISO. QSig gets its name from the Q reference point within a PBX which terminates the inter-PBX signalling and is shown in Fig. 9.8. The physical layer connection to the private telecommunications network (PTN) is designated the reference point C and is typically based on leased lines conforming to G.704 and G.704. QSig is the signalling protocol used and is transmitted in either time-slot 16 or 24 depending on the use of 2 Mbps or 1.5 Mbps connections.

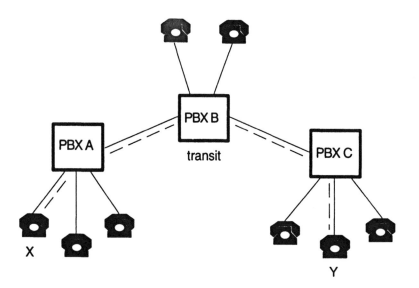

Fig. 9.9 Example of a transit call between PBXs.

QSig is based on ISDN D-channel signalling standards, and uses the same physical and data-link layers. However, the messages and functions of the layer 3 call control are extended in QSig to specifically support features and services amongst a network of interconnected PBXs. Unlike Q.931 which provides signalling between terminal equipment and the local exchange, whose capabilities and functions are quite different, QSig provides signalling between equipment that have similar functionality, in other words between PBXs. One of the differences between QSig and Q.931 is the support by QSig of transit switching. In the example shown in Fig. 9.9, a call from terminal X to terminal Y has to transit PBX B. The PBX B uses the called party number information elements in the SETUP message to determine that the call is for a terminal that is not attached to it, and thus transits the call to **PBX C**.

As QSig is based on ISDN principles, it is capable of supporting not only voice services but also data. However, as a PBX is more traditionally associated with the provision of voice services, the QSig standards refer to such equipment as **private telecommunications network exchanges** (PTNXs) that can provide a range of voice, video and data services.

Like the ISDN standards, QSig exists as a number of different standards documents. Beyond the provision of basic call services [10], PBXs from different manufacturers may not offer the same set of supplementary services. The protocol that provides the foundation on which supplementary services are implemented, known as the **generic functional protocol** [11], recognizes such cases and has the capability to provide a suitable alternative where available. QSig also has the capability to implement manufacturer specific protocols by embedding them within QSig messages. Like DPNSS, the major advantage of QSig lies in its ability to connect PBXs from different manufacturers into a network that appears to the user to deliver a single service across all PBXs regardless of location.

9.6.1 Private telecommunications networks and virtual private networks

Depending on the types of connection used to interconnect PBXs or other types of equipment, the resulting network is known either as a private telecommunications network, or a **virtual private network**. A private telecommunications network (PTN) consists of leased or owned lines that interconnect the nodes within the network. These nodes are customer premises equipment such as PBXs or data networking equipment, and provide the switching and control functions between the network nodes. Implementing such a network can be expensive, and is only worthwhile if the volume of traffic between nodes is sufficiently large that dedicated leased or owned lines are more cost effective. Where this is not the case, networking together customer premises equipment at different sites can be achieved through the public ISDN, creating what is known as a virtual private network (VPN). In this case, the normal ISDN procedures and signalling are use to create the connections between the customer premises equipment as and when required, and the QSig messages are embedded within Q.931 user-to-user signalling messages across the subscriber loop, and within the signalling messages within the network. By using the public switched ISDN, the connections between network nodes can be established as required rather than be permanently in place, and therefore may provide cost advantages to the user.

There are obviously pros and cons for each approach. Apart from the cost issue, a VPN relies on a service from the public network operator which may provide the customer with a lower quality of service and security than could be expected with a PTN.

REFERENCES

1. National Semiconductor (1994) TP3401, TP3402, TP3403 DASL digital adapter for subscriber loops. Device data sheet in *Telecommunications Databook*.
2. ITU-T (1993) Integrated Services Digital Network (ISDN), ISDN User-Network Interfaces, Primary Rate User-Network Interface – Layer 1 Specification. Recommendation I.431.
3. ETSI (1991) Integrated Services Digital Network (ISDN); Primary Rate User-Network Interface – Layer 1 Specification and Test Principles. ETS 300 011.
4. ANSI (1990) Integrated Services Digital Network (ISDN) Primary Rate – Customer Installation of Metallic Interfaces, Layer 1 Specification. T1.408.
5. ITU-T (1991) General Aspects of Digital Transmission Systems; Terminal Equipment, Physical/Electrical Characteristics of Hierarchical Digital Interfaces. Recommendation G.703.
6. ITU-T (1991) General Aspects of Digital Transmission Systems; Terminal Equipment, Synchronous Frame Structures used at Primary and Secondary Hierarchical Levels. Recommendation G.704.
7. Dallas Semiconductor (1995) E1 single-chip transceiver. DS2153Q data sheet.
8. Brooktree (1995) Bt8370 T1/E1 transceiver with physical line interface. Device overview in *Brooktree Product Selection Guide*.
9. Siemens (1994) Data Communications ICs. Overview.
10. ETSI (1994) Private Telecommunications Network (PTN); Inter-exchange signalling protocol. Circuit mode basic services. 2nd edition. ETS 300 172.
11. ETSI (1993) Private Telecommunications Network (PTN); Inter-exchange signalling protocol. Generic functional protocol for the support of supplementary services. ETS 300 239.

10

Subscriber loop technologies beyond the ISDN

In this last chapter, we examine some of the technologies that provide alternative wired subscriber loops for the delivery of digital services to users. The need to deploy new types of subscriber loops will be driven largely by user applications and the services required to support them. As a broad classification, the market for services can be divided into business and residential segments. On the business side, applications such as multi-party videoconferencing, collaborative computing or groupware, training, high-quality image communication for medical, publishing and advertising applications, and high-speed LAN interconnection will drive the need for high-bandwidth switched services. In the residential market too, it is expected that more bandwidth will be needed to support an increase in entertainment services such as interactive television or video on demand (VoD), music on demand, networked games, as well as on-line services such as home banking and home shopping.

A goal in the provisioning of new subscriber loop technologies is to deliver sufficient bandwidth with a single connection to the user for the services required, at the minimum cost of deployment, and in such a way that it gives the network operator increased flexibility in the provisioning and management of services to the user.

10.1 THE MIGRATION FROM COPPER TO OPTICAL FIBRE SUBSCRIBER LOOPS

A typical copper twisted-pair subscriber loop connection may consist of several stages of interconnected cable. Figure 10.1 shows an example of a subscriber loop topology consisting of three stages of cabling. It can be expected that the cables in the feeder network closest to the local exchange will consist of large bundles containing many pairs of fairly small gauge wire, while those in the distribution network between the distribution point and the surface cabinet will contain a smaller number of wire pairs having a larger gauge. The last stretch to the customer premises is the installation cable, and is usually the shortest distance.

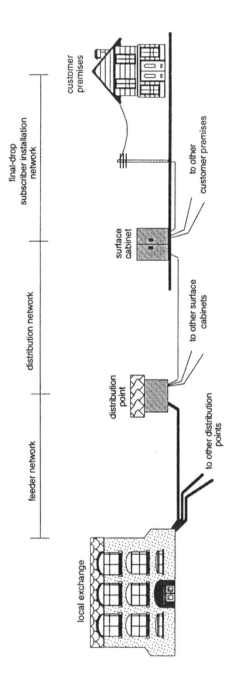

local exchange

feeder network

distribution network

final-drop subscriber installation network

to other distribution points

distribution point

to other surface cabinets

surface cabinet

to other customer premises

customer premises

Today, it is often more practical and economic to use digital multiplexed transmission systems between the local exchange and the distribution point to replace the large bundles of cables, in which case the distribution point itself assumes some of the functionality of the local exchange instead of being a simple wire patch panel. Using digital multiplexed transmission reduces the size of cabling and allows greater distances to be achieved between the exchange and the distribution point, albeit with the aid of repeaters. Newer transmission technologies based on optical fibre are being deployed today because their cost is now comparable to that of copper-based transmission systems, and as well as being able to carry significantly higher bandwidth signals, repeaters are not necessary over the distances required. Such technology is being installed as far as the surface cabinet, and is referred to generically as **fibre to the curb** (FTTC), with only the last installation cable being a copper twisted-pair. While only a minority of large enterprises are today served with optical fibre connections directly to the premises, eventually it will become cost effective to replace the last connection to the home with fibre (known as **fibre to the home**, or FTTH) such that bandwidth is no longer a major obstacle in the delivery of services available to the population at large.

Although optical fibre distribution is widely regarded as being the most cost-effective technology for the distribution of large bandwidth signals within the subscriber loop network, one cannot ignore the massive investment made to date in copper cabling[42], and overestimate the speed with which optical fibre will become pervasive in the local network. Despite its cost effectiveness, it is doubtful that many network operators have the funds available to immediately replace existing copper cabling in the distribution network. Instead, it is expected that optical fibre will be introduced into the distribution network slowly over the next 20 years. In the meantime, techniques will be developed to better use the existing copper cabling for the provision of existing and new higher bandwidth services.

10.2 EXTENDING THE LIFE OF COPPER SUBSCRIBER LOOPS

New digital subscriber loop technologies are continually being developed in order to prolong the use of existing twisted-pair copper subscriber loops. One of the first to be developed after ISDN was the high-bit-rate digital subscriber loop (HDSL) that can be used as an alternative for the provisioning of 1.5 Mbps and 2 Mbps subscriber connections. HDSL technology is essentially an extension of that used in the ISDN U transceiver, and in order to achieve higher bi-directional bandwidth, multiple subscriber loops are used. More recently, as technology has progressively improved over the years, single-line HDSL (S-HDSL) has been

---

[42]It is estimated that in North America, the investment in existing copper cabling is of the order of $100B.

demonstrated over distances of up to 3 km on 0.4 mm diameter cable. In addition, work has started on the definition of a symmetric DSL known as SDSL whose bandwidth can be adapted over a range up to 2 Mbps according to the transmission characteristics of the subscriber loop cable, and which allows co-existance of a passive analogue telephony service on the same cable.

The transmit and receive bandwidth requirements of many applications are asymmetric. For example, video on demand requires a large bandwidth from the network to the user for the video and sound information, and a smaller bandwidth in the return direction toward the network in order to convey control information that allows the user to stop, start, rewind and fast-forward the video in much the same way as a video cassette recorder (VCR). Advantage can be taken of the asymmetrical channel requirements in order to design a system that has adequate range for subscriber loop purposes despite a high bandwidth toward the user for the delivery of video and other services. The asymmetric digital subscriber loop (ADSL) can provide a bandwidth of around 6 Mbps in one direction with a bandwidth of 640 kbps in the reverse direction. Over shorter distances of several hundreds of metres, similar techniques are being used in VDSL (very-high-speed digital subscriber loop) to provide bandwidths of around 50 Mbps in one direction; it is intended for use as the final drop connection to the subscriber's premises in an FTTC system delivering broadband ISDN services.

A summary of different xDSL technologies is shown in Table 10.1 where the data rates and ranges give a broad indication of the capabilities of each system, while more details concerning HDSL and ADSL can be found in the sections that follow.

Table 10.1 Comparison of xDSL technologies

xDSL	Typical range	Data rates
HDSL	3.5 km	1.5 Mbps or 2.0 Mbps
S-HDSL	2.4 km or 2.15 km	1.5 Mbps or 2.0 Mbps
SDSL	Between 5.5 km and 1.5 km	Between 144 kbps and 2 Mbps
ADSL	3.5 km	Up to 6 Mbps downstream and 640 kbps upstream
VDSL	300 m	Up to 51 Mbps downstream and 2.3 Mbps upstream

10.3 THE HIGH-BIT-RATE DIGITAL SUBSCRIBER LOOP (HDSL)

HDSL is a replacement technology for the G.703/704 leased line or PRA ISDN 1.5 Mbps and 2 Mbps digital connections between the local exchange and subscriber premises. The G.703/704 connections usually require specially provisioned twisted-pair or coax cables. Two twisted-pairs are required per connection, one for the transmit and one for the receive direction with each pair located in a cable group having the same direction of transmission and free from any bridged taps. In addition, every 1 to 2 km a repeater is required to boost the signal-to-noise ratio of the signal. Naturally, such special provisions make these connections expensive to install and maintain. However, by using the techniques developed for the BRA ISDN U interface transmission system, and extending them to operate at higher bit rates, HDSL uses normal telephony subscriber loop cables to allow 1.5 Mbps and 2 Mbps connections.

The rationale behind the use of existing ISDN technology was that the U interface transceiver architecture has already been proven, and the design and development of HDSL transceivers would therefore follow shortly after the definition of a standard.

As discussed in Chapter 4, the main limitations that impair full-duplex digital transmissions on subscriber loop cables, are near-end crosstalk (NEXT) and impulse noise. The NEXT coupling between cables is a function of frequency, and adequate performance can be achieved on 3.5 km of 0.4 mm cable (11 500 feet of 26 AWG cable) at frequencies up to around 300 kHz. Furthermore, 2B1Q transceivers achieve a bit rate of four times the available channel bandwidth, which results in a maximum bit rate of 1.2 Mbps. To deliver 1.5 Mbps or 2 Mbps connections therefore requires more than a single subscriber loop. In most cases is not a problem as additional capacity is usually installed as a matter of course. Despite the need for multiple subscriber loops and more expensive transceiver electronics, HDSL is still a competitive technology compared to the provisioning of G.703/704 connections.

The first HDSL definition was developed in North America by Bellcore [1] as a replacement for the standard 1.544 Mbps T1 systems used for leased line and PRA ISDN connections. This was then developed further by the ANSI T1 Committee working group T1E1.4 [2], the same group that defined the 2B1Q BRA ISDN U interface standard T1.601. Using the 2B1Q line code, the system defines two full-duplex subscriber loops, each running at a bit rate of 784 kbps, to give an aggregate 1.568 Mbps capacity. The payload of a 1.5 Mbps G.704 frame maps exactly into the 1.536 Mbps payload of the HDSL system, and leaves 32 kbps for the HDSL overhead bits which include framing synchronization and maintenance functions.

The T1 standard defines operation of HDSL on subscriber loops whose characteristics lie within those defined as the carrier serving area (CSA). The CSA is a general definition used as a specification for subscriber loops that fall within an administration area. Its maximum reach is 12 000 feet (3.66 km) of 24 AWG (0.5 mm) cable that is free from any bridged taps. The distance is reduced for smaller gauge cable, for example 9000 feet (2.74 km) for 26 AWG (0.4 mm), and even further if bridged taps are included.

Standards activities in Europe within ETSI started several years behind those of North America, and therefore were able to use the T1E1.4 system as a basis for European standardization of a core HDSL system that could be applied to a range of applications, including transport of 2 Mbps G.704 frames. In addition, this time delay was sufficient for several manufacturers to demonstrate that the basic HDSL transceiver architecture could be adapted to operate on slightly shorter subscriber loops at speeds in excess of 1.2 Mbps. Consequently, the ETSI Technical Recommendation [3] defines two system configurations, one consisting of three subscriber loops, each at 784 kbps, giving an aggregate of 2.352 Mbps, and another consisting of two subscriber loops, each at 1168 kbps, giving an aggregate of 2.336 Mbps. 2.304 Mbps of this total bandwidth is available to carry payload information, and into which the standard 2.048 Mbps G.704 frame is mapped.

Like the ETSI ISDN U interface specification, transmission performance is specified in terms of dB insertion loss for different types of cable configurations. The basic loss on a single stretch of 0.4 mm cable is defined as 31 dB with noise for the three-pair system, which corresponds to approximately 4 km, and 27 dB with noise for the two-pair system. No loading coils are permitted, and up to two bridged taps are allowed, each having a maximum length of 500m.

The different subscriber loop configurations are shown illustrated in Fig. 10.2. In addition, a single regenerator is permitted on each pair in the HDSL system to extend its range, and power may be delivered equally across the pairs from the LTU to the NTU and regenerator if present.

Due to the multiple subscriber loops used by HDSL systems, each end will contain multiple transceivers as well as a core frame multiplex and demultiplex device that is required to split and combine the data streams between the application interface and the transceivers. Companies that have developed 2B1Q HDSL technology today are Level One, Brooktree, Pairgain and Metalink.

10.3.1 HDSL framing

Figure 10.3 illustrates the framing defined for European HDSL systems. The basic frame consists of four data blocks, 32 overhead bits (marked as 'H'), a 14-bit synchronization word and four optional stuffing bits that appear at the end of the frame. This frame is transmitted on each pair used within the system once every six milliseconds. The synchronization word allows both ends of the subscriber loop to become synchronized during activation of the HDSL system, while the overhead bits contain information for the control of operation and maintenance functions including a CRC value, power supply status, far-end-block-error (FEBE) occurrence, regenerator errors and alarms, an indication of a loss of signal from the far-end application interface, and an embedded operations channel (EOC).

Each of the four data blocks is made up of a number of payload blocks according to the number of subscriber loops used in the system. In the three-pair system, each payload block contains 12 bytes, while in the two-pair system each block contains 18 bytes. In order for the HDSL system to transport some other framed application bit streams, it must first map the application bit stream into

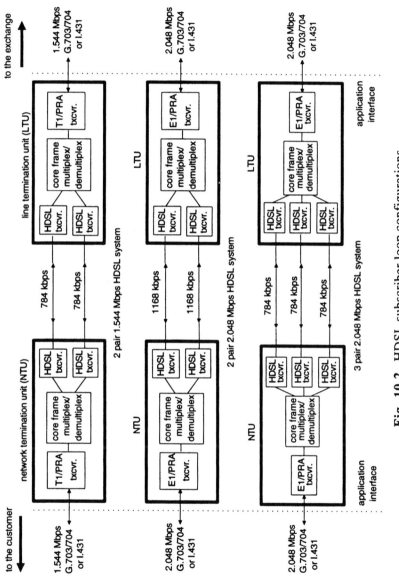

Fig. 10.2 HDSL subscriber loop configurations.

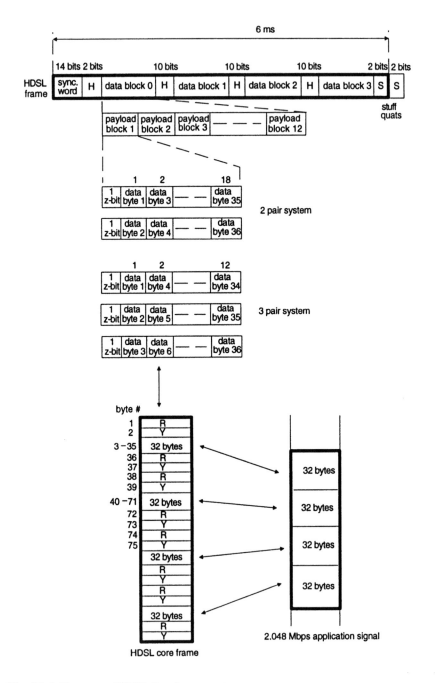

Fig. 10.3 European HDSL framing structure.

the HDSL frame by converting it into an intermediate block referred to as a 'core' frame which is 144 bytes in length. This mapping is also provides a means of rate adaptation between that of the application data stream and the aggregate payload data rate of the HDSL system by filling in unoccupied bytes with null values. Each core frame is then split on a byte-by-byte basis between the payload blocks on each subscriber loop. If, for example, the HDSL system carries a standard 2.048 Mbps payload such as a PRA stream, then successive 32 byte frames are ordered within the HDSL core frames as shown in Fig. 10.3 where the R and Y bytes are not used and set to all '1's.

As well as the H overhead bits, each payload block contains a single Z overhead bit. The first three of these on each subscriber loop are used to identify the pair in the HDSL system to allow multiplexing and demultiplexing of the core frame amongst them. Other Z-bits are reserved for the HDSL system operation as well as transparent transport of application data, and if not used, are set to '1'.

At the end of the HDSL frame, four stuffing bits may be used to provide a DC balance for the line signal. The four stuff bits, or two 2B1Q quats, are always used together such that either both quats are present or none. If no stuff quats are present, then a DC imbalance occurs from the coding of the synchronization word that results in an average +3 symbol offset per frame (the four 2B1Q quats are represented as symbols +3, +1, −1 and −3, see Chapter 4). The remainder of the frame remains balanced through a scrambling process of the frame, except the synchronization word and stuff bits, before being transmitted. Therefore, every two frames the accumulated offset can be compensated by the addition of the two stuff quats each with a value of −3.

10.3.2 Embedded operations channel (EOC)

Each pair in an HDSL system provides 13 bits per frame for use as an embedded operations channel (EOC) in which commands and data can be communicated between the line termination unit (LTU) and network termination unit (NTU) (Fig. 10.2) in order to activate loopbacks, send and notify corrupted CRCs, and read and write control and status registers in the NTU transceivers. In terms of protocol, the LTU is the master and issues commands to the NTU which operates as a slave and acknowledges correct reception of each command or message by echoing it back to the LTU. At the NTU, if the same command is received three times it is executed. Some commands have a latching effect, which means that once it is activated it remains active until another command is issued to deactivate it, and allows several commands to be active simultaneously. Non-latching commands must be persistently sent by the LTU for as long as the action is required, or alternatively a **hold state** message can be sent which has the same effect. In addition to sending commands to the NTU, the EOC protocol defines a procedure that allows the LTU to read from and write to the addressable registers in the NTU.

Each command issued by the LTU is sent in parallel on all pairs in the HDSL system, and each of the transceivers in the NTU responds independently. Such a system was chosen to accommodate EOC signalling for regenerators used for each pair in the HDSL system where necessary to extend its length. In addition, the format of the EOC message contains two address bits that allow the NTU and regenerator to be separately addressed by the LTU. In the downstream direction, that is from the LTU toward the NTU, all messages are repeated by a regenerator, even those which are addressed to the regenerator and not the NTU. In the upstream direction, that is from the NTU toward the LTU, the regenerator again passes EOC messages through unless the regenerator itself has been addressed by a message from the LTU, in which case the upstream EOC message from the NTU is overwritten by that of the regenerator. This is not a problem as NTU was not being addressed by the LTU and will only be returning a hold state message back to the LTU.

10.3.3 Start-up procedure and pair identification

As with all digital transceivers, the start-up procedure in an HDSL system involves synchronization of the receivers to the far-end transmissions. As with the 2B1Q ISDN U transceivers, this process allows the echo cancellers and DFEs (digital feedback equalizers) to train themselves to the characteristics of the subscriber loop so as to provide optimum channel compensation prior to the transmission of application information, thereby achieving similar BER (bit error rate) transmission performance of 1×10^{-7} across specified cable types.

An HDSL system can have up to three pairs, each of which must be activated and their data streams combined to create a single application interface at both ends of the connection. The maximum overall activation time allowed is therefore somewhat longer than an ISDN U transceiver, and is of the order of 28 seconds (ETSI recommendation) and double this if regenerators are used. Splitting the core frame containing the application data stream and combining the HDSL data streams is dependent on the order in which application data bytes are allocated to the HDSL pairs (Fig. 10.3). Pair ordering is achieved using the Z-bits mentioned earlier, that are present in the HDSL frame for each pair in the system. During start-up at the LTU, each pair is assigned a number that represents the order for multiplexing and demultiplexing the core frame. For example, for a three-pair system the numbers 1, 2 and 3 are assigned by setting one of the three Z-bits used to a '1', while in a two-pair system the numbers 1 and 2 are similarly assigned by setting either of the two Z-bits used to a '1'. When a transceiver in the LTU has reached an appropriate state during activation, it starts to transmit its Z-bits.

Meanwhile, each of the transceivers in the NTU are undergoing activation and await reception of the Z-bits from the LTU in order to also be able to assign a corresponding order for the individual pair data streams to its multiplex/demultiplex process. While each pair remains unassigned, the NTU transmits a '1' in each of the Z-bits back to the LTU. Eventually, each of the transceivers at the NTU receives their Z-bits from the LTU. If a valid pattern is received six consecutive times then the assignment is registered and the outgoing Z-bits from each NTU transceiver are set to match those of the incoming Z-bits. Finally, transceivers at the LTU notice the change in Z-bits from all '1's to the assigned pattern, and if received six consecutive times, the core frame multiplex/demultiplex circuitry is configured accordingly.

The functions implemented within the receiver part of HDSL transceivers using DSP, such as echo cancellation and equalization, allow it to determine the noise margin of the received signal. Should the noise margin fall below a level which gives rise to a BER in excess of 1×10^{-7}, then data reception is automatically ceased. In addition, the noise margin value at the NTU can be read by the LTU using the EOC.

10.3.4 Line Coding

2B1Q is not the only line code for use in HDSL systems. During the development of HDSL standards, there was much debate concerning the relative merits of 2B1Q and a transmission technique known as **carrierless amplitude/phase** modulation (CAP) developed by AT&T. Unlike 2B1Q which is a baseband coding scheme that occupies the frequency spectrum down to DC, CAP is more like a modem technology that occupies a passband range of frequencies above DC and allows other signals to occupy non-overlapping lower frequency bands, such as normal telephony.

In addition, CAP demonstrates some transmission performance advantages over 2B1Q and is less demanding in terms of echo canceller design. These advantages stem from the fact that the CAP frequency spectrum avoids the lower frequencies where signals experience the largest distortions due to different delays in the spectral components of the signal. Activation of CAP systems is also typically faster than 2B1Q-based systems.

Despite these benefits, CAP was a new technology at the time the North American standard for HDSL was being defined, and no working systems were available to generate sufficient confidence that CAP should be included as part of the standard. As such, the North American standard defines only 2B1Q as the line code for HDSL. However, due to the later definition of European standards, and the demonstration of CAP as a viable technology in the mean time, it has been included as an annex to the ETSI Technical Recommendation.

The ETSI specification defines a 64-CAP system. CAP is a form of quadrature amplitude modulation (QAM) that is used widely today in voice-band modems and microwave radio systems. However, unlike QAM which modulates the data with a carrier signal, CAP modulates its data by passing it through filters, and thus generates a signal for transmission which contains no carrier component. The 64-CAP system has a constellation of 64 points where each symbol point represents six bits of digital data. Once the CAP-based HDSL system is activated, trellis coding is then also used during the transmission of data to improve the performance of the system against noise. The coding is applied to consecutive groups of 11 bits of the scrambled HDSL frame before being sent as two consecutive CAP symbols at a line rate of 212.36 kbaud. After trellis coding, within each symbol 5.5 bits contain information and 0.5 bit is redundant. Only the two pair subscriber loop configuration is defined for use with CAP.

With increasing advances in DSP techniques, full-rate single-pair HDSL (S-HDSL) systems are feasible today over reduced distances of around 2.4 km, and are currently being developed. In addition, a new DSL system known as symmetric DSL (SDSL) is being defined. During its initialization, the SDSL transceivers can adjust the bandwidth of the system to one of a range of fixed bandwidths according to the transmission quality of the subscriber loop. This would provide the network operator with flexibility to deliver the bandwidth and services that make the most of the available copper subscriber connection.

10.4 THE ASYMMETRIC DIGITAL SUBSCRIBER LOOP (ADSL)

The asymmetric digital subscriber loop (ADSL) has been defined [4] to provide asymmetric bandwidth capabilities across a single subscriber loop connection. The bandwidth in the downstream direction from the network to the subscriber is designed to be substantially higher than that of the upstream direction from the subscriber to the network. This arrangement suits a number of services and applications.

10.4.1 Applications for ADSL

The delivery of entertainment video as a video on demand (VoD) service is one of the services that drove early development of ADSL. Here, the high bandwidth in the downstream direction can support a low number of digitally compressed video channels. In addition, a much lower speed control channel in the upstream direction toward the network provides the viewer with VCR type functions such as stop, play, pause, fast forward and rewind.

VoD is reliant on a suitable standard video compression technology for entertainment video, which determines the bandwidth of the downstream video channels. The ISO Motion Picture Experts Group (MPEG) have developed such standards for the storage and broadcasting of video and audio signals. Unlike the compression for interactive video communications discussed in Chapter 7, the compression and decompression of a film for entertainment requires a significantly higher quality in the decompressed picture, although the compression process need not be real-time but can take place as an 'off-line' conversion and storage prior to transmission. The MPEG standard also includes the compression of high-quality stereo sound. Using MPEG standards, an entertainment video (together with audio soundtrack) can be compressed into a digital data stream that requires a communications channel bandwidth of 1.5 Mbps, while a broadcast television channel would require of the order of 6 Mbps.

ADSL will be used to deliver a range of services in addition to VoD according to customer demand and location. Another popular application which is suited to asymmetric communications is Internet access. The Internet protocol (IP) is a packet-based, connection-oriented protocol that relies on the use of acknowledgment packets to confirm correct receipt of transmitted data packets before more data packets are sent. When a user accesses the World Wide Web (WWW) on the Internet, the volume of data is predominantly in the direction towards the user, and takes the form of WWW pages that frequently contain large amounts of graphical information, or file downloads. In the opposite direction towards the network, there exists mainly acknowledgment packets and control packets that, due to their smaller size, require only a smaller channel bandwidth. However, the upstream bandwidth must not be as small as to cause undue queuing delays to acknowledgment packets, as this may significantly reduce the throughput of packets in the downstream direction towards the user. A ratio of upstream to downstream bandwidths of the order of 10:1 is expected to be needed. In the future, the requirement for downstream bandwidth will undoubtedly increase as the content of Web pages increases through the use of video and sound.

Other services that are expected to be delivered on ADSL are interactive games, home shopping and education.

10.4.2 ADSL systems

As with HDSL, the early standardization work on ADSL has been done primarily in North America by the ANSI T1E1 committee. The channels provided by an ADSL system are described either as simplex or duplex. A simplex channel is a one-way channel, usually from the network to the user, while a duplex channel is bi-directional with the same bandwidth in both directions. The ANSI ADSL standard [5] defines two mandatory classes of ADSL transport together with additional options as to the way the bandwidth can be partitioned. This allows flexibility in provisioning services to customers.

The two mandatory classes of transport that are defined in the ANSI standard must be provided by all equipment. Of these, the highest bandwidth class available over the minimum range can provide a simplex channel of up to 6 Mbps and a 64 kbps duplex control channel. The second mandatory service is provided by a system that is capable of just a single simplex bearer channel of 1.5 Mbps together with a control channel of 16 kbps, but as a consequence of the lower bit rates, it has a longer transmission range.

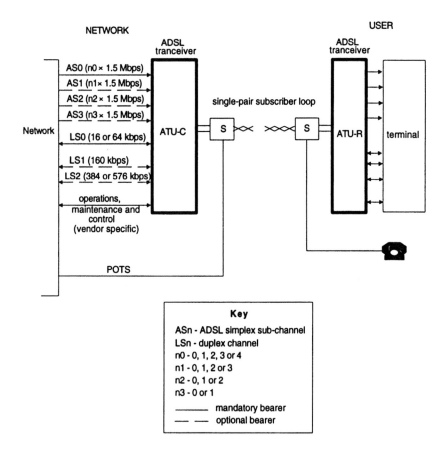

Fig. 10.4 An ADSL system and its capabilities.

Beyond these mandatory capabilities, a range of optional channels are possible depending on the capabilities of the system. As the ANSI standard is defined for North American networks, the mandatory channels are naturally based on the 1.5 Mbps hierarchy, although optional simplex channels may in addition be based on the European 2 Mbps hierarchy (ADSL is also under discussion in ETSI standards committees). Furthermore, an optional duplex ISDN H0 channel (384 kbps) may also be provided as well as an optional duplex ISDN BRA (2B+D) channel (160 kbps). Alternatively, the H0 and BRA channels can be combined to create a single 576 kbps duplex channel. Figure 10.4 illustrates the ADSL system. The available simplex bandwidth is split up into sub-channels AS0 to AS3 and can either be used as a single large channel, or optionally an assortment of smaller channels depending on application requirements and system capabilities. The duplex channel LS0 is the control channel, while LS1 and LS2 are the optional duplex channels. An embedded operations channel (EOC) is also provided for operations and maintenance. Technology in this area has progressed rapidly during the last five years, and continues to do so, with the result that some manufacturers' ADSL systems are capable of providing simplex channels of up to 8 Mbps, with duplex channels of 640 kbps.

Early ADSL systems used frequency division multiplexing (FDM) in order to separate upstream and downstream channels, as illustrated in Fig. 10.5a. This has the result that the NEXT (near-end cross talk), which for example is a major impairment in the ISDN U transmission system, is significantly reduced[43], thereby allowing a greater transmission range to be achieved for the higher bandwidth capability. In addition, the design of the ADSL receiver is simplified as it no longer needs to include echo cancellation.

Provided the size of the upstream channel remains small, the upstream and downstream channels can remain frequency separated. However, in ADSL systems that require a larger upstream channel in order to also accommodate higher bandwidth duplex services such as videoconferencing, maintaining frequency separation of upstream and downstream channels has the effect of pushing the downstream channels higher up in frequency where they suffer a higher attenuation by the subscriber loop, thus reducing the range of the system. In order to maintain an adequate range, the frequency of the downstream channel is dropped down as shown in Fig. 10.5b such that it overlaps the upstream channel, with the consequence that echo cancellation is required in the ADSL receiver in order to provide adequate separation of upstream and down stream channels.

Like other digital subscriber loop technologies, impulse noise can significantly interfere with the correct operation of an ADSL system, and for which forward error correction (FEC) techniques and data interleaving are employed to provide it with resilience against such disturbances. However, a disadvantage of such

---

[43]The performance of ADSL systems is mainly constrained by FEXT and impulse noise.

measures is that they require the data stream to pass through buffers which add to the communications delay across the ADSL system.

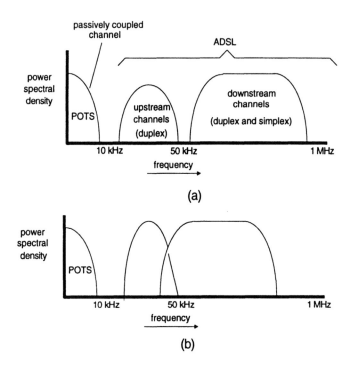

Fig. 10.5 Channel frequency allocation in ADSL. (a) Frequency separation of upstream and downstream channels; (b) overlap using echo cancellation.

10.4.3 Discrete multitone (DMT) line coding

Like HDSL, several line codes were considered during the development of the ANSI ADSL standard, the two most prominent being CAP and DMT, or discrete multitone. Although DMT has also been around for some time, it is only until recently that it has become practical for subscriber loop applications through the development of extremely powerful DSP devices.

DMT applied to ADSL divides the available bandwidth of a subscriber loop into 256 sub-channels with a spacing of around 4 kHz. During initialization of the ADSL link, the transmission capability of each sub-channel is evaluated. As subscriber loop noise and other impairments are frequency dependent, the transmission capability in some sub-channels will be better than others, and by being able to detect these impairments allows the DMT transceivers to allocate a higher number of bits per symbol in sub-channels with a high capability, while allocating a lower transmission load to sub-channels with a lower capability. Sub-channels with a capability that fall below a minimum are not used. Therefore, by splitting the available bandwidth up into smaller sub-channels, DMT isolates those areas of the frequency spectrum in which transmission performance is poor, for example due to RFI interference, and allocates a transmission load of between one and eleven bits per symbol to those sub-channels in accordance with their determined transmission capability. This technique is illustrated in Fig. 10.6. Due to its multi-channel ability, DMT is also known as a **multi-carrier** system.

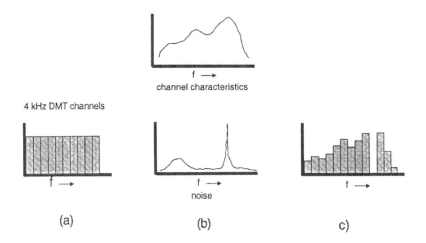

Fig. 10.6 DMT allocates more bits per frequency channel where the signal-to-noise ratio is higher. (a) DMT splits the available frequency range into fixed size (4kHz) channels. (b) The subscriber loop characteristics and noise then determine the allocation of bits per channel as shown in (c).

In 1993, the performance of prototype systems for DMT and CAP was evaluated by Bellcore, the results of which demonstrated a superior capability by DMT and an ability to deliver a 6 Mbps simplex channel over distances of 3.5 km (12 000 feet). Consequently, DMT is the chosen line code for the ANSI specification for ADSL [5], although both CAP and DMT ADSL equipment can be found in the market today for both trials and early service deployment.

10.4.4 ADSL and POTS

The ANSI draft standard defines use of a line coding that is a modulation technique which allows a baseband signal, such as telephony, to co-exist with the ADSL signal. This is an important feature, particularly in residential markets, as POTS will continue to be in use for many years yet, and in most countries is provisioned as a lifeline service that is available when mains power fails. To allow ADSL and POTS to co-exist in practice requires the addition of an ADSL–POTS splitter (the box marked S in Fig. 10.4). This POTS splitter is essentially a series of protection circuits and filters that combine and separate the POTS signal from the ADSL signal, and prevent noise that could be detrimental to the ADSL system, such dial pulses and ringing voltages, from interfering with it. Similarly, the POTS splitter prevents the ADSL signal from creating voice-band interference in the POTS system. The splitter is designed such that even if the ADSL system fails, then a basic telephony service is still available to the user.

An indirect benefit of the development of HDSL and ADSL technologies is that it has lead to a significantly better understanding of the nature and environment of the installed copper subscriber loop network, thus allowing technology to be applied to overcome the problems of noise and distortion. In addition, as the copper feeder and distribution networks age, and are eventually replaced with more modern optical fibre systems, the length of copper in the subscriber loop will be reduced to that needed for the last drop to the customer, which is typically a length of only several hundred metres or less. This reduction in length increases the available bandwidth, which together with the advances in transceiver technology, permit ever increasing data throughputs to be achieved. As an example, it has been demonstrated that asymmetric DSL techniques can be applied to deliver broadband ISDN rates of 52 Mbps from the network to the user, and are currently being standardized under the name VDSL (very-high-speed digital subscriber loop).

10.5 HYBRID FIBRE–COAX ACCESS NETWORKS

For many years, broadcast television has been delivered by networks based on coaxial cable using analogue transmission techniques as an alternative to terrestrial transmissions. Despite being more expensive than terrestrial broadcast transmitters, cable television networks are looked on today in a favourable light due to their independence from an already overcrowded frequency spectrum. Further more, the implementation of new VoD services that require bi-directional communications is cost prohibitive for terrestrial transmission based user terminal equipment such as television sets and set-top boxes, and would require large amounts of frequency spectrum that are not readily available, particularly in densely populated areas.

Because the service provided by a traditional cable television network is a broadcast service, its network infrastructure is designed to provide a one-way path from the source of the signal, referred to as the **head-end**, and the subscriber. The signal repeaters that are used at intervals to boost the signal strength are therefore typically unidirectional. Also, no switching is involved as the analogue carrier signal, which is modulated with multiple channels, is broadcast to all subscribers who then tune either their television set or set-top box to select the desired channel. Existing coax-cable network infrastructures therefore require upgrading to provide switched bi-directional capabilities in order to deliver VoD and other interactive services.

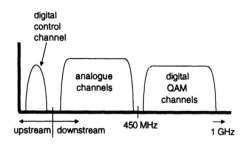

Fig. 10.7 The frequency allocation in new cable television networks.

Firstly, the repeaters in the network must be upgraded to accommodate bi-directional communications. Secondly, the uptake of the new VoD services will be gradual, with many subscribers remaining with the normal analogue cable television service. Both the old analogue cable television and the new digital VoD services must therefore be capable of co-existing on the same cable network infrastructure. This is achieved using a QAM technique that frequency multiplexes the VoD channels into a frequency band above those of the analogue television. At the lower end of the frequency spectrum, a digital back channel toward the network is created that allows subscribers to select and control the VoD and other services using signalling protocols. A typical frequency allocation is illustrated in Fig. 10.7. Thirdly, the VoD service available to subscribers is a switched service rather than broadcast. Each subscriber will be receiving and controlling his or her own movie selection or interacting with some service. Therefore, each subscriber is addressable such that the set-top box that interfaces between the television set and the cable network can identify the correct digital channel that has been allocated to that subscriber. The switched nature of a VoD service also requires that the cable network is now equipped with a switching infrastructure that allows the source of the video signal from a video server within the network to be switched to the correct digital channel for a particular subscriber.

Figure 10.8 illustrates the type of network that is being deployed to meet these service requirements. As can be seen, the digital part of the core switching and transmission network employs optical fibre transmission technologies, while the subscriber network combines both the old analogue broadcast services as well as the new digital VoD services on coaxial cables to subscriber premises. As many as 500 subscribers can be accommodated on each coax cable segment. The mixture of both optical fibre and coaxial cable has led to these types of network being known generically as **hybrid fibre–coax** (HFC) networks. In most cases, the digital video information will be stored in the video server as MPEG encoded data files. An ATM (asynchronous transfer mode) switching network will then transport and switch the MPEG stream as fixed size ATM cells across the optical fibre network until it reaches a location in the vicinity of the subscribers. There, the MPEG stream is either recovered from the ATM cells and QAM encoded in a digital channel toward the subscriber, or the ATM cell stream itself is QAM encoded and sent directly to the subscriber. A benefit of leaving the MPEG stream as ATM cells in the subscriber network is that cells from different streams can be multiplexed into a single channel, although doing so requires the ATM decoding to be performed in the set-top box.

A major difference between the new VoD and the old broadcast services is that VoD requires a back channel towards the network through which the users can select and control the delivery of the video to their television set through the set-top box. The bandwidth of this control channel is relatively small, and rather than allocate a part of the available bandwidth to each subscriber connected to the coax cable, all users share the same bandwidth which is then accessed with a multiple-access protocol that allows each set-top box to gain access to the entire bandwidth of the channel for a period of time before freeing it up to give others an opportunity to access it.

With existing broadband cable networks in place, the cable TV network companies are also in a position to be able to deliver broadband services to users. The choice of ADSL or HFC will be determined largely by cost of implementation. Studies by Bellcore have shown that for low service penetration, ADSL has an advantage over HFC due to the high capital costs associated with installing the fibre infrastructure in an HFC network. However, as the service penetration increases, the cost per subscriber in an HFC network reduces as subscriber numbers increase. In an ADSL network, the cost increases almost linearly with the number of subscribers. ADSL is therefore regarded by many as a short term solution to the delivery of high-bandwidth services, while in the longer term these services would migrate to HFC type networks.

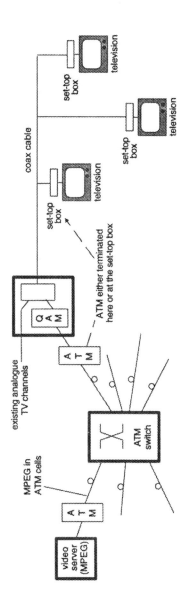

video server (MPEG)

MPEG in ATM cells

ATM

ATM switch

existing analogue TV channels

QAM

ATM

ATM either terminated here or at the set-top box

set-top box

television

coax cable

set-top box

television

set-top box

television

10.6 OPTICAL FIBRE IN THE SUBSCRIBER LOOP

The use of optical fibre in the subscriber loop, otherwise known as fibre in the loop (FITL), can occur in varying degrees from the exchange to the subscriber. The high bandwidth afforded by fibre transmission systems is most cost effective where the highest degree of multiplexing of different subscriber connections and services can be achieved using the fibre transmission system. As well as a simple point-to-point connection, the signal from a single fibre can be broadcast to several destinations by tributary fibres connected to the main fibre through passive optical splitters, and, more than a single signal can be accommodated on the fibre through wavelength multiplexing. These factors allow fibre optic subscriber loops to take on point-to-multipoint and ring architectures as well as the more usual point-to-point connections.

Several system classifications have arisen according to what point in the subscriber loop the optical fibre is terminated. The distinction between many of these classifications is minor, so a description only of the main types of system is given below.

FTTC – fibre to the curb. The optical fibre from the exchange is terminated in a road-side cabinet in the vicinity of the group of customers it will serve. The cabinet is usually mounted on a street curbside, which gives rise to its name. Once individual services have been demultiplexed from the fibre, they are delivered to the customer using existing copper cable connections. Therefore, FTTC typically services customers whose individual bandwidth and service requirements are low, for example residential and small office users. The cabinet contains the fibre termination and multiplexing/demultiplexing equipment along with the necessary interfaces to deliver the copper-based services, such as POTS and ISDN, to the customers.

FTTO – fibre to the office. Here, the optical fibre terminates at the business premises, usually in the basement of a building or large office block. The bandwidth requirements of a business user are higher than that of the residential or small office subscriber, and the fibre is used in this case to deliver a range of flexible services and higher bandwidth connections. Inside the building, services can be distributed from the termination using whatever connections are available. For multiple offices in a large leased office block, this approach is attractive as the termination can be designed with flexibility to allow the requirements of new tenants, or even the changing requirements of existing tenants, to be easily met through simple reconfiguration of the large bandwidth available from the fibre, rather than having to install new subscriber loops. The fibre termination equipment will therefore contain cross-connect devices that allow easy reconfiguration.

FTTH – fibre to the home. As its name suggests, the optical fibre terminates directly at the home, and is able to offer a range of services to individual customers. For each residential customer to be supplied with an individual fibre connection to the exchange is currently too cost prohibitive, not so much because of the cost of the fibre, but mainly due to the cost of the opto-electronic interfaces that must be provided both in the exchange and at the residence. Instead, a single fibre is used to service a group of customers, with the fibre being passively split to provide individual drops to termination units at each customer.

Although much of the justification for the use of optical fibre in the subscriber loop has been for the delivery of broadband services, it has been demonstrated that FTTH can be made cost effective even for delivery of just basic POTS. An example of such a system is TPON (telephony over passive optical networks), which has been pioneered by British Telecom and is today being trialled in several countries.

The basic TPON network consists of a single fibre from the local exchange which is then fanned out using optical splitters at two different points creating a total of 128 connections as shown in Fig. 10.9, each of which may be terminated at the customer premises. A TDM signal is broadcast from the exchange to each of the end terminations which access the assigned time-slots within the TDM frame. Each telephony channel consists of the usual 64 kbps speech PCM signal together with an 8 kbps signalling channel. These two are derived from the standard 2 Mbps G.703/G.704 interfaces that are used to interconnect the local exchange to the TPON exchange termination. The 30 PCM channels are demultiplexed and interfaced directly between the 2 Mbps interface and the TPON frame, while the 8 kbps signalling for each channel is derived from decoding the messages in the common channel signalling (CCS) within time-slot 16 of the 2 Mbps frame and creating a separate channel-associated signalling (CAS) channel for each of the 30 telephone channels. Each 2 Mbps interface from the local exchange therefore contributes towards a 2.532 Mbps signal $(30 \times (64 + 8))$ kbps plus bandwidth for overhead bits) that in turn is multiplexed with up to eight others, plus some synchronization and control bits, to form a 20.48 Mbps bit transport system (BTS).

At the customer premises, each network termination is instructed to extract the relevant information from the arriving TDM stream. The 64 kbps PCM channel and its associated 8 kbps signalling arrive contiguously and are converted into a suitable form for interfacing to a telephone. In the opposite direction, each network termination sends its $64 + 8$ kbps signal together with control at a predetermined time that ensures that they all arrive at the exchange termination in a correct order and with a precise timing such that it appears as a continuous TDM stream. In order to determine at what time each network termination should transmit, a continuous ranging process takes place which determines the network termination to exchange termination signal delay and inserts at the network termination an appropriate delay so as to bring the overall loop delay to a fixed value. For a long subscriber loop, only a small amount of compensating delay would be required, while a short loop would need close to the maximum delay. As the delay will vary with temperature and component ageing, the ranging process is carried out continuously.

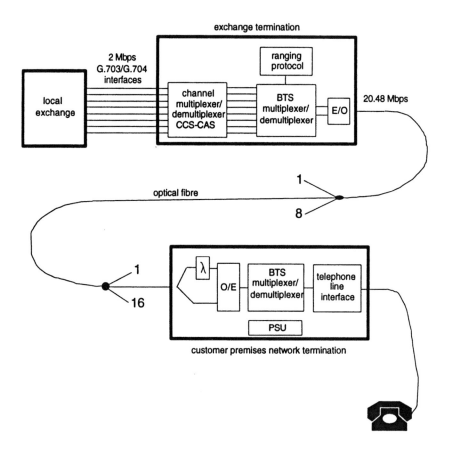

Fig. 10.9 A TPON system.

Key to the success of TPON for residential areas is the cost of the network termination which must provide all the BORSCHT (battery, overvoltage, ringing, supervision, coding, hybrid and test) features typically found in the local exchange. In particular, it must be provided with a local source of power not only to power itself but also the attached telephone, and an emergency battery backup that ensures continued service during power outages. Although these requirements may limit the viability of the provision of single-line telephony using TPON, it may also be applied to the delivery of multiple telephone or ISDN connections to small businesses, or even be used as a form of FTTC where sharing the optical infrastructure across multiple user connections makes TPON more viable. Also, like other optical network infrastructures, TPON has a benefit in that once it is installed, future broadband services may use the same fibre infrastructure by use of wavelength multiplexing and filters at the network termination that filter out all but the desired wavelength signal.

Although originally based on European digital hierarchy data rates, TPON can also interface to North American 1.5 Mbps interfaces.

10.7 THE SYNCHRONOUS DIGITAL HIERARCHY (SDH)

The life expectancy of equipment within the public telecommunications network is of the order of 10 to 15 years, during which time significant technology advances have usually taken place that allow replacement equipment to offer significant benefits in terms of reliability, cost and performance. In the transmission network, technology requirements are for high-speed transmission systems that can achieve high levels of multiplexing with a flexibility that allows many different types of traffic to be easily multiplexed and demultiplexed. For some years now, the transmission systems within telecommunications networks have been upgraded from the copper-based TDM digital systems founded on what is generically known as the plesiochronous digital hierarchy (PDH), to fully synchronous systems based on high-speed optical fibre. These latest systems use a transmission protocol and framing known as the synchronous digital hierarchy (SDH), that operate at data rates of up to 2.4 Gbps.

New subscriber network technologies such as FITL effectively extend a part of the transmission network into the subscriber network, and SDH is a good example of this. Deployment of SDH in the subscriber loop is particularly beneficial in metropolitan areas, as its flexibility and simplicity in multiplexing data streams of different rates allow it to accommodate a diverse range of services demanded by today's businesses. In addition, the availability of such high-bandwidth connections close to the customer facilitates easier deployment of broadband ISDN in the longer term.

Before describing some of the features of SDH, it is useful to understand some of the limitations of existing PDH multiplexing systems which SDH attempts to overcome. Firstly, there is the issue of cost and ease of multiplexing and demultiplexing. In a PDH system, multiplexing of several 2 Mbps or 1.5 Mbps frames into higher order multiplexes is performed on a stage-by-stage basis. In order to extract a lower bit-rate stream from a multiplex which is the next level up in the hierarchy requires it to first be completely demultiplexed into its constituent parts. Therefore, to extract a 2 Mbps stream (level 1 PDH) from a 140 Mbps multiplex (level 4 PDH) would require three stages of demultiplexing. The complexity and cost associated with PDH multiplexing therefore make its configuration, once established, rather rigid and difficult to reconfigure.

Secondly, the plesiochronous operation of a PDH network leads to the possibility of frame slips that can cause data to be lost due to the small variation in the clocks distributed amongst local exchanges. The term plesiochronous refers to the fact that each network element, such as a switching exchange, has its own clock source which is loosely synchronized to a master clock through a hierarchy of synchronization control. At the top-most level of the hierarchy, the international level, the individual national clocks are so accurate and finely tuned to one another that any differences that do exist between them do not result in any slippage. This master clock then drives the hierarchy of clocks where synchronization is achieved both with the clocks in the level above and by mutual synchronization with other clocks at the same level. The further down the hierarchy of a particular clock, the greater the inaccuracies between individual clocks are likely to be, thus leading to a small but finite possibility of slippage in the frames transmitted between exchanges. At the bottom of the hierarchy, the 125 µs network clock is generated locally in each exchange, and synchronized to other clocks by signals received from the clocks at the next hierarchical level up. In an SDH network, all network elements receive their timing from a reference clock rather than generating their own, and facilities are provided to allow data streams from a PDH network to interface and be transported by an SDH network.

Within a PDH system there is little or no support for operation, administration and maintenance functions (OA&M), a feature which is critical for the deployment of enhanced management within networks. This is remedied in SDH through provision for the communication of these functions for each part of the architecture that makes up an SDH transmission system, together with a definition of the protocol that allows the information to be exchanged between elements within the network.

10.7.1 The SDH bit rates

SDH was derived from the North American SONET (synchronous optical network) specification developed during the mid-1980s by Bellcore and the ANSI T1X1 committee. However, SONET is based only on the North American digital hierarchy rates, so when the former CCITT also started to formulate standards for a synchronous optical network, it had to consider compatibility with European rates which led to the different hierarchies between SDH and SONET that are listed in Table 10.2 below.

Apart from differences in bit rates, SONET and SDH share a high degree of compatibility with one another, the main differences being in the way the different hierarchical signals are multiplexed. The current ITU-T Recommendations that cover SDH are G.707 [6], G.708 [7] and G.709 [8], while the ANSI SONET standard is defined in T1-105 [9]. The remaining description focuses on SDH.

Table 10.2 Comparison of SDH and SONET bit-rate hierarchies

SONET	Bit rate (Mbps)	SDH
STS-1/OC-1	51.840	
STS-3/OC-3	155.520	STM-1
STS-9/OC-9	466.560	
STS-12/OC-12	622.080	STM-4
STS-18/OC-18	933.120	
STS-24/OC-24	1244.160	
STS-36/OC-36	1866.240	
STS-48/OC-48	2488.240	STM-16

STS – synchronous transport system
OC – optical carrier
STM – synchronous transfer mode

10.7.2 SDH framing

The basic frame format in SDH is known as an STM-1 (synchronous transport module, type 1) and consists of 2430 bytes transmitted every 125 µs that results in a data rate of 155.52 Mbps. The frame is represented pictorially as a matrix of 270 columns of bytes by nine rows, the transmission of which starts at the top left and is scanned on a row-by-row basis until the last byte at the bottom right corner is transmitted. Figure 10.10 illustrates the STM-1 frame structure. As can be seen, it is divided into two parts. In the first nine columns is the section overhead (SOH) information and administrative unit (AU) pointers which are described a bit later, while the remaining 261 columns contain the STM-1 payload. The payload may consist of a number of different types of channels, each of which is encapsulated within a virtual container (VC), whose different bit rates are listed as part of Fig. 10.10. The VC is the information structure which is used to support path layer connections within an SDH system. The **path** is essentially that which exists between the point in the system at which information is inserted and that at which it is extracted from the SDH transport system. The path may be made up of several multiplex sections, which in turn may be made up of several regenerator sections, as illustrated in Fig. 10.15, section 10.7.3.

Owing to different transmission path lengths, at the point in a network where a VC is inserted into the STM-1 payload, the frame synchronization of the signal within the VC may not be coincident with that of the STM-1 frame. Rather than delaying the VC until it is synchronized with the STM-1 frame, a mechanism known as the administrative unit (AU) is employed to allow the beginning of the VC to be anywhere within the payload, provided there is space within the payload for the entire AU to be accommodated, while the AU pointer contains the offset to the start position of the AU. The AU together with the AU pointer are known as an administrative unit group (AUG) which has a fixed phase relationship with respect to the STM-1 frame. For example, Fig. 10.11 illustrates how a VC-4 frame could be carried by an AU-4 within the STM-1 payload, but offset within it such that it is actually transported by two successive STM-1 frames. The AU therefore provides the VC with the ability to float inside the payload according to the synchronization offset between the STM-1 payload and the contents of the VCs which are not in frame synchronization with it. VC-3s and VC-4 are transported directly in the STM-1 payload using AU-3 and AU-4s, while lower order VCs are themselves encapsulated in tributary units (TUs) before being placed in a VC-3 or VC-4.

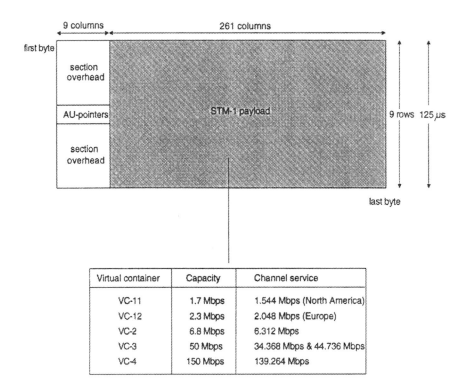

Fig. 10.10 STM-1 frame format and possible virtual containers that can be transported within the STM-1 payload.

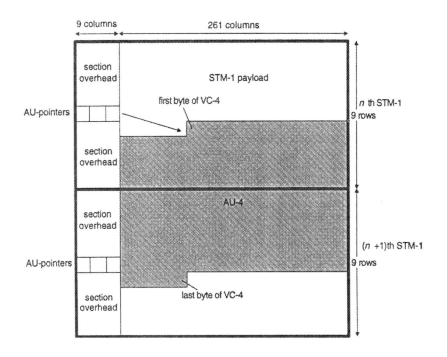

Fig. 10.11 The AU-4 transport within the STM-1 payload.

It can be expected that SDH will co-exist with PDH equipment for many years to come as transmission networks are slowly upgraded. Some data channels feeding into an SDH transmission system will therefore have originated from a PDH or other transmission system where the transmission clock may differ from that used in the SDH network. There is therefore a potential for both underrun and overrun conditions to occur, which are compensated for by bytes within the AU-pointer, known as the **justification** bytes, that are used similar to a bit-stuffing technique. Figure 10.12 illustrates the case where the data rate of the contents of the VC is actually running faster than the nominal STM-1 frame rate and results in use of the AU-pointer justification bytes to carry VC data. The overall effect of the difference in rates is to cause the end of the VC and the beginning of the next VC to be positioned earlier in the STM-1 payload such that it appears to float up the STM frame. This position shift of the AU-4 occurs occasionally in jumps of three bytes and is reflected in a corresponding change in the AU-pointer. Similarly, a slower VC would result in a delay of the VC within the STM frame, giving it the appearance of floating downwards. In this case, the justification bytes are filled with dummy data.

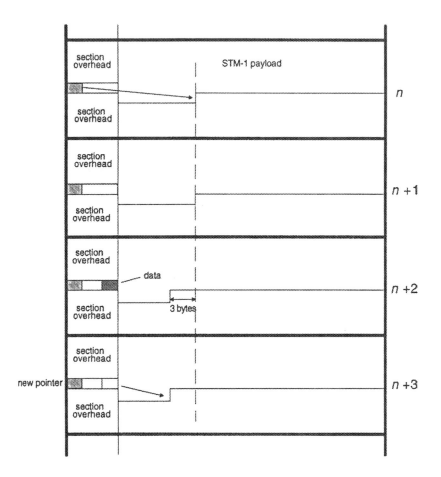

Fig. 10.12 The accommodation of VC frame slip within the STM-1 frame.

The AU-4 exactly fills the payload available in the STM-1 frame and is capable of transporting a channel of slightly less than 140 Mbps. For channels around 50 Mbps, a VC-3 is defined. Like the VC-4, the VC-3 is also capable of floating inside the STM-1 payload within its own AU-3s. Three AU-3s are accommodated in an STM-1 payload through a byte interleaving process within the payload. This means that the order of bytes within the payload follows a sequence that begins with the first byte of the first AU-3 followed by the first byte of the second AU-3, the first byte of the third AU-3, the second byte of the first AU-3, and so on. Each VC-3 is 85 columns wide and the usual nine rows high, and when multiplexed

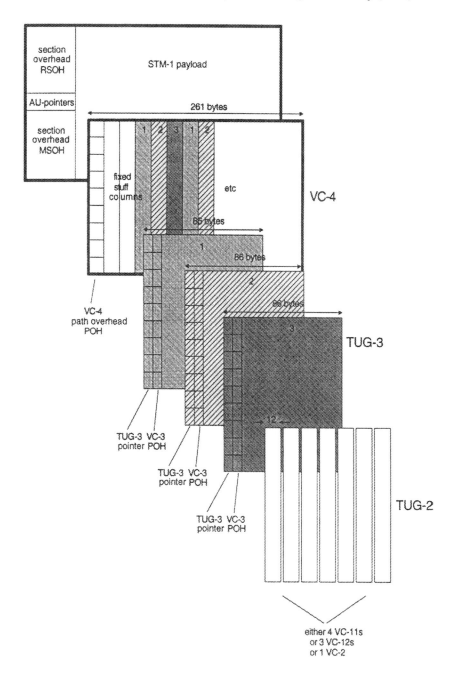

Fig. 10.13 An example of the SDH multiplexing hierarchy.

into the AU-3 is padded with an additional two columns to fill the available capacity such that three VC-3s fit into the STM-1.

As mentioned earlier, lower order VC-1s and VC-2s are multiplexed into higher order VC-3s and VC-4s through tributary units (TUs) and tributary unit groups (TUGs). A TU is similar to an AU in that it also has a pointer that allows a VC to float inside the TU. However, the purpose of the TU is to adapt information streams between the lower order and higher order path layers, whereas the AU adapts between the higher order path layer and the multiplex section layer (Fig. 10.14). One or more TUs occupy a fixed location within a higher order VC payload and collectively make up the TUG whose content may consist of different size TUs, thus allowing the transport system a degree of flexibility in how the bandwidth is used. Like in the AUG, byte interleaving is also used to multiplex the TUs within the TUG. Figure 10.13 illustrates the hierarchical multiplexing of seven TUG-2s into fixed positions within a TUG-3, and then into a VC-4.

Figure 10.14 provides an overview of the multiplexing structure within SDH and details the multiplexing of the lower order VCs into the TUG-2 structure. Note that there are two ways of multiplexing the TUG-2s into higher order structures, either into the TUG-3 as illustrated in Fig. 10.13, or into a VC-3, three of which are then combined into an AUG. Higher order STM structures, STM-4 and STM-16, are created by multiplexing multiple AUGs within its payload and AU-pointer positions using the byte-interleave process described above. The AUGs occupy fixed positions within the higher order STM frame.

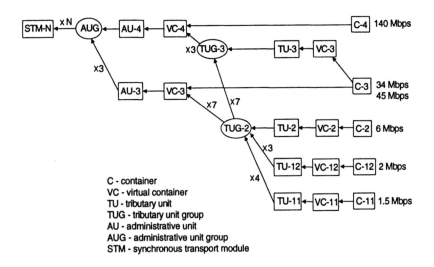

Fig. 10.14 Overview of the SDH multiplexing hierarchy.

10.7.3 The SDH system

The enhanced OA&M capabilities of SDH rely on the overhead built into the SDH frame that permits the communication of OA&M information between the entities that make up an SDH system. Two generic types of equipment can be identified within an SDH system: a multiplexer and regenerator. A multiplexer will be responsible for combining and splitting up different orders of STM signal as well as interfacing non-SDH signals to the SDH, while a regenerator will simply boost the optical signal where transmission over long distances is required. Figure 10.15 illustrates these two types of equipment within an SDH system.

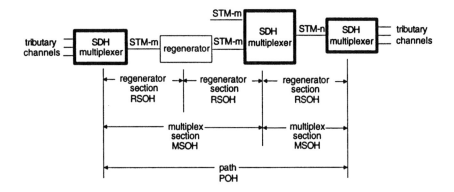

Fig. 10.15 SDH system and overheads.

There are two types of overhead: the path overhead (POH) and section overhead (SOH). As shown in Fig. 10.15, the POH is added to the payload of an STM frame as part of a virtual container, the idea being that the overhead gets added to the payload information at the point at which it is multiplexed into the SDH system, and is processed at the point at which it is demultiplexed. This defines the so called **path** to which the overhead refers. The overhead remains with the payload as it is transmitted across the path, and includes performance monitoring, alarm status indications, and signals for maintenance purposes. The POH for a VC-3 and VC-4 differ slightly from those of a VC-1 and VC-2 in that POH for the higher order structures, VC-3 and VC-4, contain additional multiplex structure indications.

The section overhead (SOH) is part of the SDH frame itself, and is separate from its payload. It contains block framing information, and information for maintenance, performance monitoring and other operational functions. As can be seen in Fig. 10.13, the SOH consists of two parts: the regenerator section overhead (RSOH) and the multiplex section overhead (MSOH). A multiplex section can be made up of several regenerator sections, and the RSOH within an STM frame is generated and terminated at the regenerator equipment within each section. The MSOH, however, passes transparently through the regenerators and is processed at the multiplex equipment. Finally, a path may typically consist of several multiplex sections.

10.7.4 SDH in the subscriber loop

While the details concerning SDH multiplexing may look complicated, the basic principles are simple and allow for a great deal of flexibility in the content of VCs and the way in which information streams can be multiplexed into and demultiplexed out of existing STM structures. SDH technology can provide flexible platforms from which services can be deployed within the subscriber network in areas with a high demand for services, for example in the inner cities and metropolitan areas. In the longer term, SDH will also provide the platform for the delivery of broadband ISDN to customer premises.

A typical method of deploying SDH in the subscriber network is through the use of equipment known as add–drop multiplexers (ADMs) that allow the SDH system to conveniently distribute bandwidth and services to points either at or close to a subscriber's premises. Such networks may be configured as a ring that interconnects each of the ADMs to a local exchange. The flexibility of SDH typically allows any of the tributary signals to be allocated to any of the VCs contained in the STM-1 streams. Rings may be interconnected through a cross-connect multiplexer (CCM). The CCM is an ADM with more than two STM-N ($N = 1$, 4 or 16) ports that allows any of the VCs that arrive at a port to be multiplexed to a VC in any of the outgoing ports, thereby allowing path connectivity between two or more rings. Figure 10.16 provides an illustration of how SDH may be used within the subscriber network.

A disadvantage of such a system is that it relies on a single fibre to distribute signals to all of the ADMs, and a failure at a particular location may render the entire system inoperable. However, protection may be provided through duplication of the optical fibre ring and transmission of VCs such that a receiver may switch between the two received VCs based on analysis of the POH. To minimize the risk of concurrent damage, each fibre may take a different physical route between ADMs.

At the local exchange, the services that are deployed across the subscriber network must be interfaced the public network infrastructure. As competition between rival network operators is introduced through deregulation of the telecommunications market, situations may occur where the subscriber networks are interconnected to another public network belonging to a rival network operator. The V5.1 interface between the subscriber network and the local exchange in Fig. 10.16 is a standard interface that creates a point of connection for any subscriber network to another network for the delivery of services such as POTS, ISDN and low-bandwidth leased lines. The V5.1 interface has been standardized by ETSI [10] as part of a program known as open network provision (ONP) that removes the technical barriers associated with the introduction of competition in the provision of public network services. V5.1, being a European standard, is based on 2 Mbps G.703/G.704 connections that includes a signalling protocol that supports both PSTN and ISDN between the subscriber network and the local exchange.

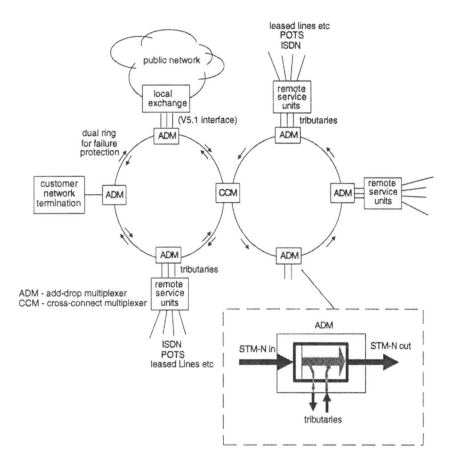

Fig. 10.16 Example of the use of SDH in the subscriber network.

The multiplexing in SDH is hardware intensive, and many of the functions associated with SDH equipment are available today as standard IC building blocks. For example, most common ICs found today are framing devices for particular STM and SONET levels, mapping devices that map different types of tributary into an STM or SONET payload, cross-connect functions, and physical layer transceiver functions for both optical and electrical interfaces. However, although the multiplexing functions are hardware intensive, many of the enhanced management functions embedded within SDH and SONET require software processing, and ICs that process the overhead bytes within frames have ports that allow access to this information by a microprocessor system that is responsible for implementing the management functions.

10.8 BROADBAND ISDN AND ATM

The integration of services in the ISDN has been at the level of the subscriber loop. Inside the network the services are not integrated and may be delivered by independent networks. For example, telephony services may be implemented by a PSTN, while data services are likely to be supported by existing packet data networks. However, through integration at the subscriber loop level, the user's viewpoint is that the services are available from a single user–network interface, regardless of how the services are implemented at the far side of the interface within the network. This type of ISDN is the narrowband ISDN (N-ISDN) which is in operation today and on which this book has focused. It is capable of supporting channel bandwidths at the user–network interface of up to 2 Mbps per subscriber connection.

However, recognizing that in the long term, a future public network will need to provide a range of switched services that require bandwidths in excess of 2 Mbps, but with the same flexibility and integrated interface that the N-ISDN does today, the ITU-T defined a broadband ISDN (B-ISDN) and recommended that asynchronous transfer mode (ATM) should be used as the mechanism to transport and switch any type of information within the B-ISDN. Unlike N-ISDN and existing networks, the B-ISDN implies the used of an entirely new network that is most likely to consist of ATM switching exchanges and SDH transmission networks. Many of the features and characteristics of B-ISDN come directly from ATM.

Many of the advances towards the availability of ATM today can be attributed to the ATM Forum, whose members include network operators, equipment vendors and users from all sectors of the computer and communications industry, working to match user requirements for ATM with the embryonic standards available from the ITU-T. A more dynamic organization than the ITU-T, the ATM Forum has moved quickly to define implementation agreements that precisely define all aspects of both user–network interfaces (UNI), and network–network interfaces (NNI) that are required when interworking between ATM and other types of network as well as between two ATM networks. These implementation agreements are based on the ITU-T standards, but go much further in defining the many aspects that are as yet not defined. In this respect, the ATM Forum works closely with existing standards bodies to ensure that the work it does can benefit existing ATM standards initiatives. The most visible results from the work of the ATM Forum are in the customer premises where ATM is being deployed as an alternative high-speed LAN technology with protocols that allow it to emulate the behaviour of an existing data LAN. However, with many public ATM trials now underway in Europe and North America, the time is fast approaching when public ATM services will also be available. ATM standards are also being developed on a regional basis by ANSI in North America and ETSI in Europe to allow multi-vendor networks to be established for the deployment of public ATM services.

10.8.1 ATM cells

All types of information are mapped into one of five ATM services. Each service transports this information in fixed sized cells consisting of 48 bytes of information and a 5 byte header that contains some addressing information together with other control and maintenance information. The cell format is illustrated in Fig. 10.17 together with a brief explanation of the different fields it contains.

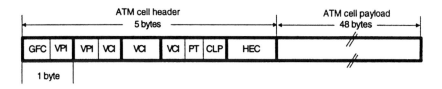

Fig. 10.17 The ATM cell and its header. Notes: GFC – generic flow control. Handles flow control across the UNI. VPI/VCI – virtual path identifier/virtual channel identifier. An address the indicates the link between two nodes in an ATM network. The address is split into a virtual channel (VC) and a virtual path (VP) in order to better manage the routing of cells within the network. The VC is equivalent to the connection itself, while the the VP is a collection of Vcs that share the same path through the network. Therefore, rather than switching based on both VPI and VCI, just the VPI can be used to switch all channels on the same path. PT – payload type. This indicates the type of information carried by the cell.

CLP – cell loss priority. Set by the user to indicate that the cell may be discarded when the network becomes congested. HEC – header error control. Provides error detection and limited error correction for the five byte header of the ATM cell only.

There are two advantages to the use of small fixed size cells in comparison to more traditional packet networks that use long variable-sized packets. First, the small size reduces latency[44] and the buffer requirements for switching elements within the network. However, the size of the cell or packet is always a tradeoff against efficiency, as for a fixed amount of overhead or header bytes, the longer the packet or cell then the greater the efficiency. As bandwidth in the B-ISDN is several orders of magnitude higher than that available when the existing packet networks were designed and built, it is not crucial that efficiency take the highest priority within a B-ISDN, and a 5 byte header is used that gives an efficiency of 90.6% as opposed to a 1000 byte LAPB frame that has an efficiency of 99.4%. The second advantage is that the cell is fixed in size, as opposed to the packet, which is variable. This makes many of the protocol functions in ATM considerably easier to implement in hardware with the consequent advantage of speed. The 48 byte size itself represents a compromise between the computer world where many of today's CPU architectures are based on 32-bit data paths, and the telecommunications world where 64 is a common factor in data rate calculations.

The 'A' in ATM stands for asynchronous, and refers to the transfer method of cells between two points in the network and not the transmission method. In this respect, ATM is similar to a packet network. ATM cells will typically be transferred in a synchronous channel, and due to the ability to statistically multiplex cells within the channel in the same way that packets are in a packet data network, it allows the channel to support multiple virtual channels, each of which may belong to different connections that have been set up across the network. However, unlike packets that must twice undergo three layers of protocol processing as it passes through a switch in the packet network, the ATM cell must only undergo simple header address translation to make sure that the cell is routed to its correct outbound port in the switch. ATM therefore has more in common with frame relay, and is sometimes referred to as **cell relay**.

One of the main advantages of ATM technology is that it is scaleable. The ATM standard is equivalent to a layer 2 protocol which describes the use of 53 byte cells, but is not constrained by being inextricably linked to any physical layer standard with fixed data rates or framing requirements. ATM cells can therefore be used in local networks, PDH or SDH transmission networks, each operating at

---

[44]For example, as reception of information is on a per-cell or per-packet basis, the delay with which any byte within the packet or cell is communicated across the physical link is determined by the length of the packet or cell.

different data rates, although the cells will be rate adapted at each interface to deliver a defined end-to-end connection bandwidth.

10.8.2 The B-ISDN reference model

The B-ISDN protocol reference model is defined in I.121 [11] and, like the N-ISDN reference model, is divided into a user plane (U-plane), a control plane (C-plane) and a management plane (M-plane) as shown in Fig. 10.18. The physical layer and ATM layer are common to all planes, while above the ATM layer, cells will be processed according to whether they contain call or connection control information (C-plane), user information (U-plane), or network management information (M-plane).

Although it adopts a layered approach, the B-ISDN protocol reference model does not strictly adhere to the OSI-RM. However, in this respect, the ATM layer and ATM adaptation layer can be considered equivalent to the OSI-RM layer 2.

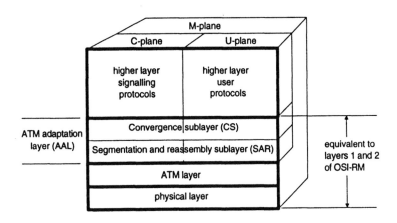

Fig. 10.18 The B-ISDN reference model.

10.8.2.1 B-ISDN physical layer

At the physical layer, the B-ISDN user–network interface as defined in ITU-T Recommendation I.432 [12] has a reference configuration, shown in Fig. 10.19, with broadband elements equivalent to those in N-ISDN, the 'B' suffix or prefix defining them as B-ISDN elements or reference points. One of two possible bit rates is allowed at the S_B and T_B interfaces: either 155 Mbps or 622 Mbps. While optical fibre cabling is preferred for both interfaces, the 155 Mbps interface can also be transmitted across short distances of coaxial cable. The ATM Forum UNI implementation agreement defines a much broader range of physical layer possibilities, including lower rates to accommodate the transfer of ATM cells across SONET and PDH links, and different media such as unshielded twisted-pair wiring.

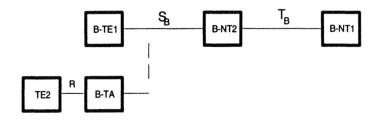

Fig. 10.19 The B-ISDN reference configuration.

ATM cells are transported either as a continuous stream of cells in a cell-based format, or as cells in an SDH- or SONET-based frame structure. In the cell-based format, the continuous stream of cells may be made up of ATM layer cells when there is information to transmit, or physical layer cells which are either idle cells or contain physical layer management information. Alternatively, cells may be transported by a synchronous transmission system such as SDH and SONET, as well as other older PDH transmission systems during the early trial phase of B-ISDN systems and other ATM-based networks. Although these different technologies will co-exist in network infrastructure for some time to come, it is expected that as the transition to SDH- and SONET-based transmission networks continues, the preferred delivery of B-ISDN to customer premises will be on SDH- or SONET-based subscriber networks, while other ATM-based networks, such as video on demand (VoD), may use other schemes.

Figure 10.20 illustrates the packing of ATM cells in an STM-1 frame at 155 Mbps. Unlike the vertical ordering of tributaries within the VCs shown in Fig. 10.13, the ATM cells are placed one after the other in a row-by-row basis within the payload. The VC payload is not an integer number of ATM cells so a cell may cross a VC boundary.

10.8.2.2 ATM layer

The ATM layer is concerned with the transmission of cells between adjacent nodes in the network and operates on the header contents of an ATM cell. Like an OSI-RM layer 2 process, the ATM layer operates on a link-by-link basis and not on an end-to-end basis across the multiple links of an entire network. As such, the information contained in the ATM cell header, including the VPI/VCI address information, is only relevant across that link, and as the cell is switched in an exchange to an outgoing transmission port, it will be assigned a new VPI/VCI address. The assignation of the VCI and VPI for each link in the end-to-end connection across the network is determined during call establishment when the optimum path through the network is chosen, and for each link a VPI and VCI are assigned. ATM switches therefore contain tables of VCI and VPI addresses that are used when switching cells between input ports and output ports.

The ATM layer also performs a cell multiplexing and demultiplexing function between the different simultaneous connections supported by the protocol stack, as well as marking each cell in the PT field (Fig. 10.17) according to the payload type of each cell. A cell loss priority (CLP) bit can be set by the user to indicate the eligibility of the cell to be discarded under heavy network congestion, while the header error control (HEC) byte allows the integrity of the header information of the ATM cell (not the information part of ATM cell) to be determined.

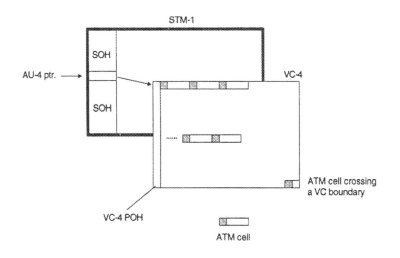

Fig. 10.20 ATM cells in an STM-1 frame.

10.8.2.3 ATM adaptation layer

The layer above the ATM layer is the ATM adaptation layer (AAL), and is responsible for adapting the many different types of user information, for example voice, video and data, into ATM cells. The fundamental differences between the types of user data lead to the implementation of different AAL services that provide different qualities of service according to the characteristics of the information. Unlike the ATM layer, the AAL operates on an end-to-end basis across a connection.

The AAL consists of two sublayers: the convergence sub-layer (CS) and the segmentation and reassembly (SAR) sub-layer. The purpose of the CS is to take user data in what ever form and break it up, or collect it, into manageable fixed size chunks. To each chunk, a header and trailer are added, as well as padding bytes as necessary. The resulting CS-PDU (CS-protocol data unit) is then passed down to the SAR sub-layer where it is split into smaller fixed size blocks and another header and trailer are added relevant to one of the AAL services appropriate for the user data. The resulting SAR-PDU is 48 bytes long such that when it is passed down to the ATM layer, it adds the 5 byte header to complete the ATM cell ready for transmission. This process is illustrated in Fig. 10.21.

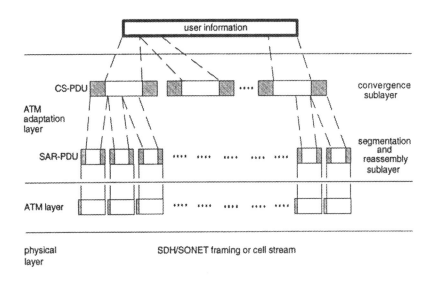

Fig. 10.21 The adaptation process of user information into ATM cells.

Five different AAL services have been defined according to the requirements of the user data. These five AALs are as follows.

AAL1: Connection-oriented, constant-bit-rate (CBR) service with end-to-end timing. This service is designed to handle traffic of an isochronous nature, for instance voice, audio and video such as that produced by H.320, that is usually transported in fixed bandwidth channels over TDM-based transmission systems. Crucial to this service is the ability to recover the clock information that was used to generate the data stream at the source (which is normally synchronized to the 8 kHz network timing) at the receiving end which allows the original information content of the data to be reconstructed.

AAL2: Connection-oriented, variable-bit-rate (VBR) service with end-to-end timing. The nature of differentially compressed video is that it generates a variable bit stream depending on the amount of motion contained within the image. The more motion that is present, the larger the difference between two successive frames, and hence the larger the amount of information produced by the compression process. This service is intended for such real-time video applications, and eliminates the stuffing and buffering processes required to generate a constant bit rate, as used for example in H.320.

AAL3/4: Variable-bit-rate (VBR) data transfer. This service was originally two proposals which have subsequently been merged into one. AAL3 provided connection-oriented packet data services intended to transfer large amounts of bursty data, for example generated by file transfer applications, while AAL4 was intended to transfer small amounts of bursty data for connectionless data services similar to that used in LANs. No timing relationship is maintained across the connection.

AAL5: Connection-oriented, variable-bit-rate (VBR) data transfer. This is similar to AAL3/4 but is more simple and efficient due to the reduced amount of error detection and correction, multiplexing and segmentation that it performs. Although AAL5 was added some time after AAL1-4 were defined, it is the most commercially advanced and used primarily in ATM LANs.

In addition to these five types of AAL for user information in the U-plane, a separate signalling AAL based on AAL5 will be used to handle signalling packets in the C-plane.

10.8.3 Signalling

The ATM UNI C-plane layer 3 protocol is derived from the N-ISDN Q.931 protocol and is known as Q.2931. It distinguishes itself from Q.931 in that eventually it will offer the ability to make not only point-to-point connections, but multipoint connections and broadcast connections, and more importantly will allow the user to negotiate with the network a quality of service for the connections being established. Owing to the size of the task to define standards for these new types of connection, and the need to establish early specifications that will allow ATM trials and initial services to begin, the development of ATM signalling specifications, by both the ITU-T and ATM Forum, has been staged starting with simple switched connections, and migrating to more complex multimedia connections at a later date. However, to ensure that the signalling protocol will be able support this migration, some minor changes to the basic Q.931 protocol are implemented in Q.2931. These consist mainly of changes to the information elements to describe attributes that are unique to ATM, and a division of the signalling into call control and bearer connection control to allow multiple connection calls to be established.

Unlike N-ISDN, there is no fixed channel in the B-ISDN user–network interface that can be used to establish ATM calls. Instead, **metasignalling** is used in a fixed VC by a terminal in order to request and be allocated a signalling VCI (SVCI) that it can then use to perform call control functions with the network. Metasignalling messages are simple and fit into a single ATM cell. The messages consist of a number of parameters that are contained in every cell, and non-relevant parameters for a particular message are coded with a null value. The metasignalling VC is shared by all terminals connected across the user–network interface. Once a terminal has been allocated an SVCI, it can proceed to establish the connection using Q.2931 protocol procedures. Inside the network, broadband extensions to the ISDN SS7 protocols will be used to provide signalling between nodes.

10.8.4 Traffic management

The ATM has an inherent method of rate adaptation up to the fixed bandwidth of the physical channels in which the cells are transported. If an application, which is associated with a connection that operates across the physical channel, has information to send, then ATM cells are generated, and can be accommodated up to the maximum bandwidth provided by the channel. When no information is being generated by the application, then no information-carrying cells are generated. The actual portion of the channel bandwidth that is used to carry ATM cells with user information therefore varies according to the rate at which information is generated by the application. ATM can therefore support as much bandwidth as required by the application up to the maximum delivered by the channel. This ability is known generically as **bandwidth on demand**.

However, the idea behind B-ISDN is that many virtual connections that carry different services and applications will be supported across the user–network interface. During the establishment of each virtual end-to-end connection across the network, a **virtual** bandwidth has to be negotiated with the network for that connection, and depending on the type of service used with the connection, other service parameters may also be defined such as to allow the instantaneous cell rate to momentarily exceed the bandwidth value while maintaining operation within other parameters such as cell loss and end-to-end delay. The danger is of course, that a terminal may exceed certain bandwidth parameters and lead to network congestion that affects other connections, which may result in either extended end-to-end delays due to increased buffering within the network, or to cell loss where cells are simply discarded by the network.

The different parameters and their values are used to define certain traffic management schemes to suit the quality of service requirements of the different services provided by ATM. The ATM Forum have currently defined several types of traffic management schemes which are described below, one of which is selected at the time a virtual connection is established across the network.

CBR: constant bit rate. This simply provides a fixed size bandwidth to allow emulation of circuit connections that carry voice, audio and constant bit rate video. It is described by peak cell rate (PCR – which is equivalent to the peak bandwidth) and cell delay variation tolerance (CDVT) parameters. If less than the PCR is being used, then the surplus bandwidth is wasted.

VBR: variable bit rate. This traffic profile is characterized by a sustained cell rate (SCR) that represents the average cell rate over time. Some statistical gain is permitted up to the PCR for periods of time determined by a burst tolerance (BT), but means that there must also be periods of time at which the cell rate falls below the SCR in order to maintain an average of the SCR value over time. Two variations of VBR are VBR-RT (VBR-real-time) and VBR-NRT (VBR-non-real-time). The real-time version has more stringent bounds on transmission delays for applications such as variable-bit-rate video. The collective bandwidth of a group of VBR connections can be managed by a method known as the **generic cell rate algorithm** (GCRA), otherwise known as the leaky bucket method [13].

UBR: unspecified bit rate. As its name suggests, there are no parameters, including cell rate, associated with the UBR management scheme. Instead, a terminal sends cells to the network when it needs to and the network makes a best effort attempt to get them to their destination. However, there are no quality of service guarantees and when the network becomes congested, the end-to-end delay increases and eventually cells are discarded. In such cases, the higher layer protocols above those of ATM are relied upon to detect and re-transmit the packets in which cells went missing, but in many cases the re-transmission just adds to further increase the network congestion. The UBR scheme was designed to allow connections to utilize the excess bandwidth not being used by CBR and VBR services.

ABR: available bit rate. The above schemes are all open-loop in that there is no feedback information from the network that indicates the current state of network congestion. The most recent management scheme to be defined, ABR, permits a reliable transport of bursty data through an indication by the network to the communicating terminals that it is congested, thereby allowing the terminals to reduce the cell rate until such time as the congestion indication is removed. In this way, the end-to-end delay is increased, but no cells should be discarded by the network. Consequently, the ABR scheme cannot guarantee end-to-end delay and is thus suitable for bursty, non-real-time applications. The ABR scheme is defined primarily by the PCR and a minimum cell rate (MCR) which the network agrees to always provide.

10.9 CUSTOMER PREMISES SUBSCRIBER LOOPS

Today, the voice and data networks within an organization are separate as they were each designed to handle two very different types of information. In the future, it is expected that ATM will be used to integrate these services within the premises network. However, ATM requires the introduction of completely new networking equipment which today represents a considerable expense on top of that already invested by an organization in its existing voice and data equipment. The expectation is therefore that ATM will be introduced gradually into enterprise networks, and initially as a backbone network where the speed and flexibility of ATM can justify its deployment. This scenario is most visible today in LANs, and ATM has only just started to be introduced in large PBX systems. Interestingly, unlike a LAN, the PBX is still very much a centralized resource, and the impact of ATM technology may well have the effect of decentralizing it into more of a distributed system, whereby smaller voice switching and processing systems that serve a specific workgroup or area are deployed locally and interconnected through ATM to each other, and to a switch that provides access to public network services.

Migration towards ATM-based customer premises networks is expected to take many years, and also depends on the availability of public network ATM services at reasonable tariffs such that the benefits of ATM can be realized on end-to-end connections across the public network. In the meantime, multimedia applications are being conceived and developed that require the delivery of voice, video and data to the desktop. Unfortunately, today's packet LAN is not suitable for the delivery of synchronous services such as PCM voice and H.320 video, while a PBX is not an optimal solution for packet data services and does not provide sufficient bandwidth to the desktop for good quality interactive video. However, a LAN standard recently adopted by the IEEE 802 committee known as IEEE 802.9a [14] brings together the most popular LAN standard, 10Base-T Ethernet, and 6 Mbps of isochronous bandwidth that allows simultaneous delivery of standard Ethernet packet services and high bandwidth ISDN services over the existing LAN twisted-pair cabling between the desktop and an IEEE802.9a-based hub. The 6 Mbps of isochronous bandwidth is available as 96 individual 64 kbps

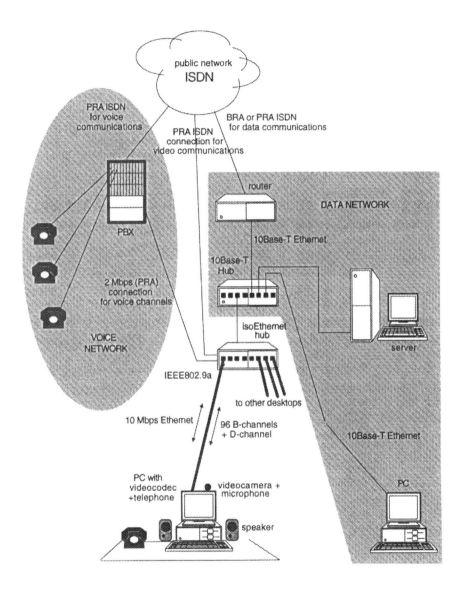

Fig. 10.22 An illustration of how IEEE802.9a (isoEthernet™) technology can be applied to existing voice and data networks and combine voice, video and data services to the desktop.

B-channels and a 64 kbps D-channel that are used in the same way as ISDN B- and D-channels. The standard also defines a call control protocol for the D-channel that is derived from the ISDN Q.931 standard. These features facilitate a simple interface between the IEEE802.9a LAN and public ISDN services.

Figure 10.22 illustrates how IEEE 802.9a, otherwise known as isoEthernet[TM][45], can be deployed in an existing enterprise network infrastructure. At the desktop, the existing 10Base-T Ethernet card in the PC is replaced with an isoEthernet card that combines the voice, video and data streams from the different sources located within the PC or connected to it. In this case, the PC is equipped with a videocodec add-in card, camera and microphone that enables it for video communication applications, while connection of the telephone to the isoEthernet card enables telephony support, and the PC has the opportunity to better manage telephony functions through its graphical user interface. This is just one example of an area that has become known as computer telephony. As for the data services, there is no change due to the support by isoEthernet of 10Base-T Ethernet, such that to the existing data network, the PC connected by isoEthernet appears to be connected a standard 10Base-T Ethernet network.

IsoEthernet was designed to run on existing 10Base-T wiring, so the same wiring will connect the PC to an isoEthernet hub located somewhere in the wiring closet within 100 m of the PC, along with the other data network equipment such as standard 10Base-T hubs. Inside the isoEthernet hub, the isochronous data streams and Ethernet streams are separated at the physical layer for each of the isoEthernet connections to the hub. Many Ethernet networks today employ repeated hubs to create the shared medium in which Ethernet operates. Alternatively, a switched technique may be used to improve the use of bandwidth. Either technique could be applied to the Ethernet streams inside the isoEthernet hub before connection to the existing Ethernet data network. Meanwhile, the isochronous streams from the different connections can either be switched locally, or to a digital connection to the PBX to allow interworking with the existing voice network, or to a PRA ISDN connection to the public network for video communications. Alternatively, another IEEE802.9 link can provide connection between isoEthernet hubs.

The impact of isoEthernet to existing networks is therefore kept to a minimum by combining their services only at the final distribution to the desktop across existing LAN wiring. Instead of two or three connections to the desktop, one for telephony, one for data, and possibly a separate ISDN connection for video, all three services can be integrated across a single connection. It can be envisaged that in the longer term, ATM will be used as a backbone technology to interconnect isoEthernet hubs to other multimedia networking equipment.

---

[45]isoEthernet is a trademark of National Semiconductor Corporation.

REFERENCES

1. Bellcore (1991) Generic Requirements for High-Bit-Rate Digital Subscriber Lines. TA-NWT-001210.
2. ANSI (1992) Draft Technical Report on High-Bit-Rate Digital Subscriber Lines (HDSL). T1E1.4/92-002R1.
3. ETSI (1993) High Bit-Rate Digital Subscriber Line Transmission System on Metallic Local Lines. DTR/TM-3017.
4. Kyees, P.J., McConnell, R.C. and Sistanizadeh, K. (1995) ADSL: a new twisted-pair access to the information highway. *IEEE Communications Magazine*, **33**(4), 52–59.
5. ANSI (1995) Asymmetric Digital Subscriber Line (ADSL) Metallic Interface. Draft T1E1.4/95-007R2.
6. ITU-T (1993) Synchronous Digital Hierarchy Bit Rates. Recommendation G.707.
7. ITU-T (1993) Network Node Interface for the Synchronous Digital Hierarchy. Recommendation G.708.
8. ITU-T (1993) Synchronous Multiplexing Structure. Recommendation G.709.
9. ANSI (1991) Digital Hierarchy – Digital Interface Rates and Formats Specifications. T1-105-1991.
10. ETSI (1994) Signalling Protocols and Switching (SPS); V Interfaces at the digital Local Exchange (LE), V5.1 interface for the support of Access Networks (AN), Part 1:V5.1 interface specification. ETS 300 324-1.
11. ITU-T (1991) Broadband aspects of ISDNs. Recommendation I.121.
12. ITU-T (1993) B-ISDN User-Network Interface – Physical Layer Specification. Recommendation I.432.
13. Onvural, R.O. (1994) <u>Asynchronous Transfer Mode Networks: Performance Issues</u>, Section 4.5.1, Artech House.
14. IEEE (1995) IEEE Standards for Local and Metropolitan Area Networks, Integrated Services (IS) LAN: IEEE802.9 Isochronous services with carrier sense multiple access with collision detection (CSMA/CD) media access control (MAC) service.

Appendix A

The seven-layer reference model for open systems interconnection

A.1 INTRODUCTION

The open systems interconnection reference model (OSI-RM) is a framework with which standardized computer communication protocols can be specified. These protocols allow any computer system to be able to communicate with any other computer system, thereby creating an open environment for the interconnection of systems that conform to these standards. Development of the OSI-RM started during the late 1970s and early 1980s by the International Standards Organization (ISO) and continues today as the model evolves to encompass many forms of computer communications.

The general trend in the public telecommunications networks is towards digitalization of equipment. This trend started in the 1960s with the introduction of digital transmission equipment, and was followed by digital switching exchanges, and finally digitalization of the subscriber loop by the ISDN. Consequently, the network equipment has become increasingly dependent on computers and their software for the control, administration and maintenance of their functions. Today, a modern digital switching exchange has all the features of a medium-sized computer system whose primary task is the control of switching functions within the exchange. The communication of signalling, administration and maintenance information is required between the computers in the exchanges of the network to connect and disconnect calls through the network, and provide the subscriber with a host of services aimed at increasing the value of the network service to the subscriber. This is achieved with a common channel signalling network which can be envisaged as a separate computer communications network that exists inside the telecommunications network.

Most of the standardization in public telecommunications networks is through ITU-T Recommendations which prior to 1984 did not fit into the OSI-RM. Consequently, basic telephony services are not covered by the OSI-RM. However, the OSI-RM is intended for data services, and so for ISDN which offers both voice and data services within a single interface, the OSI principles are applied as far as possible. The ITU-T has worked closely with ISO such that a common model for the OSI-RM is defined which can be applied both in the telecommunications environment and in computer communication networks.

The ISO international standard on OSI is known as IS7498: Information Processing Systems – Open Systems Interconnection – Basic Reference Model, while the ITU-T version is published in the X.200 series of Recommendations.

A.2 THE SEVEN-LAYER REFERENCE MODEL

The specification of a computer communication system involves the definition of the services to be provided and the protocols which are to implement them. This is generally a complex procedure, and involves the definition of a number of functions which must interact with one another in order to provide defined communication services to the user. The specification procedure is eased if the communication system architecture can be generically split into a number of more manageable parts. The reference model adopted by ISO and ITU-T comprises a hierarchical seven-layer structure as shown in Fig. A.1.

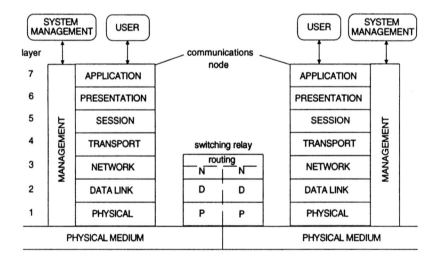

Fig. A.1 The ISO seven-layer reference model for open systems interconnection.

A key factor which determines the success with which any such model may be generally applied is that its structure must be sufficiently abstract that it remains independent of any implementation issues which determine how a system is achieved. In particular, the definition of the interfaces between the layers is important as several standards may exist for a specific layer, with the possibility that one standard may be broadly applicable for a range of applications. However, the various options available within a layer may introduce too many degrees of freedom to make universal interchangability a practical feature. This has led to the definition of standard 'profiles' which represent the selection of compatible options for standards within the seven layers for an application area.

A brief description of the physical medium and the functions of the seven layers is given in the following sections.

A.2.1 The physical medium

The OSI seven-layer model does not include the physical medium itself which provides a transparent communication channel to the other communication nodes in the system.

A.2.2 The physical layer

The physical layer is responsible for the conversion between the data stream it sends to and receives from the layer 2 and the physical signal that it transmits and receives from the physical medium. In addition, the physical layer will perform the line coding of the signal and timing synchronization between the peer physical layers, as well as the procedural requirements such as activation and deactivation that allow access to the physical medium.

A.2.3 The data-link layer

The data-link layer shields the layers above from the characteristics of the physical layer and the transmission medium, and attempts to provide these layers with an error free transmission. Standards for the data-link layer specify the way in which the data-link PDUs are to be organized, or framed, in preparation for their transfer across a connection. The beginning and end of a frame will be uniquely defined by flags, while errors that occur during transmission of a frame across the physical medium may be detected by mechanisms such as cyclic redundancy check values. The data-link layer may also provide for the correction of these errors within its protocol specification, such as the re-transmission of unacknowledged frames used in X.25 LAPB and ISDN Q.921 LAPD.

A.2.4 The network layer

The purpose of the network layer is to provide independence of the data transfer technology used at the data-link and physical layers to the layer above, the transport layer. The transport layer therefore has no knowledge of whether the underlying network technology is based on a computer local area network or a packet-switching network. In addition, the network layer provides routing and relaying functions for information which allow many different sub-networks to be combined into a single global network. The peer-to-peer connections established between the network SAPs (NSAPs) are point-to-point, and each NSAP is identified within the global network by a unique address.

The network layer is divided into four sublayers as follows.

- **Sub-network access control function.** This provides an end-to-end service across a particular sub-network.
- **Sub-network dependent convergent function,** which is responsible for mapping a particular network service on to a specific sub-network.
- **Sub-network independent convergent function.** This sub-layer is concerned with the allocation of a message to the appropriate network service.
- **Routing.** This controls the establishment and tear-down of a virtual call between NSAPs across several sub-networks.

The routing sublayer is located at the top of the network layer, while the sub-network access control function is at the bottom. Figure A.1 illustrates the use of a switching relay between communication nodes in two separate sub-networks. For clarity, only the routing sublayer in the network layer is shown.

A.2.5 The transport layer

The view of the network from the transport layer is that of a single network where each node in the network contains a unique address, regardless of how the network is physically composed and the technology with which it is implemented. A key aspect of the transport layer is that a connection may be requested from the layer above, the session layer, with a defined quality of service (QoS) which once established, must be maintained for the duration of the connection. The quality of service defines specific values for connection parameters such as the throughput of the connection and the connection establishment delay. During call establishment, one of five classes of service may be selected as follows.

- **Class 0.** This is the basic class which assumes that the service quality of the network connection is adequate and therefore provides no error correction and no multiplexing. Multiplexing allows several application sessions to share the same network connection.

- **Class 1.** Error recovery is provided, but no multiplexing. This allows the transport layer to recover from a disconnect which is indicated by the network, but cannot handle unsignalled disconnects.
- **Class 2.** This provides class 0 service but with multiplexing.
- **Class 3.** A combination of class 1 and class 2 services which allows error recovery and multiplexing.
- **Class 4.** This class provides recovery from both signalled and unsignalled disconnects, as well as multiplexing.

The transport layer is also required to optimize the use of the underlying resources based on constraints such as availability and cost, while maintaining the agreed quality of service to the session layer.

A.2.6 The session layer

The purpose of the session layer is to provide the means with which the interactions between the application processes, communicating through the presentation layer, can be structured and organized. It is responsible for the establishment of a session connection and the management of the dialogue which takes place on that connection. The session layer protocol can be negotiated during call establishment. Instead of choosing one of a selection of protocols, a sub-set of optional protocol functions within the session layer protocol is chosen for the particular session connection in order to provide the necessary services to the presentation layer above. One of the choices made will be the interaction mode between the presentation entities, which can be one of either two-way simultaneous, two-way alternate, or one way. In addition, the session layer may provide synchronization of the dialogue between the session entities, and quarantining of PDUs, whereby a session entity may send data units to another session entity which holds them until such time as the sending session entity instructs it to release them to the receiving presentation entity, thereby allowing data units to be grouped before being passed upwards to the presentation layer.

Although a one-to-one relationship exists between the corresponding session and transport connections, a transport connection may be used to support several successive session connections such that following disestablishment of one session connection, another may be initiated using the same transport connection. This process also works in reverse, such that in the case of a transport connection failing, or no longer being able to provide the negotiated quality of service, it may be disconnected and another established transparently to the operation of the communicating session entities.

A.2.7 The presentation layer

The presentation layer provides an independence to application tasks from the differences in the representation of information between the communicating application tasks. During connection establishment and through out the communication, the peer presentation entities negotiate the syntax translation that each provides.

A.2.8 The application layer

The highest layer in the seven-layer stack is the application layer which contains those processes within the application itself which are concerned with communications. Typically, an application task which resides outside the communications system and the OSI-RM, accesses it through a specific application protocol that resides within the application layer, and which provides a standardized interface based on application control service elements (ACSE). Different applications require different application services, and are chosen or defined as necessary to meet the requirements of the application and its users. A number of specific application services have been, or are currently under standardization by OSI, examples of which are the services for message handling (MHS) that support electronic mail (E-mail) and electronic data interchange (EDI), and services that support file transfer, access and management (FTAM).

A.2.9 Management

The management functions within the protocol stack provide specific management of layer resources as well as more system-oriented functions such as error monitoring, provision of alarms, performance monitoring, security, accounting, test and diagnostics. The common management information protocol (CMIP) and common management information service (CMIS) allow communication between remote management entities.

A.3 LAYERING PRINCIPLES

The principle of layering is that from the bottom upwards, successively higher layers make use of the services provided in the layer immediately below in order to provide an enhanced set of services to the layer immediately above. Application tasks which interface to the top layer are therefore presented with a powerful set of services with which to communicate information across the network.

Another reason for layering is to ensure that each layer maintains an independence from other layers by the definition of services provided to the next higher layer which are independent of the way in which these services are implemented. Therefore, changes can be made to the way a layer, or group of layers, operates while maintaining the services provided to the higher layers which need not be changed as a consequence of changes to the lower layers.

OSI standards for a particular layer will contain two important components. The first is a definition of the functionality contained within and beneath the layer and represents the services provided to the layer above. The definition of a service is expressed in terms of **service primitives** which are communicated across the layer boundaries and invoke the execution of functions within the layer which result in specific actions being taken and possibly the generation of further primitives. The second component is the definition of the protocol which fulfils the requirements of the service definition.

In OSI terminology, any layer is referred to as the (N)-layer, while the layer immediately beneath is the ($N-1$)-layer, and that above is the ($N+1$)-layer, as shown in Fig. A.2. Each layer contains one or more **entities** which can be thought of as hardware and/or software tasks within the layer that implement the protocols which process the service primitives conveyed to it via the **layer service access point (SAP)**. Again, in OSI parlance, the (N)-entities provide the (N)-service to the ($N+1$)-entities at the (N)-service access points, which are called (N)-SAPs for short.

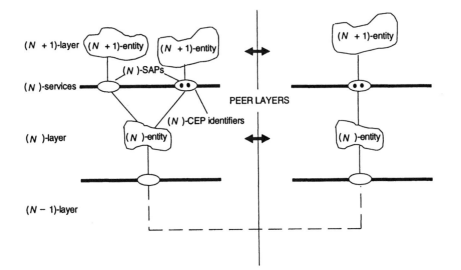

Fig. A.2 Layering principles.

A service which is offered by the network dependent layers is that connections can be created between peer SAPs, that is the equivalent (N)-SAP in a machine with which communications is to be established, in order to synchronize the peer entities and afterwards transfer information between them. In addition, several connections may be supported through the same (N)-SAP, where each connection is uniquely identified by the (N)-connection end-point identifier, or (N)-CEP identifier for short. The (N)-CEP identifier is used by both the (N)-entity and (N + 1)-entity located on either side of the (N)-SAP to uniquely identify the (N)-connection. In this way, connections are established between (N)-SAPs and not entities, which allows the independence of layers to be maintained.

Layers may function is one of two ways, either in connection-oriented mode, or in connectionless mode. In the connection-oriented service, a connection must be established between the SAPs used by the communicating entities before any communication can take place, rather like the establishment of a connection in a phone call before the two parties can talk to one another. The communication transaction therefore consists of three phases: a connection establishment, data transfer, and disconnection. The establishment of a connection effectively allocates resources within a layer to that connection for its duration. For the lower layers in a communications node that operates in the connection-oriented service, an (N)-service requires that an (N)-connection be established between the peer (N)-SAPs before any communication between the peer (N + 1)-entities can occur. If this connection is not available, then it must first be established before communications between the (N)-SAPs can take place. However, this assumes that an (N − 1) connection is already available in the layer below, over which the (N)-connection can be established, which if not the case, must be established prior to the connection above it. Connections are therefore established sequentially in a bottom-up fashion. When the connections are fully established between peer (N)-SAPs, data transfer may take place between the peer entities associated with the (N)-SAP, and the entities situated on either side of a SAP.

In the connectionless mode, prior establishment of a connection is not necessary, and the information communicated is self-contained in that it has everything required by a layer to process it, including its source and destination identification. This type of operation is typically found in the lower layers of local area networks, although the higher layers will typically operate through a connection-oriented service.

As information is passed from an application task at the top of the seven layers, down through successive layers to the bottom layer and eventually over the physical communications link to its destination, it is processed by the functions in each layer and information added to that which already exists in order to control or conform to the way the information is processed by the corresponding peer layer in the receiving node. At the receiving end, the converse process takes place whereby, from the bottom up, successive layers process the information added by their peer layer which then is discarded, so that only the application information is eventually passed upwards to the application task. Information which is communicated across a layer boundary is referred to as a **service data unit** (SDU). For information which is passed downwards across the $(N+1) - (N)$-layer boundary, the functions within the (N)-layer then add any necessary **protocol control information** (PCI) for peer layer processing in the form of a header, and the combined unit SDU + PCI known as a **protocol data unit** (PDU) is passed across the next boundary to the $(N-1)$-layer. In doing so, the PDU becomes the SDU for the $(N-1)$-layer and the process is repeated. The relationship between SDUs, PCIs and PDUs is illustrated in Fig. A.3.

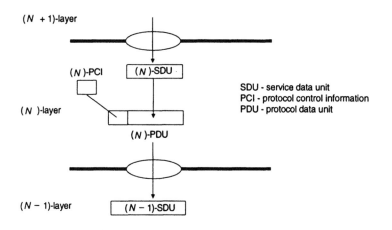

Fig. A.3 An illustration of the addition of protocol control information as data units are passed down through layers.

A4 SUMMARY

The OSI-RM provides a standardized framework with which standards for communication protocols can be defined for open systems that allow software products and communications equipment from numerous suppliers to communicate with one another. However, the OSI-RM has its roots in the world of data communication networks, and is therefore currently evolving further to encompass the multi-service nature of telecommunication networks of today and the future.

Appendix B

List of ITU-T I-Series Recommendations

This appendix contains a list of those ITU-T I-Series Recommendations, including those of the former CCITT, that are in publication. The date of publication is given in brackets after the description of the Recommendation. Where the term **Blue Book** is used, this means that the Recommendation was included in the last series to be published as the Blue Books when they were updated once every four years. The publication date of the Blue Books was June 1989. Since then a policy of more frequent updating has been adopted by the ITU-T, and the dates given represent the last known update for that Recommendation.

Part I – General structure – I.100-Series

ISDN – General structure – terminology
I.110   Preamble and general structure of the I-Series Recommendations (Blue Book).
I.111   Relationship with other Recommendations relevant to ISDNs (Blue Book).
I.112   Vocabulary of terms for ISDNs (Nov. 93).
I.113   Vocabulary of terms for broadband aspects of ISDNs (Nov. 93).
I.114   Vocabulary of terms for Universal Personal Telecommunications (Oct. 93).

Description of ISDNs
I.120   Integrated services digital networks (ISDNs) (Nov. 93).
I.121   Broadband aspects of ISDNs (Jul. 91).
I.122   Framework for frame mode bearer services (Oct. 93).

General modelling methods
I.130   Method for the characterization of telecommunication services supported by an ISDN and network capabilities of an ISDN (Blue Book).

Telecommunications network and service attributes
I.140   Attribute technique for the characterization of telecommunication services supported by an ISDN and network capabilities of an ISDN (Mar. 93).
I.141   ISDN network charging capabilities attributes (Blue Book).

General description of asynchronous transfer mode
I.150   B-ISDN asynchronous transfer mode functional characteristics (Nov. 95)

Part II – Service capabilities – I.200-Series

Service capabilities – scope
I.200   Guidance to the I.200-Series Recommendations (Blue Book).

General aspects of services in ISDN
I.210   Principles of telecommunication services supported by an ISDN and the means to describe them (Mar. 93).
I.211   B-ISDN service aspects (Mar. 93).

Common aspects of services in the ISDN
I.220   Common dynamic description of basic telecommunication services (Blue Book).
I.221   Common specific characteristics of services (Nov.93).

Bearer services supported by an ISDN
I.230   Definition of bearer service categories (Blue Book).
I.231   Circuit mode bearer service categories (Blue Book).
I.231.1 Circuit-mode 64 kbits/s unrestricted, 8 kHz structured bearer service (Blue Book).
I.231.2 Circuit-mode 64 kbits/s, 8 kHz structured bearer service, usable for speech information transfer (Blue Book).
I.231.3 Circuit-mode 64 kbits/s, 8 kHz structured bearer service, usable for 3.1 kHz audio information transfer (Blue Book).
I.231.4 Circuit-mode alternate speech/64 kbits/s unrestricted, 8 kHz structured bearer service (Blue Book).
I.231.5 Circuit-mode 2×64 kbits/s unrestricted, 8 kHz structured bearer service (Blue Book).
I.231.6 Circuit-mode 384 kbits/s unrestricted, 8 kHz structured bearer service (Jul. 96).
I.231.7 Circuit-mode 1536 kbits/s unrestricted, 8 kHz structured bearer service (Jul. 96).
I.231.8 Circuit-mode 1920 kbits/s unrestricted, 8 kHz structured bearer service (Jul. 96).
I.231.9 Circuit-mode 64kbits/s 8 kHz structured multi-use bearer service (Mar. 93).
I.231.10 Circuit-mode multiple-rate unrestricted 8 kHz structured bearer service (Mar. 93).
I.232   Packet-mode bearer service categories (Blue Book).
I.232.1 Virtual call and permanent virtual circuit bearer service category (Blue Book).
I.232.2 Connectionless bearer service category (Blue Book).
I.232.3 User signalling bearer service category (USBS) (Mar. 93).
I.233   Frame mode bearer services (Oct. 91).
I.233.1 ISDN frame relaying bearer service (Oct. 91).
I.233.1 Annex F Frame relay multicast (Jul. 96).
I.233.2 ISDN frame switching bearer service (Oct. 91).

Teleservices supported by an ISDN
I.240   Definition of teleservices (Blue Book).
I.241   Teleservices supported by an ISDN (Blue Book).

I.241.1 Telephony (Blue Book).
I.241.2 Teletex (Blue Book).
I.241.3 Telefax 4 (Blue Book).
I.241.4 Mixed Mode (Blue Book).
I.241.5 Videotex (Blue Book).
I.241.6 Telex (Blue Book).
I.241.7 Telephony 7 kHz teleservices (Mar. 93).
I.241.8 Teleaction stage one service description (Oct. 95).

Supplementary services in ISDN
I.250 Definition of supplementary services (Blue Book).
I.251 Number identification supplementary services (Blue Book).
I.251.1 Direct-dialling-in (Aug. 92).
I.251.2 Multiple Subscriber Number (Aug. 92).
I.251.3 Calling Line Identification Presentation (COLP) (Aug. 92).
I.251.4 Calling Line Identification Restriction (COLR) (Aug. 92).
I.251.5 Connected Line Identification Presentation (Feb. 95).
I.251.6 Connected Line Identification Restriction (Feb. 95).
I.251.7 Malicious Call Identification (Aug. 92).
I.251.8 Sub-addressing supplementary service (Aug. 92).
I.251.9 Calling name identification presentation (Jul. 96).
I.251.10 Calling name identification restriction (Jul. 96).
I.252 Call offering supplementary services (Blue Book).
I.252.1 Call Transfer (CT) (Blue Book).
I.252.2 Call Forwarding Busy (Aug. 92).
I.252.3 Call Forwarding No Reply (Aug. 92).
I.252.4 Call Forwarding Unconditional (Aug. 92).
I.252.5 Call Deflection (Aug. 92).
I.252.6 Line Hunting (LH) (Blue Book).
I.253 Call completion supplementary services (Blue Book).
I.253.1 Call waiting (CW) supplementary service (Jul. 90).
I.253.2 Call Hold (Aug. 92).
I.253.3 Completion of calls to busy subscribers (Jul. 96).
I.253.4 Completion of calls on no reply (Jul. 96)
I.254 Multiparty supplementary services (Blue Book).
I.254.1 Conference calling (CONF) (Blue Book).
I.254.2 Three-Party Supplementary Service (Aug. 92).
I.255 Community of interest supplementary services (Blue Book).
I.255.1 Closed User Group (Aug. 92).
I.255.2 Support of Private Numbering Plans (Jul. 96).
I.255.3 Multi-level precedence and preemption service (MLPP) (Jul. 90).
I.255.4 Priority service (Jul. 90).
I.255.5 Outgoing call barring (Aug. 92).
I.256 Charging supplementary services (Blue Book).
I.256.2a Advice of charge: charging information at call set-up time (AOc-S) (Mar. 93).
I.256.2b Advice of charge: charging information during the call (AOc-D) (Mar. 93).
I.256.2c Advice of charge: charging information at the end of the call (AOc-E) (Mar. 94).

I.256.3 Reverse charging (Aug. 92).
I.257 Additional information transfer supplementary services (Blue Book).
I.257.1 User-to-User Signalling (Oct. 95).
I.258.1 Terminal portability (TP) (Oct. 95).
I.258.2 In-call modification (IM) (Feb. 95).
I.259.1 Address screening (Jul. 96).

Part III – Overall network aspects and functions – I.300-Series

Network functional principles
I.310 ISDN – Network functional principles (Mar. 93).
I.311 B-ISDN general network aspects (Aug. 96).
I.312 (Q.1201) Principles of intelligent network architecture (Oct. 92).

Reference models
I.320 ISDN protocol reference model (Nov. 93).
I.321 B-ISDN protocol reference model and its application (Apr. 91).
I.324 ISDN network architecture (Oct. 91).
I.325 Reference configurations for ISDN connection types (Mar. 93).
I.326 Functional architecture of transport networks based on asynchronous transfer mode (Nov. 95).
I.327 B-ISDN functional architecture (Mar. 93).
I.328 (Q.1202) Intelligent network - Service plane architecture (Oct. 92).
I.329 (Q.1203) Intelligent network - Global functional plane architecture (Oct. 92).

Numbering, addressing and routing
I.330 ISDN numbering and addressing principles (Blue Book).
I.331 (E.164) Numbering plan for the ISDN era (Blue Book).
I.333 Terminal selection in ISDN (Mar. 93).
I.334 Principles relating ISDN numbers/sub-addresses to the OSI reference model network layer addresses (Blue Book).

Connection types
I.340 ISDN connection types (Blue Book).

Performance objectives
I.350 General aspects of quality of service and network performance in digital networks, including ISDNs (Mar. 93).
I.351 Relationships among ISDN performance Recommendations (Mar. 93).
I.352 Network performance objectives for connection processing delays in an ISDN (Mar. 93).
I.353 Reference events for defining ISDN and B-ISDN performance parameters (Aug. 96).
I.354 Network performance objectives for packet-mode communication in an ISDN (Mar. 93).
I.355 ISDN 64 kbits/s connection type availability performance (Mar. 95).

I.356 B-ISDN ATM layer cell transfer performance (Nov. 93).
I.357 B-ISDN semi-permanent connection availability (Aug. 96).

Protocol layer requirements
I.361 B-ISDN ATM layer specification (Nov. 95).
I.362 B-ISDN ATM adaptation layer (AAL) functional description (Mar. 93).
I.363 B-ISDN ATM adaptation layer (AAL) specification (Mar. 93).
I.363 Add. B-ISDN ATM adaptation layer (AAL) specification (Nov. 93).
I.363.1 Types 1 and 2 AAL (Aug. 96).
I.363.3 Types 3/4 AAL (Aug. 96).
I.363.5 Types 5 AAL (Aug. 96).
I.364 Support of the broadband connectionless data bearer service by the B-ISDN (Nov. 95).
I.365 B-ISDN ATM adaptation layer sublayers (Nov. 93).
I.365.1 Frame relaying service specific convergence sublayer (Nov. 93).
I.365.2 Service specific coordination function to provide the connection-oriented network service (Nov. 95).
I.365.3 Service specific coordination function to provide the connection-mode transport service (Nov. 95).

General network requirements and functions
I.370 Congestion management for the ISDN frame relaying bearer service (Oct. 91).
I.371 Traffic control and congestion control in B-ISDN (Aug. 96).
I.372 Frame relaying bearer service network-to-network interface requirements (Mar. 93).
I.373 Network capabilities to support universal personal telecommunications (UPT) (Mar. 93).
I.374 Framework Recommendations on "network capabilities to support multimedia services" (Mar. 93).
I.376 ISDN network capabilities for the support of teleaction service (Mar. 95).

Part IV – ISDN user-network interfaces – I.400 Series

General
I.410 General aspects and principles relating to Recommendations on ISDN user–network interfaces (Blue Book).
I.411 ISDN user–network interfaces – Reference configurations (Mar. 93).
I.412 ISDN user–network interfaces – Interface structures and access capabilities (Blue Book).
I.413 B-ISDN user–network interface (Mar. 93).
I.414 Overview of Recommendations on layer 1 for ISDN and B-ISDN customer accesses (Mar. 93).
I.420 Basic user–network interface (Blue Book).
I.421 Primary rate user–network interface (Blue Book).

Layer 1 Recommendations
I.430   Basic user–network interface – Layer 1 specification (Nov. 95).
I.431   Primary rate user–network interface – Layer 1 specification (Mar. 93).
I.432   B-ISDN user–network interface – Physical layer specification (Mar. 93).
I.432.1   General characteristics (Aug.96).
I.432.2   155 520 kbits/s and 622 080 kbits/s operation (Aug.96).
I.432.3   1544 kbits/s and 2048 kbits/s operation (Aug.96).
I.432.4   51 840 kbits/s operation (Aug.96).

Layer 2 Recommendations
I.440 (Q.920)   ISDN user–network interface data link layer – General aspects (Mar. 93).
I.441 (Q.921)   ISDN user–network interface – Data link layer specification (Mar. 93).
(Q.921 bis)   Abstract test suite for LAPD conformance testing (Mar. 93).
(Q.922)   ISDN data link layer specifications for frame mode bearer services (Aug. 92).

Layer 3 Recommendations
I.450 (Q.930)   ISDN user–network interface layer 3 – General aspects (Mar. 93).
I.451 (Q.931)   ISDN user–network interface layer 3 specification for basic call control (Mar. 93).
I.452 (Q.932)   Generic procedures for the control of ISDN supplementary services (Mar. 93).
(Q.933)   Signalling specification for frame mode basic call control (Mar. 93).

Multiplexing, rate adaptation and support of existing interfaces
I.460   Multiplexing, rate adaptation and support of existing interfaces (Blue Book).
I.461 (X.30)   Support of X.21, X.21bis and X.20bis based data terminal equipment (DTEs) by an integrated services digital network (ISDN) (Mar. 93).
I.462 (X.31)   Support of packet mode terminal equipment by an ISDN (Nov. 95).
I.463 (V.110)   Support of data terminal equipments with V-Series type interfaces by an integrated services digital network (ISDN) (Sept. 92).
I.464   Multiplexing, rate adaptation and support of existing interfaces for restricted 64 kbits/s transfer capability (Oct. 91).
I.465 (V.120)   Support by an ISDN of data terminal equipment with V-Series type interfaces with provision for statistical multiplexing (Sept. 92).

Aspects of ISDN affecting terminal requirements
I.470   Relationship of terminal functions to ISDN (Blue Book).

Part V – Interwork interfaces – I.500-series

Interwork interfaces
I.500   General structure of ISDN interworking Recommendations (Mar. 93)
I.501   Service interworking (Mar. 93)
I.510   Definitions and general principles for ISDN interworking (Mar. 93).

I.511   ISDN-to-ISDN layer 1 interwork interface (Blue Book).
I.515   Parameter exchange for ISDN interworking (Mar. 93).
I.520   General arrangements for network interworking between ISDNs (Mar. 93).
I.525   Interworking between networks operating at bit rates less than 64 kbits/s with 64 kbits/s-based ISDN and B-ISDN (Aug. 96).
I.530   Network interworking between an ISDN and a public switched telephone network (PSTN) (Mar. 93).
I.540 (X.321)   General arrangements for interworking between circuit switched public data networks (CSPDNs) and integrated services digital networks (ISDNs) for the provision of data transmission (Blue Book).
I.550 (X.325)   General arrangements for interworking between packet-switched public data networks (PSPDNs) and integrated services digital networks (ISDNs) for the provision of data transmission (Blue Book).
I.555   Frame relaying bearer service interworking (Nov. 93).
I.560 (U.202)   Technical requirements to be met in providing the international telex service within an integrated services digital network (Mar. 93).
I.570   Public/private ISDN interworking (Mar. 93).
I.571   Connection of VSAT based private networks to the public ISDN (Aug. 96).
I.580   General arrangements for interworking between B-ISDN and 64 kbits/s based ISDN (Nov. 95).

Part VI – Maintenance principles – I.600-Series

I.601   General maintenance principles of ISDN subscriber access and subscriber installation (Blue Book).
I.610   B-ISDN operation and maintenance principles and functions (Nov. 95).

Part VII – ATM – I.700-Series

ATM equipment
I.731   Types and general characteristics of ATM equipment (Mar. 96).
I.732   Functional characteristics of ATM equipment (Mar. 96).

Management of ATM equipment
I.751   Asynchronous transfer mode management of the network element view (Mar. 96).

Appendix C

Differences in ISDN services and signalling implementations

The following sections in this appendix provide a brief description of some of the differences to be found between the ISDN subscriber loop signalling and implementations in Europe and North America.

C.1 SERVICES

To start with, we take a look at the difference in service provision between North America and Europe. In both cases, these services are designed to provide support for existing applications aswell as new ones, so many of the differences arise from the regional popularity of various applications. In general, many of the services offered in North America support a variety of voice and both circuit-switched and packet-switched data terminals, while in Europe a number of optional teleservices have been defined to support facsimile and videotex, as well as telephony and data applications such as file transfer. Despite the wider range of defined services available in Europe, most are optional, and only a small basic set of services are mandatory. In America, NI-1 also requires uniform capability across the services delivered by the regional service providers with cooperation from the long distance carriers. However, the target set for deployment of NI-1 services is higher than the minimal set of Euro-ISDN services.

NI-1 represents the first stage of national ISDN roll-out in North America, and the services it provides lay the foundations on which further services are built in subsequent revisions known as NI-2 and NI-3. NI-1 defines five bearer services as follows.

NI-1 bearer services
- Circuit mode, 64 kbps, 8 kHz structured, speech
- Circuit mode, 64 kbps, 8 kHz structured, 3.1 kHz audio
- Circuit mode, 64 kbps, 8 kHz structured, unrestricted digitial information (64 kbps)

- Circuit mode, 64 kbps, 8 kHz structured, unrestricted digital information, rate adapted from 56 kbps to 64 kbps
- Packet mode, unrestricted digital information (B- and D-channel)

In contrast, Euro-ISDN defines the following two mandatory bearer services.

Euro-ISDN bearer services
- Circuit mode, 64 kbps, 8 kHz structured, unrestricted digital information
- Circuit mode, 64 kbps, 8 kHz structured, audio

In addition to these, the following optional services are also defined:

- Circuit mode, 64 kbps, 8 kHz structured, speech
- Packet mode, unrestricted digital information (B- and D-channel, X.31 Case B)

The supplementary services offered by Euro-ISDN and NI-1 are significantly different, as illustrated by the following lists. The first is a list of services defined for NI-1. These services are available on voice- and circuit-switched data bearer services except the electronic key telephone system (EKTS), call pickup, flexible calling and automatic callback which are only available for voice.

NI-1 supplementary services

Call forwarding
- Subfeatures – Call forwarding variable
                Call forwarding interface busy
                Call forwarding don't answer
                Reminder notification
                Redirecting number
                Redirecting reason
                Courtesy call

Automatic callback
- Subfeatures – Automatic callback intra-switch

Call hold
- Subfeatures – Hold and retrieve

Additional call offering
- Subfeatures – Additional call offering, unrestricted
                Notification busy limit

Flexible calling
- Subfeatures – Consultation hold
  - Add on
  - Implicit and/or explicit transfer
  - Threeway conference calling
  - Six party conference calling
  - Conference hold and retrieve
  - Drop last call on conference
  - Add previously held call to conference

Calling number identification services
- Subfeatures – Calling party number privacy
  - Network provided number delivery
  - Redirecting number
  - Redirecting reason

Message service
- Subfeatures – Message waiting indicator

Display service
- Subfeatures – Protocol and procedures

Electronic key telephone system (EKTS)
- Subfeatures – Multiple directory number per terminal
  - Multiple directory number appearances
  - Hold/retrieve
  - Bridging/directory number (DN) bridging
  - Intercom calling
  - Membership in a multiline hunt group
  - Abbreviated and delayed ringing
  - Automatic bridged call exclusion
  - Manual bridged call exclusion
  - Call appearance call handling
  - Station message detail recording

Multiline hunt groups
- Subfeatures – Linear hunting
  - Circular hunting
  - Uniform hunting
  - Stop hunt
  - Make busy
  - Analog members in hunt group

Basic business group
- Subfeature – Simulated facility groups for in and out Calls

Business group dial access features
- Subfeatures – Business group dialing plan
                Intecom dialing
                Abbreviated dialing
                Dial access to private facilities
                Dial access to ARS
                Customer access treatment code restrictions
                Code restriction and diversion
                Direct outward dialing
                Direct Inward dialing

Call Pickup
- Subfeature – Call pickup

In addition to these voice- and circuit-switched data supplementary services, NI-1 also defines a range of packet mode bearer services [1]. However, packet-switched data services are restrictive in that B-channel connections between the terminal and the packet handler are permanently connected like a leased line connection, and means that the particular B-channel can only be used for packet-switched data calls. This service restriction is overcome in NI-2. In addition, within NI-1, the provision of primary rate services is only very basic, with most implementations being supplier specific. These are also enhanced in NI-2 along with the access, call control and signalling protocols for PRA to conform to a uniform set of standards. At the basic rate interface, due to technical limitations in NI-1, only two 'B-channel' terminals can be connected at the user–network interface, and each service supported by a terminal must be accessed through a separate directory number rather than a single number being used for any type of call. Again, these shortcomings are addressed in NI-2 [2].

In keeping with the two-tiered approach adopted by Euro-ISDN, a small number of supplementary services are mandatory, together with a range of optional services. Therefore, while all national networks conforming to Euro-ISDN will provide a basic set of services, networks in different countries may well support different services beyond these. For contrast with those of NI-1, the Euro-ISDN services are listed below.

Euro-ISDN supplementary services

Mandatory services
- Calling Line Identification Presentation (CLIP)
- Calling Line Identification Restriction (CLIR)

- Direct Dialling In (DDI)
- Multiple Subscriber Number (MSN)
- Terminal Portability (TP)

Optional services
  Number identification services
- Subfeatures – Connected line identification presentation (COLP)
                   Connected line identification restriction (COLR)
                   Malicious call identification (MCID)
                   Sub-addressing (SUB)

  Call offering services
- Subfeatures – Explicit call transfer (ECT)
                   Call forwarding busy (CFB)
                   Call forwarding unconditional (CFU)
                   Call forwarding no reply (CFNR)
                   Call deflection (CD)

  Call completion services
- Subfeatures – Call waiting (CW)
                   Call hold (HOLD)
                   Completion of calls to busy subscribers (CCBS)

  Multiparty services
- Subfeatures – Conference call, add-on (CONF)
                   Three party (3PTY)
                   Meet-me conference (MMC)

  Community of Interest Services
- Subfeatures – Closed user group (CUG)

  Additional Information Transfer
- Subfeature – User-to-User Signalling (UUS)

  Others
- Subfeature – Freephone (FPH)

  Charging Related Services
- Subfeatures – Advice of Charge – charging info. at call setup time (AOC-S)
                   Advice of Charge – charging info. during a call (AOC-D)
                   Advice of Charge – charging info. at the end of a call
                   (AOC-E)

Other differences between NI-1 and Euro-ISDN are evident from a comparison between the signalling protocols of both the European and American networks, and those defined in the generic ITU-T standards.

C.2 PHYSICAL LAYER DIFFERENCES

At the BRA S interface, very little difference exists between the ITU-T I.430 (1993) [3] Recommendation and the corresponding American ANSI T1.605-1991 [4] and European ETSI ETS 300 012 [5] specifications. The differences which do appear are mainly concerned with clarification, and variations to detailed parameters rather than overall functionality. In fact, the European prETS 300012:1991 document refers in whole to the ITU-T I.430 specification, but contains a detailed list of specific changes to the text of I.430. Many of the changes noted in the 1991 ETSI document referred to the 1988 Blue Book I.430 Recommendation which was subsequently updated in 1993, incorporating many of these changes. Specific examples of where differences exist between European and American S interface standards are in the use of the Q- and S-channels, and in loopback configurations.

At the BRA U interface, American ISDNs use the 2B1Q line code system, while in Europe there is a mixture of both 2B1Q and 4B3T systems, although the trend is towards the use of 2B1Q. Within 2B1Q, American ISDNs make more use of the test and maintenance facilities than in European ISDNs due to the fact that the NT1 in America is considered customer premises equipment and is the property of the subscriber and not the network operator. There is therefore a greater need to remotely access the NT1 for test and maintenance purposes.

Differences in power-feeding capabilities across the user–network interface were summarized in Table 3.3. A typical power-feeding configuration in Europe is for an NT1 to be powered from the network, while the S interface power source 1 under normal operating conditions is derived locally from a mains power supply within the NT1. If mains power fails, then a restricted primary power source is derived from the power feed to the NT1 from the network. Activation of the backup is automatic on mains power failure such that the current status of any voice calls is maintained. The situation in America is different in that the NT1 can be expected to provide its own S interface primary power source backup using batteries as a suitable supply arrangement from the network is not available.

For the subscriber loop of American ISDNs, the TE1s and NT1s are self activated instead of using the on-demand activation and deactivation procedures found in European ISDNs. This means that as soon as a TE1 is connected to the S bus and powered-up it automatically initiates activation and remains permanently activated until it is physically disconnected from the subscriber loop.

C.3 DIFFERENCES IN DATA-LINK LAYER (LAPD) STANDARDS

Only minor differences exist between the ITU-T Q.920/1 layer 2 LAPD specifications and those adopted by different countries for their ISDN implementations. This is mainly due to the fact that LAPD is a derivative of the well established LAPB protocol, and was available early as a CCITT Recommendation in the 1984 Red Books.

The ETSI standard defining LAPD for use in the subscriber loop of European ISDNs is ETS 300 125 [6]. This standard is identical to the ITU-T standard apart from additional text used to assist in the description of the procedures and their application in European ISDNs, and the following clarification of optional items.

- The TEI value 0 is reserved for the point-to-point signalling connection between an NT2 and the network. This signalling connection will also support the broadcast data-link connection.
- The optional procedure for the re-transmission of REJ frames by a data link while still in a reject exception condition is not permitted.

The American standard defining LAPD is the Bellcore Technical Reference TR-TSY-000793 [7], and further clarifications are made in the Bellcore Special Report SR-NWT-001953 [8] for its application in the NI-1 network. The main points are listed below.

- It is recommended that a terminal maintains its data-link connection in a Multiple-frame Established state even during periods of call inactivity. When the data link is deactivated, the TEI is lost. This is in contrast to Euro-ISDN which allows the call control protocol stack to be deactivated when there are no calls present, but retain the current TEI value.
- Should a power failure condition occur over a large area serviced by an ISDN exchange switch, then the re-establishment of power will cause all those terminals affected to activate their physical layers and data-link layers to the Multiple-frame Established state. Consequently, those terminals with automatic TEIs will all request a new TEI from the network exchange at about the same time with the danger that it becomes overloaded. To overcome this problem, a T-WAIT timer is defined based on the count-down of a random number, after which the terminal may proceed with TEI assignment procedures. The duration of the timer is in the range of 1 second to 5 minutes, unless a user attempts to make a call while the T-WAIT timer is active in which case TEI assignment procedures are started at that instant.
- The default value of the link monitor timer T203 is 30 seconds.
- An optional counter, d1, is defined to reflect the maximum number of calls that are allowed to be established at one time. This feature is particularly useful in the network exchange to prevent it from overload conditions.

C.4 DIFFERENCES IN NETWORK LAYER (CALL CONTROL) STANDARDS

Differences between the signalling network layers of different ISDNs can be expected to be high due to the different functional requirements of the networks that arises from their implementation. This may result in different options being chosen for the implementation of a specific function, and although signalling will usually be based on the core messages and information elements defined in Q.931, additional messages and information elements may also be used implement other functions not yet covered by Q.931. The differences at this level between today's ISDNs are too numerous to list, so we confine the discussion here to only the main functional differences. For further reference, the layer 3 implementation features for NI-1 are described in reference 8 while the basic call control specifications for Euro-ISDN are contained in reference 9.

The NI-1 layer 3 protocol control defines several non-standard messages required to support electronic key telephone system (EKTS) installations and a number of national and network specific information elements. The specific support EKTSs is unique, and is provided through special messages such as KEY HOLD, KEY RELEASE, KEY SETUP and KEY SETUP ACKNOWLEDGE. These messages are supported according to the method described in Q.931 whereby the first byte of the message type in the information element field is encoded with the escape sequence '0' followed by a second byte that identifies the national specific message. Similarly, the national specific information elements, such as Display Text and Operator System Access, make use of codeset 5, while the network specific information elements such as Call Appearance, Redirecting Subaddress and Redirection Number use codeset 6.

Although not a part of NI-1, reference 5 defines support for network controlled soft keys to support a type of terminal known as a display phone that allows a high degree of interaction with services provided by the network. Due to the infancy of standards in this area, and the need to provide multi-vendor compatibility, these services are intended for deployment as part of NI-2.

Euro-ISDN departs only slightly from the messages and information elements defined by Q.931. In particular, it makes optional use of a SEGMENT message and procedures for circuit-mode calls[46] that segment and reassemble layer 3 messages that are too long to fit in a single layer 2 data-link frame.

Supplementary services are very different between Euro-ISDN and NI-1. Although the description of some services may appear similar, their implementations are significantly different. NI-1 makes use of the feature key management as a means of invoking and controlling service features, while Euro-ISDN defines both feature key management and functional protocol

---

[46] Q.931 defines the use of the SEGMENT message only for the control of non-call associated temporary signalling connections, for example in the control of supplementary services while no call is present, and the transfer of user-to-user information.

procedures (Chapter 5), although it is expected that most services will be implemented in the longer term using the functional protocol. American ISDNs conforming to NI-1 invoke supplementary services through the a combination of the HOLD and RETRIEVE groups of messages and the feature key management procedures. A service that involves a second call will make use of the HOLD messages to place one of the calls on hold while the other is active or being processed using the feature key management protocol, as for example in the establishment of a conference call. While exchanges conforming to NI-1 are required to support the service invocation methods, they are not required to support the uniform operation of the services themselves, thus leading to differences in the way a user perceives the service to operate. A uniform service operation is implemented in NI-2.

The feature key management protocol involves storage in the local exchange of a terminal service profile (TSP) that is unique to the users terminal, and lists those services the user is allowed to access. It also provides a mapping between the activators and indicators (buttons and lamps) on a users terminal and the functions they perform in activating a service or indicating the status of a service to the user. As part of NI-2 it is possible for this list to be downloaded to a terminal from the exchange rather than being manually entered into the terminal.

C.5 REGIONAL DIFFERENCES WITHIN EUROPE

Due to the successful early regional deployment of ISDN in some European countries prior to the establishment of a unified Euro-ISDN standard, those countries are now faced with the problem of forward migration toward Euro-ISDN now that the standard exists and commitments have been made by network operators to implement it. In some cases, where deployment and uptake of a regional ISDN has not been widespread, a quick migration to Euro-ISDN may be possible due to the low installed base of existing ISDN lines. However, in some cases, a prolonged migration period will be necessary, and may result in the network operator having to support both regional and Euro-ISDN solutions over a period of time. It is therefore useful to comment here on some of the differences between regional ISDN standards that illustrate some of the deviations to Euro-ISDN. The two examples chosen here are the UK DASS 2 and the German 1TR6 protocols.

C.5.1 UK DASS 2

In the UK, an early form of BRA ISDN known as single-line IDA (integrated digital access) was put in place as early as the middle of 1985, and provided the subscriber with a 64 kbps user channel for voice or data, and an 8 kbps channel for low-speed data. Another 8 kbps channel is used for signalling. In total, the aggregate channel bandwidth supported by the subscriber loop amounted to only 80 kbps full-duplex due to lack of technology and uncertainty of installed subscriber loop characteristics at that time. Instead of U-transceivers employing echo-cancellation techniques, burst mode transceivers were used with a line rate of 250 kbps. The signalling protocol employed was known as DASS (digital access signalling system), which was later developed into the DASS 2 protocol [10] that was employed in the early versions of the UK PRA ISDN service known as multi-line IDA, and later as ISDN 30 launched in 1988. In 1990, a BRA ISDN service known as ISDN 2 was made available whose features, as well as signalling, conformed to ITU-T specifications as far as they were defined at that time, in other words, to the CCITT Blue Book Recommendations. Since 1992, BRA Euro-ISDN services have been available to which all existing subscribers were upgraded. However, the DASS 2 protocol used for PRA ISDN has proved to be successful and is still supported today, although most users are migrating to Euro-ISDN standards.

DASS 2 is entirely different from the ISDN call control protocols. At the data-link layer, although HDLC framing is used, an entirely different and more simple link access protocol (LAP) is defined. Call control information is communicated in UI command frames between peer LAPs, and acknowledged with a UI response frame with no information field. Error recovery is provided by a time-out, which if it occurs, or a bad response frame is received before the time-out, then a persistent mode of communication is employed by the sending LAP where the original UI command frame is persistently transmitted until an appropriate UI response frame is received. A limit of 64 is placed on the maximum number of re-transmissions allowed, after which any queued-up frames are discarded and the link is reset.

The DASS 2 protocol is time-slot oriented; that is to say, whereas in ISDN a message is associated with a particular call function through the SAPI, TEI and call reference values, in DASS 2 the association is made through reference to a particular time-slot in the PRA interface, and this convention applies at both the data-link and call control layers. The equivalent layer 3 protocol is totally different to Q.931 although it is less complex, and employs a different state machine and a different set of messages.

C.5.2 German 1TR6

The regional call control protocol defined in Germany is 1TR6 [11] for both BRA and PRA ISDN. Although existing 1TR6 installations continue to be supported, all new installations conform to Euro-ISDN. The physical and data-link layers of 1TR6 are similar to ITU-T Recommendations, while the network layer shows considerable deviation. It is made up of three processes as follows:

Transaction process: Basic call control functions.

Editor process: Provides status of supplementary services, and was intended as platform on which further record processing functions could be implemented.

Global process: Processes broadcast messages before passing them either to the transaction process or the editor process.

Two different protocol discriminators are used to identify messages as belonging to either the transaction process (0x41) or the editor process (0x40) instead of the usual layer 3 message protocol discriminator (0x08). As well as the usual Q.931 messages, 1TR6 defines a number of specific messages and information elements, in particular for the editor process. Handling of supplementary services is also different, and messages such as FACILITY and STATUS within the 1TR6 network layer are also treated differently to that in Q.931. Finally, no restart procedure is defined in 1TR6.

Unlike Euro-ISDN, 1TR6 provides a semi-permanent connection service with a fixed price tariff applied for calls to pre-defined numbers, which is suited for data applications such as LAN-to-LAN interconnection or leased-line backup.

REFERENCES

1. Bellcore (1991) National ISDN-1. Special Report SR-NWT-001937, Issue 1.
2. Bellcore (1992) National ISDN-2. Special Report SR-NWT-002120, Issue 1.
3. ITU-T (1993) Basic User-Network Interface – Layer 1 Specification. Recommendation I.430.
4. ANSI (1991) Integrated Services Digital Network (ISDN) – Basic Access Interface for S and T Reference Points (Layer 1 Specification). T1.605-1991.
5. ETSI (1991) Integrated Services Digital Network (ISDN); Basic user-network interface Layer 1 specification and test principles. ETS300012.
6. ETSI (1991) Integrated Services Digital Network (ISDN); User-Network interface data link layer specification. Application of ITU-T Recommendations Q.920/I.440 and Q.921/I.441. ETS 300 125.
7. Bellcore (1988) ISDN D-channel Exchange Access Signalling and Switching Requirements (layer 2). TR-TSY-000793.
8. Bellcore (1991) Generic Guidelines for ISDN Terminal Equipment on Basic Access Interfaces. Special Report SR-NWT-001953.
9. ETSI (1990) Integrated Services Digital Network (ISDN); User–network interface layer 3. Specifications for basic call control. ETS 300 102-1.
10. British Standards Institution (1991) Apparatus for connection to public telecommunications systems using the Digital Access Signalling System No.2 (DASS 2) via a 2048 kbps CCITT Recommendation G.703 interface. Part1. General Requirements. BS 7378.

11 FTZ (1988) FTZ-Richtliniensammlung. Technische Forderung an digitale Endgerate mit So-Schnittstelle. FTZ-RichtlS 1TR3, Band III, 1TR6, D-Kanal Protokoll (Schicht 2 und 3). 1TR6 Ausgabe 1.90.

Bibliography

Bellamy, J. (1982) *Digital Telephony*, John Wiley and Sons.

Bocker, P. (1992) *ISDN. The Integrated Services Digital Network - Concepts, Methods, Systems*, 2nd edn, Springer-Verlag.

Brewster, R. L. (1993) *ISDN Technology*, Chapman and Hall.

Deniz, D. Z. (1994) *ISDN and it's Applications to LAN Interconnection*, McGraw Hill.

Freeman, R. L. (1989) *Telecommunication System Engineering*, 2nd edn, John Wiley and Sons, Wiley Series in Telecommunications.

Goldstein, F. R. (1992) *ISDN in Perspective*, Addison-Wesley.

Griffiths, J. M. (1992) *ISDN Explained*, 2nd edn, John Wiley and Sons.

Hardwick, S. (1991) *ISDN Design, A Practical Approach*, Academic Press.

Helgert. H. J. (1991) *Integrated Services Digital Networks, Architectures/ Protocols/ Standards*, Addison-Wesley.

Pujolle, G., Seret, D., Dromard, D. and Horlait, E. (1988) *Integrated Digital Communications Networks*, Volumes 1 and 2, John Wiley and Sons.

Reeve, W. D. (1995) *Subscriber Loop Signalling and Transmission Handbook - Digital*, IEEE Press, IEEE Telecommunications Handbook Series.

Ritchie, W. K. and Stern, J. R. (eds) (1993) *Telecommunications Local Networks*, Chapman & Hall, BT Telecommunication Series.

Ronayne, J. (1991) *Introduction to Digital Switching*, Pitman.

Stallings, W. (1992) *Advances in Integrated Services Digital Networks (ISDN) and Broadband ISDN*, IEEE Computer Society Press.

Stallings, W. (1992) *ISDN and Broadband ISDN*, 2nd edn, Macmillan.

Winch, R. G. (1993) *Telecommunication Transmission Systems*, McGraw-Hill.

Contacts

In several places throughout this book, examples have been taken from 'real' devices and numerous references have been made to comparable devices from a number of manufacturers. The purpose of these examples has been to illustrate the principles discussed in the text rather than to act as a detailed reference guide for a particular device. Consequently, as manufacturers choose to modify their designs in subsequent revisions of devices, it is likely that with time there will be some discrepancies between current revisions of devices and some of the details explained or tabulated here. If a reader is looking for the latest information of a particular device, then this can be gained from the manufacturers' databooks. Alternatively, many companies now publish their datasheets on the World Wide Web as a convenient source of reference. Below there is a list of Internet World Wide Web sites for the manufacturers of those devices mentioned in this book. In addition, a section has been added for other relevant Web sites on ISDN and related material that the reader may find useful.

ISDN DEVICE MANUFACTURERS

Advanced Micro Devices:
http://www.amd.com

AT&T Microelectronics (now known as Lucent Technologies):
http://www.attme.com

Brooktree Corporation:
http://www.brooktree.com

Dallas Semiconductor:
http://www.dalsemi.com

Level One:
http://www.level1.com/products.html

Mitel:
http://www.semicon.mitel.com/Products.html

Motorola:
http://www.mot.com/SPS/MCTG/MDAD/isdn.html

National Semiconductor:
http://www.nsc.com

NEC:
http://www.ic.nec.co.jp/index_e.html

SGS Thomson:
http://www.st.com

Siemens Semiconductor:
http://www.siemens.de/Semiconductor/products/products.htm

OTHER USEFUL CONTACTS

ADSL Forum:
http://www.adsl.com/adsl

Bellcore ISDN:
http://www.bellcore.com/ISDN/ISDN.html

Broadcast ISDN Users Guide and Directory:
http://www.sms.co.uk/isdn/home.htm

Dan Kegel's ISDN Page:
http://www.alumni.caltech.edu/~dank/isdn/isdn.html

ECMA:
http://www.ecma.ch

ETSI:
http://www.etsi.fr

European ISDN Users Forum:
http://www2.echo.lu/eiuf/en/eiuf.htm

IETF RFC documents:
http://www.cis.ohio-state.edu/htbin/rfc

ISDN for LAN Interworking:
http://www.acc.com/White/isdn.html

ITU:
http://www.itu.ch

North American ISDN Users Forum:
http://www.nist.gov/misc/niuf.html

Telechoice (information and news on xDSL technologies):
http://www.telechoice.com/xdslnewz

Glossary

2B1Q – two bits in one quaternary symbol. This is the line code use for the basic rate access U interface transmission system in North America and in parts of Europe.

3B2T – three bits in two ternary symbols. A line code used in early pre-standardized ISDN U transmission systems.

4B3T – four bits in three ternary symbols. A line code used in parts of Europe for the basic rate access U interface transmission system.

A

AC – alternating current. A current whose direction periodically reverses and whose amplitude varies with time according to a sinusoid. The average current over time is zero.

A/D – analogue to digital. Conversion of an analogue signal into a digital signal.

address – an identification which allows a device or a piece of data to be uniquely located or identified in a group of similar objects. In a telecommunications network, the number which is used to call the terminal can be thought of as its address. Similarly, in a microprocessor system, every memory or external register location has a unique binary number address which the microprocessor uses to access the information within the memory or register.

ADPCM – adaptive differential pulse code modulation. An encoding/decoding method that provides compression of a standard PCM speech channel to 32 kbps.

ADSL – asymmetric digital subscriber loop. A subscriber loop technology that delivers a high-bandwidth channel in one direction and a low-bandwidth channel in the opposite direction. The use of such subscriber loops was originally intended for the delivery of video on demand (VoD) services to residential subscribers where a TV/video channel is provided from the network to the subscriber, and control channel from the subscriber to the network in order to select and control the TV/video. Additional bi-directional services such as ISDN and POTS can also be delivered.

AM – amplitude modulation. A modulation technique where the amplitude of a carrier signal is varied proportionally to the amplitude of a modulating signal.

AMI – alternate mark inversion. A three-level line code consisting of a positive and negative line voltages, and a zero line voltage (high impedance). Either successive '1's or '0's in a digital stream are alternately encoded as positive and negative line voltages, while the remaining value takes on a zero line voltage. AMI is used in the ISDN S transmission system.

ANI – automatic number identification. ANI is the name given to the service in North America which provides the calling user's number to the user receiving the call. This service is similar to the CLI (calling line identification) services in Europe.

ANSI – American National Standards Institute. The main standards body in North America for ISDN standards.

API – Application program interface. An abstract set of definitions of function calls used by an application to a service function such as a communications protocol stack or an operating system. An API divorces the application from the implementation dependencies of the service.

ASCII – American Standard Code for Information Interchange. The code used to represent alphanumeric and terminal control characters for information exchanged between a computer and a terminal.

Asynchronous transmission – a mode of communication which is characterized by transmissions which start and stop with undefined time intervals between them. The serial clocks used for the transmission and reception of data at each end of the communication link are not synchronized with one another.

ATM – asynchronous transfer mode. The communication technique chosen for the implementation of Broadband ISDNs, and is also used in local high-speed data networks. Different types of data, such as voice, video and data, are split into groups of 48 bytes to which a 5 byte header is added to create an ATM cell. These cells can be transmitted at a number of rates up to 2.4 Gbps and can be switched using fast hardware techniques. A major benefit of ATM is that it is scaleable and allows flexibility in the way the bandwidth is used.

Attenuation – the relative amount by which the energy of a signal is reduced. A common way of expressing attenuation is in terms of decibels (dBs). A decibel is defined as $10\log_{10}(P_1/P_2)$, where P_1 is the reduced signal power and P_2 is the normal signal power.

AU – administrative unit. A data structure within an SDH frame that allows a virtual container (VC) to be offset to allow for varying transmission path lengths, and for payloads that are slightly out of synchronization with the SDH.

AWG – American wire gauge. A standardized system for the classification of wire conductors according to their thickness.

B

Balanced – a transmission method on a pair of wires where the signal voltage and polarity on one wire is referenced to that of the other rather than ground potential. The electrical path for both wires is balanced such that any common mode signals (common to both wires) do not affect the transmission signal.

baud – the unit of symbol rate used in digital transmission systems. Using different coding schemes, one or more data bits are encoded as a symbol for transmission. A transmission rate of one symbol per second is equal to one baud.

BC – bearer capability. An information element within ISDN call control messages that defines the channels to be used for a call connection.

BER – bit error rate. The rate at which bits are corrupted by a transmission system such that they are received incorrectly. ISDN systems generally specify bit error rates of the order of one bit in 10^7.

B-ISDN – broadband ISDN. The next generation of ISDN that can provide high-bandwidth switched connections with speeds greater than 2 Mbps. All types of information, voice, video and data, are transported and switched in fixed size ATM (asynchronous transfer mode) cells.

bps – bits per second. Terminology used to represent a unit of digital transmission speed. Other units frequently used are kbps (kilo–bits per second) for a unit of a thousand (10^3) bits per second, Mbps (mega–bits per second) for a unit of one million (10^6) bits per second, and Gbps (giga–bits per second) for a unit of one thousand million (one US billion) (10^9) bits per second.

BRA – basic rate access. An ISDN subscriber loop that provides the user with access to two individual 64 kbps B-channels for calls, and a 16 kbps D-channel for call signalling (often referred to as 2B+D).

BRI – basic rate interface. The user–network interface at which the user can access the 2B+D channels.

Bridged tap – extra pair of wires connected in a shunt configuration to the main subscriber loop cable pair. The provision of this extra wire is for future connection of a new subscriber premises to the main cable, and in the meantime may be left unconnected. While the bridged tap connection has no affect on analogue voice frequency signals, its unterminated open circuit connection causes echoes that can seriously degrade the transmission of higher speed digital signals.

C

CAP – carrierless amplitude/phase modulation. A form of QAM (quadrature amplitude modulation) developed by AT&T and used in high-speed digital subscriber loops.

C/R – command/response. A bit within the control field of an HDLC frame that marks the frame as a command or a response.

CCITT – Comité Consultatif International Télégraphique et Téléphonique (International Telephone and Telegraph Consultative Committee). The organization that is responsible worldwide for defining telecommunication standards known as **Recommendations**.

CCSS 7 – Common channel signalling system number 7. This is the common channel signalling used within the network of the ISDN.

Centrex service – the provision by the network operator of services to a group of user terminals which make them appear as if they were connected to a PABX. A group of users with the Centrex service could therefore dial each other directly using only their extension number, forward calls, place calls on hold, connect to waiting calls, and so on.

CEPT – Conference Européenne des Administrations des Postes et des Télécommunications. An organization of representatives from European national network operators and administrations that define European telecommunications policy.

Channel associated signalling – a method of signalling in which the signalling information associated with a channel is carried in the channel itself, or in a separate signalling channel which is permanently associated with it.

Circuit switching – the manner in which an end-to-end connection is established across a network that provides the user with a dedicated channel for the duration of the call (connection).

Clk – clock. A continuous digital signal which drives digital synchronous circuits. The transitions in a clock signal cause an update in the state of the circuit based on its input conditions. In a serial digital communications system, the transition of a clock signal delimits the transmission and reception of a unit of information.

CO – central office. North American terminology for a local telephone switching exchange.

Codec – coder/decoder. A device which provides bi-directional conversion of signal coding format.

Common channel signalling (CCS) – a method of signalling in which the signalling information associated with a number of user channels is carried in a channel which is common to those user channels.

CPE – customer premises equipment. A term used to refer to equipment connected to a public telecommunications network and which is the responsibility of the customer. The boundary of the public network is at the customer premises.

C-plane – control plane. A term used to describe the set of protocols that perform signalling and control. It is used in conjunction with the U-plane, or user plane, which describes those protocols which operate on the user data within the communications channel. The distinction between C-plane and U-plane is convenient when describing a network such as ISDN in which the end-to-end call control signalling is entirely out-of-band, and may be considered to be performed by a separate network within the ISDN.

CPU – central processing unit. The main computation and control unit of a computer. These functions may be implemented on a single IC as a microprocessor.

CRC – cyclic redundancy check. A check word appended to a data string which is used to detect simple errors caused by transmission of the data string on noisy channels. The check word is calculated as the remainder obtained when the data is divided by a generating polynomial. The process is also applied at the receiving end using the same generator polynomial, and if the remainder matches that sent as the CRC, then there is a high probability that no errors were introduced into the data during transmission.

Crosstalk – the cross-coupling of signals between communications channels in close proximity with one another.

CSTA – computer-supported telecommunications application. A protocol defined by ECMA (European Computer Manufacturers Association) and ETSI (European Telecommunications Standards Institute) for the exchange of call control information between a computer and call processing equipment such as a PBX.

CTI – Computer telephony integration. A generic name for applications that involve the control of telephony functions by a computer.

CTR – Common Technical Regulation. A test standard used in Europe that defines tests and test procedures for the connection of terminal equipment to the public networks, including ISDN.

D

D/A – digital to analogue. The conversion of signals from a digital representation to an analogue representation.

DASS – Digital access signalling system. The call control signalling protocol used in early non-standard ISDNs in the UK. A later revision, DASS2 is still supported on some PRA connections.

dB – decibel. See **attenuation**.

DC – direct current. A constant current whose polarity and amplitude do not vary with time.

DCE – data communication equipment. Equipment that provides the functions required to establish, maintain and terminate a connection. In addition, the DCE also performs the signal conversion and coding required to interface to the subscriber loop cable.

DCT – digital cosine transform. A transform used to encode a pixel image in terms of its spatial frequency components.

DDI – direct dialling in. A supplementary service that allows the individual extensions of a PBX to be dialled directly by an external call.

DFE – decision feedback equalizer. An adaptive equalizer that continually compensates for the frequency distortions introduced as a pulse propagates down a subscriber loop in order to provide the optimum sampling instant for the received signal.

DISC – disconnect message. A LAPD/LAPB control message used to terminate multiple frame operation of the data-link layer.

DLCI – data-link connection identifier. A combination of the LAPD data-link SAPI and TEI addresses in an HDLC frame that uniquely identifies the relevant data link for the message.

DM – disconnect mode. A LAPD/LAPB control response message which is sent to indicate that the data-link layer is not in a multiple frame operation state.

DMI – digital multiplexed interface. A specification defined by AT&T for terminal adaptation. DMI was developed prior to ISDN and several of its features are included in the ISDN terminal adaptation specifications.

DPCM – differential pulse code modulation. A form of PCM in which the difference between the current sample and the previous sample of an analogue signal is encoded and sent as a PCM value. Using the difference signal rather than the absolute value results in a reduction in the number of bits required to encode the signal.

DPLL – digital phase-locked loop. A circuit used in digital transceivers to recover the clock information encoded as part of the received data stream. This clock signal is then used to sample the received signal to recover the digital information.

DPNSS – digital private network signalling system. An inter-PBX signalling system defined prior to ISDN which allowed the features of a PBX to appear uniformly across a network of PBXs from different manufacturers.

DPSK – differential phase shift keying. A modem modulation technique in which the difference in phase between successive line symbols is used to represent digital information.

DRAM – dynamic random access memory. A type of RAM whose contents is retained through a continuous refresh process. Its advantage is that a DRAM memory cell is small and very simple allowing high-density memory devices to be realized. However, because of the overhead associated in refresh circuitry, it is mostly suitable in systems with large RAM requirements.

DTMF – dial-tone multifrequency. A series of dual tones that correspond to the keys on a telephone, and which are used as an alternative to pulse-dialling and for in-band end-to-end control of terminal equipment such as answering machines.

DTE – data terminal equipment. Equipment that provides the source and destination for information communicated across the network.

E

ECMA – European Computer Manufacturers Association. A European association whose purpose it is to facilitate the standardization of information processing and telecommunication systems.

ECSA – Exchange Carriers Standards Association. An North American association established after the split-up of AT&T in 1984 to administer the establishment of standards for telecommunications networks. ECSA sponsored the formation of the T1 committee that formulates many of the North American standards that are published by ANSI. ECSA is now known as the Alliance for Telecommunications Industry Solutions (ATIS).

EDI – electronic data interchange. The transfer of structured data, in an agreed format, from computer to computer by means of electronic communications.

EMI – electromagnetic interference. Interference radiated by circuits containing rapidly changing signals, for example logic switching circuits.

EOC – embedded operations channel. A channel embedded within a framed bit stream of a communications link that is used to convey information concerned with operation and maintenance functions associated with the communications link.

EPROM – a shortened form of UV-EPROM meaning UV-erasable electrically programmable read-only memory. A semiconductor memory whose contents can be electrically programmed, and erased by exposure to UV light.

EEPROM – electrically erasable programmable read-only memory, also known as E^2PROM. A semiconductor memory whose contents can be both electrically programmed and erased.

ETSI – European Telecommunications Standards Institute. The standards body charged with defining standards for public telecommunications networks in Europe.

F

FCC – Federal Communications Commission. The main body in North America that determines regulatory policy on all aspects of communications.

FCS – frame check sequence. A method of appending a codeword to a data string or block, which is calculated from the data itself and used at the receiver to determine if errors occurred during transmission. See also CRC.

FDM – frequency division multiplexing. A process that allows two or more signals to be transmitted over a common connection through the use of a different frequency band for each signal.

FEBE – far-end block error. An error that occurs in the reception of a block of data at the far-end receiver in a communications system. Often, a counter is used to record the occurrence of errors, and the value of the counter is included in data sent back towards the near-end transmitter. This allows the transmitter to directly monitor the quality of the communications channel.

FEXT – far-end crosstalk. In a duplex communications link, the far-end crosstalk is that induced at the far-end transmitter from another adjacent far-end transmitter.

FIFO – first-in first-out. A buffer store whose output preserves the order in which information was placed at the input.

FITL – fibre in the loop. A broad classification of technologies that employ optical fibre in the subscriber loop network.

FM – frequency modulation. A modulation technique where the frequency of a carrier signal is varied according to the amplitude of a modulating signal.

FRMR – frame reject message. A LAPB/LAPD message used by indicate that a receiver has rejected the previous frame sent due to an unrecoverable error.

FS – frame synchronization signal. A digital pulse used in a serial time division multiplexed bus to delineate periods of 125 µs that are normally synchronized with the network 8 kHz clock.

FTAM – file transfer, access and management. An OSI applications layer protocol for the access, management and transfer of files across an OSI-based communications network.

FTTC – fibre to the curb. A technology which uses optical fibre to deliver subscriber access channels from a local exchange to a point close to a group of users, typically a curb side cabinet, and from where traditional copper subscriber loop pairs or other established technologies distribute services to the users premises.

FTTH – fibre to the home. The use of optical fibre subscriber loops deployed between the local exchange and customer premises in a residential environment for the delivery of a range of services from low-bandwidth telephony to high-bandwidth television.

Firmware – the name given to executable code derived from software that is embedded within a microprocessor system in fixed memory, typically some kind of read-only memory (ROM).

G

GCI – general circuit interface. A standardized synchronous serial TDM bus used to interconnect devices or sub-systems within a piece of equipment. The bus was defined by Siemens (Siemens refer to it as IOM-2), Alcatel, GPT and Italtel. As well as conveying synchronous data streams, such as PCM coded voice, in the time-slots within the TDM bus, a control channel is defined that allows devices connected to the GCI bus to be controlled from a master GCI control device.

GOB – group of blocks. A data structure that defines the organization of picture information within an H.261 encoded video stream.

H

HDLC – high-level data-link control. A ISO standardized OSI layer 2 data-link protocol originally used for communications between mainframe computers, but whose framing and elements of procedure have subsequently provided the basis for many telecommunication data-oriented data-link protocols such as LAPB and LAPD.

HDSL – high bit-rate digital subscriber loop. An extension of ISDN echo-cancellation transceiver technology on multiple subscriber loops to deliver 1.544 Mbps or 2.048 Mbps connections across PSTN subscriber loop cables up to distances of 3.5 km (12 000 feet).

Hz – hertz. A unit of frequency representing one cycle per second. Higher order units are the kilohertz (kHz) of one thousand (10^3) cycles per second, megahertz (MHz) representing one million (10^6) cycles per second, and gigahertz (GHz) representing one thousand million (one US billion) (10^9) cycles per second.

I

I/O – input/output.

IA5 – International Alphabet number 5. A mapping between alpha-numeric characters and digital numbers that allows text to be communicated in a standard format. IA5 corresponds to the common ASCII (American Standard Code for Information Interchange) codes used for the same purpose.

IC – integrated circuit. A sub-miniature electronic circuit consisting of semiconductor devices created on a single piece of semiconducting material.

IDN – integrated digital network. A generic name that is given to a core telecommunications network capable of transporting and switching circuit and packet information, and whose switching, transmission and internal signalling uses digital technology.

IDL – interchip data link. A four-wire serial TDM bus developed by Northern Telecom used to interconnect semiconductor devices or sub-systems within a piece of equipment.

IEC – interexchange carrier. A company in North America, for example AT&T or MCI, that provides long-distance services between local exchange areas.

I frame – The LAPB and LAPD information frame used to transfer information between peer network layer protocols. I frames use a numbering sequence to provide error-free, in sequence delivery of frames across the data link.

IOM-2 bus – ISDN Oriented Modular (revision 2) bus (trademark Siemens). Siemens' version of the GCI bus, a serial TDM bus used to interconnect semiconductor devices or sub-systems within a piece of equipment.

IN – intelligent network. A digital telecommunications network with an advanced management architecture that allows quick and flexible provisioning of services to customers.

ISDN – integrated services digital network. A digital network and access that provides multiple services through a single connection to the user.

ISI – intersymbol interference. An effect caused by the spreading of pulses, or symbols, in time as they are transmitted down a cable, and that eventually overlap with adjacent pulses resulting in interference. The pulse spreading itself is caused by its frequency components travelling with different speeds down a cable.

ISO – International Standards Organization. An international organization which specifies standards for data communication network protocols. It is most commonly known for its reference model for open systems interconnection (OSI-RM).

ISUP – ISDN user part. Part of the signalling system number 7 that provides network signalling support for ISDN services beyond basic telephony. For example, supplementary services as well as support for teleservices require ISUP.

ITU - International Telecommunication Union. This is a specialized agency of the United Nations that is responsible for telecommunications. The ITU-T, a part of the ITU, is charged with the formulation on a world-wide basis, of non-mandatory recommendations for standards in the area of telecommunications.

IVDT - integrated voice and data terminal. A single terminal equipment that integrates access to both data and voice communication services.

J

JPEG - Joint Photographic Experts Group. An ISO standards group that defines standards for the digital encoding and compression of high-quality colour images for storage and transmission in a digital format.

K

kbps – kilobits per second. See **bps**.

kHz – kilohertz. See **Hz**.

KTS – key telephone system. A small PBX that is typically used for small offices with up to 50 extensions. The operation of a key system is simpler than a PBX in that no operator is required, and a user has access to external lines and other users through simple operation of a single key. Hence, the handsets of a key telephone system have a large number of keys, which give the system its name.

L

LAN – local area network. A network used within the customer premises for packet data communications. LANs operate at moderately high data rates, typically between 1 Mbps and 100 Mbps, and are used to interconnect all forms of data processing equipment including PCs and workstations, printers and file servers. LANs are interconnected to each other and other types of network through routers. ISDN is becoming one of the most popular ways of interconnecting LANs across the public network.

LAPB – link access procedure balanced. The data-link protocol used in X.25 that defines a set of procedures used in the communication of HDLC frames between two pieces of equipment, typically a terminal and the access point to the network.

LAPD – link access procedure for the D-channel. The data-link protocol used in the D-channel of an ISDN user–network interface. An extension of LAPB, the LAPD also uses HDLC frames to carry information generated at higher layers of the protocol stack, and a set of procedures that ensure their correct delivery.

LE – local exchange. The first level of switching within a public switched telephone network. The local exchange is that to which a subscriber's telephone or terminal is connected.

LEC – local exchange carrier. A name used to refer to the network operator in North America that operates the local exchanges and transmission network amongst them within a defined geographic area.

LME – layer management entity. That part of the management entity whose functions are specific to a particular layer within the protocol stack.

LSB – least significant bit. Either the first of last bit within a collection of bits whose individual value has the least significance to the overall value when the bits are considered as a binary number.

LT – line termination. The function at the network end of a subscriber loop at which the subscriber line or cable is physically terminated.

M

Mbps – megabits per second. See **bps**.

MCU – multipoint conference unit. Equipment that establishes and controls a conference between participants connected to the MCU through simple point-to-point connections.

MHz – megahertz. See **Hz**.

µs – microsecond. One millionth (10^{-6}) of a second.

MMU – memory management unit. A device that manages the logical access by a CPU to a particular physical memory implementation.

Modem – modulator/demodulator. A modulator is a device that transforms a signal into one that is more suitable for transmission on a particular medium. The process of modulation encodes the information from a signal onto a carrier signal that is then transmitted. Demodulation is the reverse process, and a modem is a device that combines both modulator and demodulator to provide full-duplex communications.

MPEG – Motion Picture Experts Group. The ISO standards group that has defined compression and decompression schemes for the storage and transmission of broadcast quality video. The compression is complex and not intended to be done in real-time, and MPEG is therefore used for video playback applications.

MSB – most significant bit. Either the first or last bit within a collection of bits whose individual value has the most significance to the overall value of the bits when considered as a binary number.

ms – millisecond. One thousandth (10^{-3}) of a second.

MSN – multiple subscriber number. A supplementary service that allows a single user-network subscriber connection to support more that one number in order to address multiple connected terminals.

N

N(R) – receive sequence number. Used in LAPD and LAPB in transmitted frames to indicate the number of error free frames received by the end of the data link that originates the transmit frame.

N(S) – send sequence number. Used in LAPD and LAPB in transmitted frames to indicated the number of frames transmitted by the end of the data link that originates the transmit frame.

NEBE – near-end block error. A block error that occurs in the reception of a frame at a local receiver. A counter is usually implemented in the receiver and locally read by a management system that monitors transmission performance.

NET – Normes Européenes de Télécommunications. A qualification standard used in Europe for the approval of the use of a particular terminal equipment for connection to public networks within the European Community.

NEXT – Near-end crosstalk. Noise induced locally at a receiver from adjacent transmitters at the same end of a transmission system.

NRZ – Non-return to zero. A serial digital format in which, during a bit period, a '1' is represented by the presence of a line voltage level, and a '0' by zero level.

NRZI – Non-return to zero inverted. A serial digital format in which the transitions of the line signal during a bit period represent a '1' while no transition represents a '0'.

NT – network termination. The network side termination of a user–network interface. The NT marks the boundary between the public network and the customer premises.

NT1 – network termination of type 1. An simple NT that performs only physical signal translation between the customer premises wiring and the public network wiring that connects the NT to the local exchange.

NT2 – network termination of type 2. An NT that performs a degree of local switching and concentration, for example a PBX.

O

OA&M – operations, administration and maintenance. A generic name used for the management functions and information that form part of a communications system.

OSI – open systems interconnection. An initiative that will allow different types of communication systems to be built from multi-vendor hardware and software products through the definition of standardized open interfaces (Appendix A).

OSI-RM – OSI reference model. The reference model for a communications system based on seven layers, and which also defines the function of each layer and the interfaces between them. The OSI-RM is frequently used as a framework to describe any type of communications system regardless of its conformance to OSI protocols (Appendix A).

P

PAD – Packet assembler–disassembler. Equipment at the periphery of an X.25 network that allows simple character-based terminals to interface to the network. The PAD receives characters from a connected terminal and assembles them in to packets which are then sent into the network, while in the reverse direction, packets from the network are disassembled and their information content sent as characters to the terminal.

Pairgain – The use alternative technologies to increase the single voice-channel capacity of a subscriber loop pair. For example, it can be doubled to handle two separate voice calls through the use of ISDN BRA transceivers, and quadrupled if both ISDN and ADPCM technologies are used.

PAM – pulse amplitude modulation. A form of sampling and modulation where an analogue signal is sampled and the height of the pulse at the sampling instant is made equal to the amplitude of the signal.

PBX – private branch exchange. Customer premises voice call processing and switching equipment that allows switching between local extensions and access to external public network connections.

PC – personal computer.

PCM – pulse code modulation. A form of PAM and A/D conversion where an analogue signal is sampled and the amplitude of the signal at that instant is converted into a digital code. As part of this process, the signal is also companded in order to reduce distortion. This results in a serial digital stream at a data rate equivalent to the sampling rate multiplied by the number of bits generated per sample.

PDH – plesiochronous digital hierarchy. A transmission and multiplexing system in which each node within a network generates its own clock which is synchronized to a master clock through a synchronization hierarchy. The resulting chain of synchronization produces clocks that may deviate slightly in frequency from one another that gives rise to frame slips in the transmission and reception of frames and the potential loss of data.

PDU – protocol data unit. Within the OSI-RM a layer will process information from the layer above by adding any necessary control information to control the and coordinate its peer layer. The resulting frame passed down to the next layer is known as a PDU (Appendix A).

pel – picture element. An image is sampled by placing a grid or matrix on top of it such that it consists of a number of rows and columns. Each element within the matrix is known as a pel whose characteristics can be encoded (for example chrominance and luminance) in order to create an electronic version of the image suitable for processing and transmission.

P/F – poll/final. A bit used within the control field of LAPB and LAPD frames to indicate if the frame is issued as a command (poll) or a response (final). The terminology originates from the HDLC protocol on which LAPB and LAPD are based, where the communication may be between a master and several slave devices on a bus network.

PID – protocol identifier. A first byte within the information field of a D-channel message that identifies the protocol that is to be used to process the message.

PLP – packet level protocol. The network layer in X.25.

PM – phase modulation. A form of modulation where information is encoded as phase variations in the carrier signal.

POTS – plain old telephone service. A term used to refer to the basic analogue telephony service provided by a PSTN.

PRA – primary rate access. An ISDN subscriber loop that provides the user with access to either 30 (European) or 23 (North American) individual 64 kbps B-channels for calls, and a 64 kbps D-channel for call signalling (often referred to as 30B+D or 23+D).

PRI – primary rate interface. The user–network interface at which the user can access the 30B+D or 23B+D channels.

PSE – packet-switching exchange. A node within an X.25 network responsible for routing packets through the network to their correct destination.

PSK – phase shift keying. A form of phase modulation where the '1's and '0's in a serial digital stream are represented as phase shifts in a carrier signal.

PSPDN – packet-switched public data network. A public data network, the subscriber access to which is through X.25-based connections or through the PSTN to a PAD.

PSTN – public switched telephone network. The telephone network that uses analogue subscriber loops to connect subscribers to a basic telephony service, and in many cases facsimile, videotex and other low-speed data services through the use of voice-band modems.

PTT – post, telephone and telegraph. A general name give to network operators in Europe. In the past, each country in Europe possessed its own PTT, which was usually state owned and operated the telephone network as well as both postal and telegraph services. In many European countries today, the postal services and telecommunication services have been split, and many of the telecommunication operators have been privatized in order to be able to react to competition introduced through deregulation.

PVC – permanent virtual circuit. A path through a packet network between two users that is permanently established. No call establishment or tear-down is therefore required.

Q

QAM – quadrature amplitude modulation. A form of modulation that combines both amplitude and phase variations in the carrier signal in order to transmit multiple bits per baud.

QPSK – quadrature phase shift keying. A form of phase modulation in which four phases of the carrier signal, separated by 90°, are each used to represent two bits of digital information.

R

RAM – random access memory. A memory device that can be written to and read from, and may therefore be used to store data variables. The contents of a RAM device are usually volatile, meaning they are lost when power is removed.

RBOC – Regional Bell Operating Company. A public network operator in North America that operates over a defined geographic region. The seven RBOCs were created as a result of the divestiture of AT&T in 1984.

REJ – reject message. A reject message is used by LAPD and LAPB to request a re-transmission of frames. The reject message occurs during the transmission of numbered I frames, and will contain a reference to the frame number from which re-transmission is requested.

RNR – Receiver Not Ready. A message used in LAPD and LAPB to indicate that the side of the communications data link that transmitted the RNR message is busy and not able to receive any further I frames.

ROM – read-only memory. A memory device that once programmed with its contents, can only be read. This type of device also tends to be non-volatile such that when power is removed, it retains its contents. They are typically used for storing constant data such as programs.

ROSE – remote operations service elements. Entities within the public network that are responsible for processing and implementing supplementary services.

RR – Receiver Ready. A message used in LAPD and LAPB to indicate that the side of the communications data link that transmitted the RR message is ready to receive further I frames.

S

SABME – set asynchronous balanced mode extended. A supervisory frame in LAPD that initializes the data-link for numbered I frame operation. A modulo-128 numbering sequence is used.

SAP – service access point. A logical interface between entities within adjacent layers in the OSI-RM across which information is passed between the entities.

SAPI – service access point identifier. More that one entity exists at the data-link layer of ISDN, and the SAPI is used to identify which entity must process the information. The SAPI is implemented as part of the address of a LAPD frame.

SB-ADPCM – sub-band ADPCM. A compression scheme used to give a wideband, 7 kHz speech channel instead of the usual 3 kHz voice across a standard 64 kbps channel.

SCP – Serial Communications Port (trademark Motorola). A serial control port used to control peripheral devices connected to a microcontroller or micorprocessor.

SDH – synchronous digital hierarchy. A standardized transmission system based on high data rate (155 Mbps to 2.4 Gbps) optical fibre systems that provides flexible multiplexing of many types of lower bit-rate channels while maintaining synchronization with a single master clock.

SDL – specification and description language. A formal method defined by the ITU-T for the unambiguous specification and description of the behaviour of telecommunications systems.

S-frame – supervisory frame. A frame used in LAPD and LAPB to coordinate and control the operation of the data-link.

SLD bus – Subscriber Line Datalink Bus (trademark Intel). A serial TDM bus used to link devices in a system that process information within the synchronous channels in the TDM bus.

SLIC – subscriber line interface controller. A device that provides the electrical interface to the subscriber line at the local exchange. For analogue subscriber loops it is also responsible for control and supervision functions such as off-hook detection, battery feed, dial detection and ringing.

SNR – signal-to-noise ratio. A ratio that indicates the amount of noise present in a signal. SNR is usually given in units of decibel (dB) defined as $SNR(dB)=10\log_{10}(P_s/P_n)$ where P_s is the signal power and P_n is the noise power.

SONET – synchronous optical network. The transmission network defined by Bellcore and ANSI for use in North America. SONET provided much of the foundation for SDH which embodies many of its techniques although it is based on a superset of the SDH hierarchy.

SPID – service profile identifier. A number used in the feature key management protocol for the control of supplementary services that identifies a subscriber number with a set of services to which the user has subscribed.

SRAM – static random access memory. A type of RAM whose operation relies only on the presence of power in order to retain its contents. An SRAM memory call is based on a bistable circuit which is significantly more complicated than its DRAM counterpart, making it most suitable for systems with low RAM and low-power requirements.

Statistical multiplexing – the ability to establish multiple logical links in a single channel. The link is identified by the address of the frame or packet communicated in the channel. During its transmission, a frame or packet will occupy all of the channel bandwidth, but due to the fact that they are communicated with sufficient interframe intervals so as to be equivalent to the required link rate (which is less than the channel bandwidth), frames or packets associated with other links can occupy the interval. The amount of bandwidth occupied by a link is determined by the rate and length of frames transmitted, thus giving rise to the name statistical multiplexing.

SUB – sub-addressing. A supplementary service that allows a sub-address, for example an extension number on a PBX, to be included in the call setup messages across the network, that allows the call to be automatically directed by the PBX to a specific extension.

SVC – switched virtual circuit. A connection in a packet data network that is established on-demand using call setup procedures that defines the end-to-end route through the network for the duration of the call. Other connections may share parts or all of the same physical route, but due to the statistical multiplexing nature of packet data networks, the connection appears to the user as a circuit connection.

Synchronous transmission – a form of transmission in which the serial clock used to recover data at the receiver is synchronized with that of the transmitter. The clock information may be communicated between the receiver and transmitter separately from the data, or is more usually encoded within the data stream itself.

T

TA – terminal adapter. A device that adapts a non-ISDN terminal to the user–network interface of the ISDN.

TCM – time compression multiplexing. A digital transmission technique used to achieve full-duplex communications on a single twisted-pair cable without the need for a hybrid circuit and echo cancellation. The technique relies on using twice the bit rate required and alternately reversing the direction of transmission such that within a specified time information is transmitted in one direction and then in the reverse direction.

TDM – time division multiplexing. A technique in which a repeated frame is partitioned into defined slots in time within the frame. Usually, the slots are equal in size and are referenced by their position from the beginning of the frame with respect to time. The information in the time-slots are independent from one another such that the composite frame can be considered a multiplex in the time domain of the individual time-slots or channels. If each channel has a bandwidth of N bps, then multiplexing M channels will result in an aggregate transmission bit rate of $N \times M$ bps. In addition, in transmission networks, frame delineation may be added as well as other overhead bits that will slightly increase this value.

TDMA – time division multiple access. An access technique for shared media networks where many terminals share a common communications channel. Each terminal is given access to the entire bandwidth of the communications channel for a specific period of time, after which another terminal has access, and so on. The order of access to the channel by terminals remains constant for the duration of a connection between terminals.

TE1 – terminal equipment of type 1. An ISDN terminal that has the ability to connect to the user–network S interface, uses one or more of the channels at the S interface to communicate information, and is capable of call control signalling.

TE2 – terminal equipment of type 2. A non-ISDN terminal that does not have the ability to connect to the user–network S interface but may be connected to it by a terminal adapter (TA) through some other common interface.

TEI – terminal end-point identifier. A seven-bit address used in the HDLC frame of LAPD to identify to which terminal connected at the user–network interface the message is addressed. Up to eight terminals may be connected to the S bus, and each has a unique TEI either pre-assigned (for example through the use of a switch on the terminal) or is assigned a TEI by the local exchange when the terminal is newly connected to the S bus and either makes or receives a call.

TSI – time-slot interchange. A circuit used to assign information from one time-slot to another time-slot in a TDM frame.

TU – tributary unit. A data structure within an SDH frame similar to an administrative unit (AU) that allows data streams from tributaries feeding into and out of the SDH to be multiplexed into higher order virtual containers (VC) for transport across an SDH network. A TU has relevance over an end-to-end path within the SDH network, while the AU is relevant across a multiplex section within the SDH, several of which may make up a path.

TUP – telephone user part. That part of common channel signalling system number 7 that is used to support basic telephony functions.

Transcoder – a device which converts information from one encoding scheme to another.

U

UA – unnumbered acknowledge. A LAPD supervisory frame used to acknowledge receipt of an unnumbered message.

UART – universal asynchronous receiver transmitter. A serial data communications device that formats and sends individual bytes of information using an asynchronous transmission scheme, that is, no clock information is transferred with the transmitted data. Clocks at the transmitter and receiver run independently of one another. However, the receiver clock synchronizes itself to the receipt of a start bit that indicates the beginning of a byte. Although the clocks are independent of one another, they have the same nominal frequency, and as only 8 to 11 bits are transmitted (7 or 8 data bits, an optional parity bit, and either 1 or 2 stop bits) the likelihood that drift between the two clocks would result in the receiver sampling the received line data incorrectly, is small.

U frame – unnumbered frame. On of three types of frames used in LAPD, the unnumbered frame service transmits and receives frames without sequence numbering.

UI – unnumbered information. A U-frame used in LAPD to transfer information between peer layer 3 entities.

UUI – user–user information. A supplementary service that allows terminal or NT-2 equipment to exchange information across the ISDN during call control. The information is embedded as information elements within call control messages and can be used, for example, to configure the operation of the application at both terminals prior to the exchange of user information.

U-plane – user-plane. A conceptual segmentation of protocols that operate on user or application information as opposed to call control information. See also C-plane.

V

VAN – value-added network. A network which provides a service to the user that adds value above that of the basic transmission and switching of information between end-points by the provision or processing of information itself such that it is in a more suitable form for the user.

VC – virtual container. Part of an SDH frame that contains the payload information that is transported along a path within the SDH network. As well as the payload, the VC contains some overhead that relates to control of the path connection.

VCI – virtual channel identifier. An identifier within the header of an ATM cell that identifies the virtual channel with which the cell is associated, and which will form part of the logical end-to-end connection between terminals. Cells belonging to the same virtual channel will have the same VCI.

VLSI – very large-scale integration. The integration of the components of a circuit on a single semiconductor substrate that produces a significant reduction in space and increase in reliability. VLSI devices typically integrate upwards of 100 000 components.

VPI – virtual path identifier. An identifier within the header of an ATM cell that identifies the virtual path between two nodes within an ATM network to which the virtual channel associated with the cell belongs. A virtual path therefore consists of a collection of virtual channels between two points within the network.

X

XID – exchange identification. An LAPD supervisory message which causes the peer LAPD entities to exchange operating parameters and to adjust their operation to the lowest common capability.

Index